Lethal Arrogance

HUMAN FALLIBILITY AND DANGEROUS TECHNOLOGIES

Lloyd J. Dumas

St. Martin's Press

ISBN 0-312-22251-3

Library of Congress Cataloging-in-Publication Data

  Lethal Arrogance : human fallibility and dangerous technologies / Lloyd J. Dumas.
   p.   cm.
  Includes bibliographical references and index.
  ISBN 0-312-22251-3 (cloth)
  1. Technology—Risk assessment. 2. Technology—Social Aspects.
I. Title.
T174.5.D86   1999
303.48'3—dc21   99-31677
                                                              CIP

Book design by ACME Art Inc.

10   9   8   7   6   5   4   3   2   1
08 07 06 05 04 03 02 00 01 99

Lethal Arrogance

HUMAN FALLIBILITY AND DANGEROUS TECHNOLOGIES

Also by Lloyd J. Dumas

The Overburdened Economy:
Uncovering the Causes of Chronic Unemployment,
Inflation and National Decline

The Socio-Economics of Conversion from War to Peace

Making Peace Possible: The Promise of Economic Conversion

The Conservation Response:
Strategies for the Design and Operation of Energy-Using Systems

TABLE OF CONTENTS

In celebration of
the creativity that makes us powerful
and the
imperfections that make us human

ACKNOWLEDGMENTS

This book is a tapestry, woven from many strands spun by many people. Some have been sources of information, others have been sources of encouragement and inspiration. Over the quarter century since I began to explore the connections among human fallibility, technical failure and security, I have had so many discussions and debates, been given so many useful comments, spent so many hours reading works drawn from so many literatures that I cannot even list all of those whose contributions to this work should be recognized. For this, I hope I will be forgiven.

Among those physical and behavioral scientists, computer scientists, engineers, medical doctors, mental health professionals and military officers whose comments and insights have contributed directly to the genesis of this work are Carl Sagan, Robert Lifton, Herbert Abrams, Warren Davis, Georgia Davis, Victor Sidel, Severo Ornstein, Lester Grinspoon, Bruce Blair, Herbert York, Bernard Feld, James Thompson, Margaret Ballard, Robert Karasek, Mikhail Milstein, James Bush and Eugene Carroll. Among those who have been important sources of support and encouragement through many years of researching and writing about what often has been a frightening and upsetting subject are Abdi Yahya, Alice Barton, Lynne Dumas, Roger Kallenberg, Martha Hurley, Randy Riddle, Amy Riddle, Sarah Flynn and Phyllis Hersh.

There is no way to express my gratitude for all the help my parents Marcel and Edith Dumas have given me in this and every other project I have undertaken, or for the encouragement and inspiration of my friend and colleague, Seymour Melman. I am thankful also for the assistance of my longtime friend and intellectual co-conspirator Janine Wedel, who came through for me once again when I most needed her help. No one could ask for more constant enthusiasm or support than my dearest friends and confidants, Dana Dunn and Yolanda Downey, have so freely given me over the years. To be immersed for so long in such a difficult and often depressing subject and still remain certain that these obstacles too can be overcome, there must be a deeply felt sense of optimism, joy and belief in the future. This has been the priceless gift of my intellectual partner and soulmate, Terri Nelson Chastain.

PREFACE

It was astonishing news. For the first time in billions of years, the earth had a new partner in near space. And it was created and put in place, not by the great forces of the cosmos, but by the power of the human mind. In the fall of 1957, the steady beep . . . beep . . . beep of the first artificial space satellite heralded the dawn of an exciting new era in which the otherworldly dreams of science fiction began to seem within reach. But the satellite bore the name *Sputnik I,* and so that excitement was tempered by fear. The Soviet Union had achieved this remarkable breakthrough. It meant that their missile and space technology had surpassed that of the United States. America was now in grave danger, or so it appeared.

To a young boy just beginning the process of becoming a man the challenge was unmistakable. In science and technology lay the future, the destiny of humankind. As a child, I had been a "polio pioneer," a participant in the first mass public trial of Jonas Salk's killed-virus vaccine. The miraculous conquest of this dreaded, crippling disease had already nurtured the seed of my early fascination with the wonders of science. But *Sputnik* raised this interest to a new level. Here was not only an object lesson in the power of science and technology, but in the nuclear age, a clear and compelling link between science and human survival.

By the time I entered college, I aspired to be a nuclear physicist, as my high school yearbook had dutifully recorded. After years of physics and math my aspirations shifted—as happens with so many students—first in the direction of applied science (leading to a master's degree in engineering), then in the direction of social science (leading to a doctorate in economics). During these years, the escalating American involvement in Vietnam drew my attention sharply back to issues of national security. It did not take long before this renewed interest began to focus more broadly on the arms race in general and its most dangerous technology—nuclear weapons. Equipped now with both advanced technical training in science and engineering and advanced training in the social sciences, I became more and more intrigued by the interface between technology and human behavior, especially as it related to matters of security and survival.

In 1974, as a newly minted professor at the School of Engineering and Applied Science at Columbia University, that interest led me to sign up as a student in a special summer course held in the former ducal capital Urbino, birthplace of the Italian Renaissance painter Raphael. Day after day in that lovely setting, I listened to lectures on aspects of the nuclear arms race delivered by the likes of political scientist Hans Morgenthau (Henry Kissinger's mentor), Herbert York (former director of defense research and engineering at the Pentagon and of Lawrence Livermore Laboratory), and Victor Gilinsky (soon to be appointed one of the first commissioners of the Nuclear Regulatory Commission). Increasingly concerned that something vital was being overlooked, I rose from the audience to request time to make an unscheduled presentation on the importance of human error and technical failure to the issues at hand. Within a few days, I presented my first thoughts on the subject in that august company, fully expecting to be patiently but firmly told why the problems I had raised were not really a matter of concern. When I saw instead how feeble the response was, I knew I was on to something.

In a sense, then, this book has been more than 25 years in the making. In May 1976, my earliest version of this analysis was published in the *Bulletin of the Atomic Scientists,* later supplemented by an article focused particularly on human fallibility in the November 1980 issue of the same journal. Begun in my years as an engineering professor at Columbia, most of the work has been done since I became a professor of social science at the University of Texas. It may have initially been the result of a passionate interest in exploring the linkage between national security and nuclear weapons, but the work ultimately broadened to encompass the interactions between human fallibility and the whole range of technologies that threaten our well-being and even our physical survival.

My work on this set of problems continued alongside research in a series of other arenas, all sharing the common characteristic of lying at the boundary between technology and social science. Two hours sitting in my car waiting in line to get gas during the OPEC oil embargo convinced me that there was something to the energy problem. I began to teach a doctoral seminar at Columbia on energy conservation, a subject then only infrequently discussed, but key to simultaneously solving both the energy and environmental pollution problems. My first book, *The Conservation Response,* resulted from these efforts and detailed engineering strategies and public policy changes that could reduce energy use in the United States by 30 to 50 percent without negatively affecting the standard of living.

I also spent many years researching and writing about the largely technological roots of American industry's competitiveness problems and their connection to public policy. Among other things, in *The Overburdened Economy* I argued that the replacement of many highly paid jobs with a future by what a colleague of mine once called dead-end "McJobs" was a direct consequence of these problems and the public policies that underlay them. And I have done extensive work on how to smoothly convert military-sector engineers and scientists to efficient civilian activity as military budgets decline. But through all of this, I remained fascinated by and continued to work on the problem of human fallibility and its implications for the control of technology.

For all of its complexities, the essence of this problem lies in a very simple fact. We all make mistakes. That is at once one of the most compellingly obvious and most profound statements that can be made about human behavior. It is obvious because we see it all around us all the time. If we ever come to doubt it for even a moment, our own day-to-day experience soon makes it clear once again that it is true. It is also profound because the implications of this simple fact run through everything that people do—every relationship we have, every social or political system we construct, every technology we create.

Though we rarely deny our fallibility (in the legal sense) when it is put to us directly, we often deny it (in the psychological sense) when it inconveniently interferes with some grand social or technological vision that captures our imagination. Over the years, we have poured enormous energy and resources into trying to compensate for and ultimately overcome the limits imposed by this simple fact of being human. We have found many ways to mitigate the problem of human error, but we have not and cannot overcome it completely.

The simple observation that we all make mistakes lies at the core of this book. For the inevitability of human error—and the broader, more complex and profound problem of human fallibility—has placed unavoidable limits on our ability to control and safely operate the increasingly powerful technologies our advanced brain and millennia of accumulated technical knowledge have led us to create. We ignore these ultimate and inconvenient limits at great peril—not just to ourselves as individuals, but to our societies and even to the continued existence of our species.

There is much in this book that is frightening. So many things can go so wrong in so many different ways. Yet the multisided problems of human fallibility and technological failure must be explored before we can understand both why we cannot avoid these problems and, more importantly, what positive steps we can take to prevent disaster. These solutions are analyzed in the final chapter.

Keep in mind as you read this book that there is a way out. If you come to feel hopeless or helpless as you read on, remember that you are neither. You are, in fact, the key to finding a solution. For it is only when we as citizens in a free society take the time and trouble to really understand the threats we face that we can shift direction and find a more viable path to the future.

Lloyd J. Dumas,
Carrollton, Texas

PART I

Threatening Ourselves

Will Twentieth-Century Incidents Become Twenty-First-Century Nightmares?

NUCLEAR DISASTER: WAITING IN THE WINGS?

Suddenly, tons of highly toxic nuclear waste erupt like a volcano. Hundreds of people are killed outright, many thousands more exposed to dangerous ionizing radiation and forced to relocate. Close to 400 square miles of land becomes so contaminated with deadly radioactivity that it is rendered uninhabitable.

A scenario from the fertile if paranoid mind of a television screenwriter? No, a real-life disaster in the former Soviet Union, covered up for more than thirty years. In the late 1950s, during the height of the Cold War, the buildup of heat and gas in stored nuclear weapons waste caused an explosion that devastated the countryside near Kyshtim in the Ural Mountains. Think it couldn't happen here?

On July 23, 1990, more than 30 years after disaster at Kyshtim, the report of a U.S. government advisory panel warned that something very much like it could indeed happen here. The panel was headed by John Ahearne, a physicist who was formerly a high official in the Departments of Defense and Energy and chair of the U.S. Nuclear Regulatory Commission.

There are millions of gallons of highly radioactive waste, accumulated during four decades of producing plutonium for nuclear weapons, stored in 177 tanks at the Department of Energy's Hanford nuclear reservation in Washington state. The study argued that heat generated inside the tanks or a shock from the outside could cause one or more of them to explode. Although the explosion would not itself be nuclear, it would throw a huge amount of radioactivity around. How imminent is the danger?

According to the panel, "Although the risk analyses are crude, each successive review of the Hanford tanks indicates that the situation is a little worse." According to the *New York Times*, as of July 1990, "Experts outside the Department of Energy say the risk of explosion is now so high that the department has imposed a moratorium on all activity at one tank because of fear that a jolt or spark could detonate the hydrogen that has built up inside the tank."[1]

Hanford's long-term plan to solve the problem is to get the liquid wastes out of the sometimes leaky tanks, turn them into a glass-like solid form and bury them. But some engineers are afraid that parts of this very process that is intended to make waste storage safer will actually make an explosion more likely.

THE TERROR UNDERGROUND

Like characters in a grade-B horror film, they emerged from under the ground by the thousands, choking and gasping for air, some with blood streaming from the nose or mouth. But this horror was all too real. In a carefully timed attack, terrorists had set off at least five canisters of deadly nerve gas at almost the same time in the tunnels of three different subway lines beneath the streets of Tokyo.

More than 5,500 people were injured in the attack, which came at the peak of the Monday morning rush hour, 12 of them fatally. It is something of a miracle that many more did not die. Tokyo subway trains are packed at rush hour, and sarin, the Nazi-era nerve gas released, is 500 times more toxic than cyanide, the gas used for executions in the United States. Half a milligram is a lethal dose.

Leaders of the wealthy Japanese end-time religious cult known as Aum Shinrikyo were later arrested and charged with the attack. It seems that the ghastly assault on Tokyo was to be just the beginning. When thousands of police officers raided dozens of Aum Shinrikyo buildings around Japan, they found gas masks, tons of chemicals and sophisticated chemical manufacturing equipment. They also found evidence that the cult might have been trying to develop biological weapons. And they found a cult newsletter warning that by the end of 1996, 90 percent of the people living in major cities like Tokyo would be killed by earthquakes, epidemics—or poison gas.

Sarin was specifically mentioned by name.

"HOLES IN THE FENCES EVERYWHERE"

Enriched uranium is standard fuel for the Russian nuclear navy. If it is enriched enough, it is also the stuff of nuclear weapons.

In the early afternoon of November 27, 1993, two guards patrolling the Sevmorput shipyard, one of the Russian navy's main nuclear fuel storage depots, saw a discarded padlock lying on the ground. They noticed that the door of a nearby storehouse was open. When they looked inside, they found that 4.5 kilograms (10 pounds) of enriched uranium was missing.

Six months later, three men were arrested and the stolen fuel recovered. The man charged with breaking the padlock on the door and stealing the uranium was the deputy chief engineer at the shipyard. The man accused of hiding the stolen uranium was a former naval officer. And the alleged mastermind of the operation was the manager in charge of the refueling division at the shipyard!

Although this theft had been an inside job, stealing uranium from Sevmorput was apparently not that big a challenge. According to Mikhail Kulik, the chief investigator, "'On the side [of the shipyard] facing Kola Bay, there is no fence at all. You could take a dinghy, sail right in—especially at night—and do whatever you wanted. On the side facing the Murmansk industrial zone there are . . . holes in the fences everywhere. And even in those places where there aren't holes, any child could knock over the half-rotten wooden fence boards.'"[2] In Kulik's view, if the amateur thief had not made the bush-league mistake of leaving the door open, the theft at Sevmorput "could have been concealed for ten years or longer."

BAD YEAR AT BANGOR

At the end of the Cold War, Bangor, Washington, was one of the most heavily armed cities in the world. With some 1,700 nuclear weapons at the Bangor Submarine Base and more stored at the Strategic Weapons Facility nearby, there was enough nuclear firepower in Bangor to destroy any country on earth many times over.

More than a thousand military personnel in the area were on active nuclear duty, certified by a special Pentagon program as physically and mentally reliable. Yet just how reliable were they?

One of those certified as reliable was an 18-year-old marine, Lance Corporal Patrick Jelly. Claiming to be a reborn soldier killed in Vietnam, he punctured his arms with a needle and thread. For weeks he had threatened to kill himself. Still he was kept on active duty. At 9:30 P.M. on January 14, 1989, while standing guard over the fearsome nuclear arsenal stored at the Strategic Weapons Facility, Jelly finally made good on his threat. The troubled young marine aimed an M-16 rifle at his head and pulled the trigger. Jelly had remained certified as reliable until the night he died.

Tommy Harold Metcalf was also certified as physically and mentally fit. He was a fire-control technician aboard the Trident submarine *Alaska,* part of the team directly responsible for launching the ship's nearly 200 city-destroying nuclear warheads. On July 1, 1989, Metcalf suddenly went to the home of an elderly couple and murdered both of them by suffocation.

William Pawlyk had been commander of Submarine Group 9 at Bangor and had served aboard the nuclear submarine *James K. Polk* for five years. In early August 1989, Commander Pawlyk was arrested after stabbing a man and a woman to death. He was head of a naval reserve unit in Portland, Oregon, at the time of the murders.

Shyam Drizpaul was assigned to duty aboard the nuclear submarine *Michigan.* Like Tommy Metcalf, he was part of the missile launch team, certified as physically and emotionally healthy. On January 15, 1990, he shot and killed a fellow crew member in the lounge at his living quarters, then another in bed. Afterward, while attempting to buy a pistol at a pawnshop, Drizpaul grabbed the gun from the clerk, shot her to death and critically wounded her brother. Fleeing the scene of that crime, he checked into a motel near Vancouver and used the same weapon to kill himself. A subsequent Navy investigation discovered that Drizpaul drank heavily, carried an unregistered hand gun, and boasted of having been trained as an assassin.

All of these incidents occurred between mid January 1989 and mid January 1990. It had been a bad year at Bangor.[3]

A WARNING

If you are thinking that there is something unreal, something unique about this set of stories, think again. Not only are these stories real, they are anything but unique. They turn on problems of human fallibility and technical failure that are embedded in the fabric of everything we do.

Our brilliant technological accomplishments have made us too complacent, too arrogant about our ability to control even the most dangerous technologies we create and permanently avoid disaster. But we will not, because we cannot.

We humans are fallible. We are not perfect, and never will be. Understanding the many-sided nature of our unavoidable fallibility and how it interacts with the most dangerous technologies we create is the key to the fundamental change that can lead us away from disaster. It is what this book is all about.

If we do not fundamentally change the way we do things, these frightening twentieth-century incidents may just be the forerunners of still more horrifying twenty-first-century nightmares.

If you think that I am exaggerating, that it can't really be that bad, read on.

CHAPTER ONE

Technology,
Human Fallibility and Survival

We humans are creatures of paradox. When we direct our talents toward positive ends, we compose music that makes the spirit soar, create artworks of timeless beauty, design and build devices that extend the reach of our unaided senses and connect us to the wider world. Bent to different ends, we despoil our environment, debase and degrade each other and create weapons so destructive that they may yet make us the first species responsible for its own extinction.

Of all the ways we affect the physical world for better or for worse, the development and application of technology is the most powerful. On the positive side, the progress of technology has allowed us to make more and better goods available at prices low enough to put them within the reach of much of the earth's population. We can now travel faster, exchange information more readily, enjoy a wider variety of foods and entertainments, have more protection from the elements, and access far better medical care than was available to even the richest kings and queens of centuries gone by.

Yet at the same time we have fouled the air, land and water and driven thousands of living species into extinction. We have put into our own hands the power to extinguish human society in a matter of hours and trigger a process that could eventually turn our bountiful planet into yet another lonely, lifeless sphere spinning aimlessly through space. This too is the "gift" of technology. Its two-sided nature is simply a reflection of the brighter and darker sides of the species that created it.

Much has been written about the power of technology. There has been more than a little rhapsodizing about the peaceful and prosperous Eden to which it will carry us some day, a day that never quite seems to arrive. Then too, there have been those who have ranted about its evils and urged us to return to a more natural lifestyle. This book is neither a paean to the glories of technology nor a cautionary tale about the price we have paid for its bounty. It is instead a passport to reality and a guide to making wiser choices. And it is warning, not about technology as such, but about technological arrogance—the unthinking assumption that we can always control the technologies we create, no matter how powerful or dangerous, and permanently avoid disaster. It is a particularly lethal form of arrogance.

We need to take the simple fact of our fallibility much more seriously than we now do in deciding what technological paths to follow. We cannot allow our fascination with the power of what we can do to blind us to how easily it can all blow up in our faces.

TECHNOLOGY AS MAGIC: THE TECHNOLOGICAL FIX

Technology reaches into the very fabric of society, reshaping the economic, social, political and cultural world in which we live. But at the same time it affects these dimensions of our lives, it is greatly affected by them. Technology is not self-propelled.

For all the mystique that surrounds it, technology is nothing more than the application of the biological, chemical and physical laws of nature to the design and development of useful products and processes. Scientists extend and deepen the store of knowledge by applying the scientific method to the exploration of these laws. Engineers apply this knowledge to the design of actual products and processes.

Yet to many of us, technology is a kind of magic. Why it works is a mystery, but there is no doubt that it does remarkable things. Flip a switch and a room lights up, push a few buttons and talk to a friend on the other side of town or the other side of the world, sit inside an enormous metal tube and roaring engines catapult you through the sky. It's all pretty amazing.

How some technology works is easy to understand. It is obvious why a bicycle moves faster when you peddle harder or changes direction when you turn the handlebars. You can see what is happening and why. But most modern-day technology is much more opaque: you can drive a car for years without having the slightest idea why it moves faster when you press the accelerator pedal or turns when you turn the steering wheel.

Engineers have worked hard to make increasingly sophisticated technology more "user friendly." Because of their efforts, it isn't necessary to understand how telephones, videocassette recorders or computers work in order to use them. And so, these technologies have become accessible to millions. But for all of its advantages, the user-friendly approach has one unfortunate side effect: it allows the technology in common use to advance far beyond even the vague comprehension of the vast majority of those who use it. For most of us, more and more of the products of everyday life have become what engineers call "black boxes"—we know what goes in and what comes out, but not what goes on inside. This has reinforced the unconscious image of technology as magic, an image that has very serious consequences for making socially intelligent decisions.

The problem is that in magic, anything can happen. There are no limits, no tradeoffs, no impossibilities. So, if technology is magic, it has no limits, no tradeoffs, no impossibilities. Whatever the problem, if scientists and engineers are clever enough and work hard enough, they can eventually find a way to solve it, they can eventually find a technological fix.

In the 1950s, it was widely believed that the new agricultural technologies known as the "green revolution" would finally put an end to world hunger. These technologies were, in fact, enormously successful in increasing agricultural yields. Yet there are more chronically hungry people in the world today than there were in the 1950s. The reason is simple: world hunger was not and is not the result of inadequate food production. It was and is mainly the result of economic, social and political factors that affect food distribution and use.[1] The problem of world hunger is not solvable by a technological route because it is not a technological problem.

When nuclear weapons were first created, many believed they were so terrifying that they would finally put an end to the age-old scourge of war. Yet since the dawning of the atomic age, we have fought more than 150 wars, taking the lives of 23 million people, two-thirds of them civilians.[2] This technological fix did not work either, and for the same reasons. Neither war itself nor the broader issue of security is a technological problem. Security and war are political, social and economic problems. We will put an end to war and find more humane and reliable paths to security when we become committed to less threatening means of resolving even our most contentious conflicts. No brilliant technological innovation will save us from the difficult task of learning how to live with each other.

There is not always a technological fix even for problems that are mainly technological. In the real world, there *are* limits, tradeoffs and impossibilities. Often enough, the technology that solves one problem creates others. As hard as it may be to believe today, in the late nineteenth and early twentieth centuries, the automobile was promoted in part as a technological fix for urban pollution. Horse-drawn transport had polluted the cities with an accumulation of animal waste that threatened public health. The automobile neatly solved that problem. Yet today, automobiles are one of the largest sources of health-threatening urban air and noise pollution. History certainly has its ironies.

The fact is, technology is not magic. It is the result of a systematic creative process of thinking and experimentation carried out by fallible human beings who inescapably embed the essential imperfection that makes them human into every technology they develop. Technology is very powerful, but it is not omnipotent.

Although technological knowledge itself is neither good nor bad, it is not true that all technologies can be applied in ways that are beneficial. Some are extremely difficult if not impossible to put to constructive use. Given our fallibility and the darker side of the human psyche, there may even be some things we would be better off not knowing. There is simply too great a chance that they will be used in ways that turn out to be highly destructive, intentionally or unintentionally.

Despite the dangers, few of us would want to give up the benefits of modern technology. But the choice is not to embrace or reject at once the full range of technologies of modern society. We can *and should* pick and choose among existing technologies (and research directions) those that have the greatest net benefit. And we should firmly reject those that have the potential for doing us grave harm. While it is impossible to eliminate risk and danger from technologically advanced society, there is no reason why we cannot avoid subjecting ourselves to the very real possibility of technology-induced catastrophe.

DANGEROUS TECHNOLOGIES

Dangerous technologies are those capable of producing massive amounts of death, injury and/or property damage within a short span of time. Some technologies are designed to be dangerous. Their intentional use, accidental use or failure can do enormous damage. Nuclear weapons are a prime example. Accidental nuclear war, triggered by human error, miscalculation or technical failure, would be no less devastating than deliberate nuclear attack. Even the accidental explosion of a single nuclear weapon in the wrong place at the wrong time could take millions of lives.

Other technologies designed for benign purposes can still produce devastating damage if enough goes wrong. Nuclear power plants supply electricity to consumers, business and industry. Yet the catastrophic human or technical failure of a nuclear plant has the potential to kill or injure hundreds of thousands of people and contaminate huge areas of the surrounding landscape. A so-called maximum credible accident at a single nuclear power plant could do billions of dollars worth of damage. Therefore, nuclear power must also be classified as a dangerous technology.

Weapons of Mass Destruction

Weapons of mass destruction are the outstanding example of dangerous technologies. Although it is technically possible to kill millions of people with conventional explosives (or even with rifle bullets or knives), what makes weapons of mass destruction different is that they can kill large numbers of people *with a single use.*

Nuclear weapons are, in turn, the pre-eminent example of a weapon of mass destruction. They have enormous destructive power relative to bomb size or weight, and a reach that extends far beyond their moment of use: people were still dying from the Hiroshima bombing decades after the attack. Used in large numbers, nuclear weapons assault the environment, tearing at the ecology, at the very fabric of life.[3]

Weapons that disperse nerve gas or other deadly chemicals are also capable of killing thousands or even millions of people. Biological weapons that release virulent infectious organisms can be still more destructive, because the organisms are alive and can reproduce. Some pathogens (such as anthrax bacilli) are very resistant to external conditions and can remain dormant but infectious for many years.[4] As frequent flu epidemics clearly indicate, once a chain of infection begins, it can spread rapidly through large numbers of people. Wartime conditions of disruption and stress would make the epidemic worse.

Despite the advanced state of modern medicine, some deadly infections can be extremely difficult to treat, let alone cure. This is especially true when the nature of the infectious organism itself is unusual, as the AIDS epidemic has illustrated with frightening clarity.[5] Recent remarkable advances in biotechnology have made it far easier to produce genetically altered and therefore unusual versions of known pathogens, as well as wholly artificial new infectious life forms.[6]

First cousins to nuclear weapons, radiological weapons kill without blast and fire by dispersing radioactive materials. They can contaminate large areas with high levels of radiation that can remain deadly for many centuries. Nuclear accidents like that at Chernobyl have already illustrated how easily radioactive materials can be dispersed over a wide area. In a sense, any nation that has nuclear power plants or waste storage sites within its borders already has "enemy" radiological weapons on its own soil. A military or terrorist attack against those facilities could unleash enormous amounts of dangerous radiation.

Given the ingenuity of weapons designers and military strategists, there will undoubtedly be more to come in the way of methods for killing large numbers of people. Perhaps we will come up with weapons that alter the climate, shifting rainfall patterns and temperatures so as to create widespread starvation or ecological devastation.[7] In the 1980s, computer simulations of nuclear war indicated that the large-scale use of nuclear weapons might produce a "nuclear winter."[8] We have already tried small-scale, primitive ecological weapons such as Agent Orange, used to defoliate the jungles of Vietnam.[9] What about weapons that would produce hurricanes, tornadoes or earthquakes on demand? At least as far back as the 1960s, research was already underway on the possibility of using underwater nuclear explosions to generate tsunamis (tidal waves).[10]

How about "ethnic weapons," designed to target one particular ethnic group? Small genetic variations in the gene pools of different ethnic groups (such as the predisposition of blacks to sickle cell anemia) might make it possible to find chemical agents, bacteria or viruses to which that group would be particularly vulnerable. As grisly and bizarre as this sounds, it was discussed long ago in the public military literature.[11] There is no doubt that the remarkably creative human mind directed to such dark purposes can add still unimagined means of annihilation to this already chilling list.

Nuclear Power and Nuclear Waste

Nuclear power is a dangerous technology not only because the consequences of a major malfunction are so potentially catastrophic, but also because it generates large amounts of highly toxic radioactive waste. Despite periodic announcements of breakthroughs, the problem of developing a safe, practical way to treat and/or store long-lived nuclear waste is still far from being solved.[12] The combination of inadequate nuclear waste treatment and storage technology and large amounts of very toxic, long-lived nuclear waste is a nightmare, a disaster waiting to happen.

Nuclear waste is often corrosive and physically hot, as well as radioactively "hot." This makes it difficult to store safely and isolate from the biosphere, especially when it remains dangerous for extremely long periods of time. A single half-life of plutonium—a critical raw material in nuclear weapons and a byproduct of nuclear power reactors—is 24,000 years. That is far longer than all of recorded human history. Considering all the political, social, economic and cultural changes that have occurred from a time more than 150 centuries before the pyramids of ancient Egypt up to the space age, what reason is there to believe that we can safely contain, isolate or even keep track of materials that remain so deadly for so long? Yet until the problem of safe storage is solved (if ever), nuclear waste storage sites are ecological time bombs waiting to explode in the faces of this or future generations. This is hardly a legacy of which we can be proud.

Highly Toxic Chemicals

The spectacular progress of chemicals technology during the twentieth century, especially since World War II, has created its own set of problems. Some chemicals routinely used in industry are extremely toxic themselves or are manufactured by processes that produce highly toxic chemical waste. The worst industrial accident in history occurred at a chemical plant—the Union Carbide pesticide plant at Bhopal, India. On December 3, 1984, water entering a methyl isocyanate storage tank triggered a reaction that resulted in the release of a cloud of deadly gas that drifted over the city, killing at least 2,000 people and injuring 200,000 more.[13] The isocyanates used in industry are related to phosgene, a chemical used as poison gas as far back as World War I.[14]

The Bhopal disaster may have been the worst to date, but there have been many accidents during manufacturing, transportation, storage and use of highly toxic chemicals. Still, not all technologies involving toxic chemicals are "dangerous." The chemicals must be toxic enough and in quantities large enough for a major release to produce disaster. Fortunately, most noxious chemicals in common use are not that toxic. Unfortunately, all too many are.

Only in the last few decades have we begun to understand just how much toxic chemical waste we have created as a result of our fascination with the "magic of modern chemistry." Highly toxic chemical waste may be less complicated to treat and store than nuclear waste, but it is still a very serious problem. And it is very widespread. Toxic waste dumps are scattered all over the United States. There are in fact few, if any, parts of the world where this problem can be safely ignored.

Despite the fact that they kill or injure large numbers of people, some technologies are not classified as "dangerous" because the damage they do is not concentrated. Automobiles, for example, kill or maim tens of thousands every year. But the victims are injured in thousands of separate crashes, each involving a relatively small number of people. Because it is dispersed in time and place, this carnage on the roadways has neither the same social impact nor the same social implications as concentrated disasters like Bhopal or Hiroshima.

PROLIFERATION

Dangerous technologies have spread all over the globe. Sometimes they were purposely propagated in pursuit of profit or other self-interested objectives;

sometimes they spread despite determined efforts to prevent others from acquiring them. The laws of nature cannot be patented. Once it has been shown that a technology works, it is only a matter of time before other well-trained and talented people will figure out how to do it too.

Nuclear Weapons

It took years for a huge, well-orchestrated program led by an unprecedented collection of the world's most talented physical scientists to develop the first atomic bomb in 1945. Intensive efforts to keep the technology secret did not succeed. The American monopoly on nuclear weapons did not last long. In 1949, the Soviet Union successfully tested an atomic bomb of its own. Britain followed in 1952, and France in early 1960. By the end of 1964, China too had joined the "nuclear club." A decade later, India detonated a "nuclear explosive device," but continued to insist that it was not a nuclear weapons state for almost 25 years. All such pretenses were blown away in May 1998, when India exploded a series of five nuclear weapons in three days.[15] Arch rival Pakistan then followed suit within a few weeks.[16] To date, then, these 7 nations have publicly declared themselves to be nuclear weapons states. But they are by no means the only countries capable of building nuclear weapons. All 20 countries listed in Table 1-1 have (or had) a significant nuclear weapons capability. Of them, 11 are known to have had nuclear weapons on their territory (the 7 "declared" nuclear states plus Belarus, Kazakhstan, Ukraine and South Africa). At least one more, Israel, is widely believed to have a substantial secret nuclear arsenal today. Almost 20 countries in addition to those listed have the technical capacity to develop nuclear weapons, but are believed not to have yet tried to do so (14 European countries plus Australia, Canada, Japan and South Korea).

It is also possible that both nuclear materials and important elements of nuclear weapons technology have already proliferated to revolutionary or subnational terrorist groups (see Chapter 2). There is no hard public information on the extent of this proliferation nightmare, but there is no doubt that it must be taken seriously.

The longer government arsenals of nuclear weapons and government-sanctioned weapons-development programs continue, the more likely both types of proliferation become. Renegade governments, revolutionaries, criminals and terrorists will have more opportunities to steal the weapons themselves or acquire the critical materials, blueprints and equipment needed to build them. Just as important, there are also more people with experience in designing, building, maintaining and testing nuclear weapons with every passing day. Is it reasonable to assume that none of them will ever be convinced, coerced or bought off by rogue governments, revolutionaries, criminals or terrorists?

TABLE 1-1
NATIONS THAT HAVE (OR HAD)
ACTIVE NUCLEAR WEAPONS PROGRAMS

NATION	NUCLEAR WEAPONS STATUS	SIZE OF ARSENAL
United States	Declared Nuclear Weapons	8,500 Strategic & 7,000 Tactical Warheads
Russia	Declared Nuclear Weapons	7,200 Strategic & 6,000-13,000 Tactical Warheads
United Kingdom	Declared Nuclear Weapons	100 Strategic & 100 Tactical Warheads
France	Declared Nuclear Weapons	482 Strategic & No Tactical Warheads
China	Declared Nuclear Weapons	284 Strategic & 150 Tactical Warheads
Belarus	Part of former Soviet Union, all nuclear weapons on its territory are believed transferred to Russia by mid 1990s	None
India	Exploded "peaceful" nuclear device in 1974, and series of admitted nuclear weapons in 1998	Estimated to be more than 60 nuclear warheads
Israel	Has not acknowledged possession of nuclear arsenal	Estimated to be more than 100 nuclear warheads
Pakistan	Exploded series of nuclear weapons in 1998	Estimated to be 15-25 nuclear warheads
Iran	Despite denials, Western intelligence sources believe active nuclear weapons program underway	Estimated to be 7-10 years away from building any nuclear weapons
Iraq	Unveiled in 1995 by defection of high-level official, covert nuclear bomb development program operating since early 1970s	Estimated to have been a few years from building nuclear warheads in 1991; status of program now unclear

Testifying before Congress in January 1992, CIA Director Robert Gates estimated that some one to two thousand of the nearly one million people involved in the former Soviet Union's military nuclear programs were skilled weapons designers.[17] Add to that the numbers in the American, French, British, Chinese, Pakistani, Indian, Israeli and South African programs, and it is clear that many thousands of people have skills and experience critical to developing nuclear weapons.

Gates emphasized the grave danger posed by the possibility that the economic, social and political distress in the former Soviet republics could lead to a large-scale

TABLE 1-1 CONTINUED

NATION	NUCLEAR WEAPONS STATUS	SIZE OF ARSENAL
Libya	Despite denials, Western intelligence sources believe active nuclear weapons program underway	Weapons program believed to be embryonic, despite 25-year effort
North Korea	Suspected active nuclear weapons program believed frozen, as result of 1994 agreement with United States	Believed close to capacity to make dozens of nuclear weapons per year
Algeria	Once suspected of having embryonic program, Algeria renounced any attempt to develop nuclear weapons in 1995	None
Argentina	Suspected of having begun nuclear weapons program in 1970s, renounced any attempt to build nuclear weapons in mid 1990s	None
Brazil	Secret nuclear weapons program run by previous military rulers in 1980s was ended in 1990 by newly elected government	None
Kazakhstan	Part of former Soviet Union, all nuclear weapons on its territory were transferred to Russia by April 1995	None
Ukraine	Part of former Soviet Union, all nuclear weapons on its territory were transferred to Russia by June 1996	None
Romania	Secret nuclear weapons program during Ceausescu regime ended with 1989 overthrow of Ceausescu	None
South Africa	Successful secret nuclear weapons program ended by 1989 order of President de Klerk	6 nuclear weapons built; weapons and related facilities dismantled by 1991

Sources: Spector, Leonard S., McDonough, Mark G., and Medeiros, Evan S., *Tracking Nuclear Proliferation* (Washington, D.C.: Carnegie Endowment for International Peace, 1995) and accompanying "Tracking Nuclear Proliferation: Errata and Essential Updates" sheets (June 7, 1996); Albright, David, "The Shots Heard 'Round the World" and "Pakistan: The Other Shoe Drops," *Bulletin of the Atomic Scientists* (July-August 1998); Hamza, Khidhir, "Inside Saddam's Secret Nuclear Program," *Bulletin of the Atomic Scientists* (September-October 1998).

emigration of nuclear, chemical and biological weapons designers. Where would they go? Perhaps to places like Cuba, India, Syria, and Libya, where they might already have contacts from the Soviet era. In late 1998, it was reported that Iran had already recruited some, and was actively trying to recruit more, Russian biological weapons experts for its own germ warfare program.[18] The desperate scientists might also find work with the prosperous criminal underground in their own or other countries, or perhaps with a well-funded revolutionary or terrorist group. By the mid 1990s, the wealthy Japanese Aum Shinrikyo doomsday cult (which seems to have had a penchant for the technologies of mass destruction) boasted of having thousands of members in Russia.

While the former Soviet Union is the greatest present concern, none of the countries that have nuclear weapons are immune to this problem. The United States today has a growing extremist militia movement whose vicious, racist, antigovernment rhetoric leaves plenty of room for more and more brutal acts of violence. Indeed, America experienced increasingly destructive acts of terrorism on its own soil in the 1990s. Britain and France have both been plagued by terrorist activities for years. In China, tensions continue to build between the repressive old-guard regime in Beijing and a rapidly modernizing urban economy whose young people (at least) yearn for a greater degree of personal and political freedom. This, along with the frequently troubled relationship between mainland China and the descendants of the heavily armed Chinese opposition forces on Taiwan, could someday lead to an explosion.

India and Pakistan, the newest nuclear powers, have yet to settle the differences that have led them to war with each other three times since 1947 and nearly triggered a full-scale accidental war in January 1987 (see Chapter 5). Both countries have also had internal ethnic violence and secessionist movements. And the region around Peshawar, Pakistan (a staging area for rebel supporters during the Soviet-Afghan war) later became a hotbed of international terrorism.

Israel, a still undeclared nuclear nation, has long been a favorite target of Arab terrorists. Israel has recently seen the rise of extremist violence within its own citizenry. The murder of Itzhak Rabin in the mid 1990s shocked the nation not only because it was the first assassination of a prime minister in Israel's history, but also because the assassin was an Israeli. It was not the first act of murder carried out by a right-wing Israeli fanatic. Although South Africa dismantled its fledgling nuclear arsenal in the early 1990s, it is unlikely that those who designed and built those weapons have disappeared. It is a terrible thought, but should South Africa's remarkably smooth shift toward a racially integrated society collapse before it has become irreversible—perhaps with the death of Nelson Mandela—long-nurtured racial hatreds could cause this near nuclear nation to descend into chaos.

Nuclear Power

In addition to increasing the chances of diversion of materials and know-how to the development of nuclear weapons (see Chapters 2 and 3), nuclear power poses many dangers of its own. Nuclear plants are always subject to catastrophic accident, especially as they age. Even when operating normally, the reactors and

TABLE 1-2

NUCLEAR POWER PLANTS IN OPERATION
(DECEMBER 31, 1995)

COUNTRY OF PLANTS	NUMBER
United States	109
France	56
Japan	51
United Kingdom	35
Russia	29
Canada	21
Germany	20
Ukraine	16
Sweden	12
South Korea	11
India	10
Spain	9
Belgium	7
Bulgaria	6
China, Taiwan	6
Switzerland	5
Czech Republic	4
Finland	4
Hungary	4
Slovak Republic	4
China, Mainland	3
Argentina	2
Lithuania	2
Mexico	2
Netherlands	2
South Africa	2
Armenia	1
Brazil	1
Kazakhstan	1
Pakistan	1
Slovenia	1
TOTAL	437

Source: International Atomic Energy Agency, Nuclear Power Reactors: Power Reactor Information System Database (January 26, 1999) http://www.iaea.org/worldatom/inforesource/pressrelease/prtable.html)

TABLE 1-3

DISTRIBUTION OF
OPERATING POWER REACTORS BY AGE CATEGORY
(DECEMBER 31, 1995)

AGE (YEARS)	NUMBER OF REACTORS
0-2	10
3-5	19
6-8	35
9-11	90
12-14	73
15-17	51
18-20	50
21-23	53
24-26	30
27-29	13
30-40	18
TOTAL	442

(No explanation is provided by IAEA for the discrepancy in total numbers between these data and those on number of reactors.)

Source: International Atomic Energy Agency, Nuclear Power Reactors: Power Reactor Information System Database (January 26, 1999) (http://www.iaea.org/worldatom/inforesource/pris/age.gif).

the waste storage areas that service them are a tempting target for terrorists or hostile militaries bent on mass destruction. Yet, as shown in Table 1-2, this dangerous technology too has spread around the world. By the end of 1995, 437 nuclear power reactors were operating in 31 countries on five continents.

Along with the normal problems of aging, nuclear reactors are subject to embrittlement caused by exposure to high levels of radiation over extended periods of time. Because this can seriously shorten their safe operating life, it is important to consider the age distribution of the world's nuclear power reactors. Table 1-3 gives the number of operating reactors of each indicated age category. By the end of 1995, nearly half the reactors were 15 years old or older; more than 25 percent were greater than 20 years old, and 18 were at least 30 years old.

Something on the order of 70,000 kilograms of plutonium are produced each year in the normal course of operation of these 400-plus power reactors. Reactor-grade plutonium can be used to make many reliable, efficient military nuclear weapons or a smaller number of crude, easily transported, terrorist nuclear bombs (see Chapter 2). The technology required to extract plutonium from spent power plant fuel has been widely accessible for decades. Indeed, roughly 200,000 kilograms of plutonium has already been separated, and is currently stored in about a dozen countries. That is an enormous amount of plutonium, about the same amount used to produce *all* of the tens of thousands of devastating weapons in the world's nuclear arsenals today.[19]

Chemical and Biological Weapons

The technology of designing and building chemical weapons of mass destruction for military use is more complex than the technology of manufacturing toxic chemicals. In addition to producing the deadly nerve gases and the like, there must be ways of delivering the chemicals to the target, as well as shielding one's own forces against them. The delivery system must protect the lethal chemicals against heat, vibration, contamination, shock and whatever other problems of premature release or deterioration they might encounter. Nevertheless, some two dozen countries are known or believed to have significant chemical weapons arsenals.[20]

Of course, for terrorist use, deadly chemicals require little if any sophisticated "weaponization." Terrorists can inject toxic chemicals into ventilating systems using something as simple as a pesticide sprayer. They can release nerve gas into subway tunnels with the crudest of devices (as the Aum Shinrikyo cult did in Tokyo in 1995), especially if they are willing to die in the process.

Biological weapons pose similar problems. The knowledge and facilities required to breed highly infectious, lethal microorganisms are widespread and are bound up with the ordinary medical and biological technologies of modern life. But the attacker's problems of weaponization and of protecting its own forces and population are much greater.

Biological weapons are even more dangerous than chemical weapons. On a weight basis, they have much more killing power than the most deadly chemicals. Inhaling a hundred millionth of a gram (eight thousand spores) of anthrax bacillus causes illness that is almost 100 percent fatal within five days. The lethal inhaled dose of the highly toxic nerve gas sarin is nearly a hundred thousand times larger (one milligram).[21] Beyond this, when living organisms are released into the environment, there is always the chance that they will mutate or otherwise behave in unpredictable ways, particularly if they are novel, genetically engineered life forms.

Historically, illness has been a much more devastating part of warfare than many of us realize. In Vietnam, American soldiers were hospitalized twice as often for infectious diseases as for wounds inflicted by enemy fire. In World War II and the Korean War, disability from illnesses cost the U.S. Army five times as many soldier-days as did combat injuries.[22] Fortunately, the difficulties of controlling deadly biologicals have so far made them less attractive to militaries. But they keep trying. As many as a dozen countries reportedly have or are seeking a biological (or toxin) warfare capability.[23] Unfortunately, as with lethal chemicals, the difficulties of control would not stop an irrational terrorist group or doomsday religious cult from using them.

Missiles

While neither missiles themselves nor missile technology is classified as danger-ous, their proliferation makes effective military delivery of weapons of mass destruction much easier. In 1987, seven nations with advanced missile technol-ogy (United States, United Kingdom, Canada, France, Germany, Italy and Japan) enacted the Missile Technology Control Regime (MTCR) to coordinate and control export of missiles and related technology. By the mid 1990s, more than two dozen other nations had agreed to abide by MTCR restrictions, including Russia and China—both major missile exporters.[24]

The agreement seems to have helped stem the proliferation of long-range ballistic missile technology, a real advantage to the industrialized countries out of range (like the United States). But it was too little and too late to prevent spread of short-to-medium-range missile technology, making many of the world's nations vulnerable to attack from hostile neighbors. North Korea's 1998 launch of a missile that overflew Japan dramatically punctuated that point.[25]

Table 1-4 lists those countries deploying missiles that they bought or manufactured, along with those countries that had active missile R & D or space launch vehicle programs in the 1990s. They include nearly every country that has, or once had, an active nuclear weapons program (as shown in Table 1-1).

A potential missile builder can learn the theory of missile guidance and control systems—among the most difficult parts of missile development—from publicly available books, magazines and university courses. The hardware required is also not that difficult to come by. Furthermore, it is now possible to purchase hand-held receivers that access the Global Positioning System (GPS) satellite network. They could tell a missile where it is anywhere on earth, at any point in time, with an accuracy of a few yards.[26]

Does the proliferation of missile technology also raise a terrorist threat? Beyond using an antiaircraft missile to bring down a civilian airliner, missile attack is likely have little appeal for subnational terrorist groups. Without the dedicated resources of a government, it would be very difficult to secretly build and successfully launch a missile of any substantial range. Even more important, there is no need. Successful actions by terrorists to date show that they are

TABLE 1-4
COUNTRIES WITH BALLISTIC MISSILES
OR MISSILE PROGRAMS, 1990s

COUNTRY	NATURE OF MISSILES OR PROGRAM	MISSILE RANGE (MILES)
U.S.	Currently Deploys Long Range Missiles	8,100 (missile)
Former USSR	Currently Deploys Long Range Missiles	8,100 (missile)
U.K.	Currently Deploys Long Range Missiles	3,000 (missile, purchased)
France	Currently Deploys Long Range Missiles	2,175 (missile)
China	Currently Deploys Long Range Missiles	8,100 (missile)
Japan	Advanced Space Launch Vehicle Program	6,200 (space launch vehicle)
Israel	Advanced Missile/Space Launch Vehicle Program	900 (missile) 3,100* (space launch vehicle)
India	Advanced Missile/Space Launch Vehicle Program	150 & 1,500* (missiles) & 8,000* (space launch vehicle)
North Korea	Missile Program	175, 375 & 1,250* (missiles)
Egypt	Missile Program	175 (missile) and 750* (missile, abandoned)
Iran	Missile Program and Purchased Missiles	80, 100, 800* (missiles) & 175 (missile, purchased)
South Korea	Missile Program	150 (missile)
Brazil	Missile R & D	2,500* (space launch vehicle)
South Africa	Missile R & D	900* (missile)

perfectly capable of delivering any weapon of mass destruction they might construct in simpler, cheaper and more reliable ways: by boat, truck, airplane— or suitcase.

Toxic Chemicals

Toxic chemical technologies are thoroughly integrated with the nearly ubiquitous operation of the chemicals industry on which so much of modern industrial and agricultural output depends. There is little doubt that industry and agriculture can become much less dependent on highly toxic chemicals. But the knowledge of chemistry and the techniques of chemical production required to make toxic chemicals are so widespread, so available, and such a normal part of modern life that the potential for producing dangerous chemicals will always be with us.

COUNTRY	NATURE OF MISSILES OR PROGRAM	MISSILE RANGE (MILES)
Argentina	Missile R & D	750 (missile, abandoned)
Taiwan	Missile R & D	375 (missile)
Pakistan	Missile R & D	175 (missile, purchased) & 375* (missile)
Indonesia	Missile R & D	65* (missile)
Saudi Arabia	No Indigenous Program	1,675 (missile, purchased)
Bulgaria	No Indigenous Program	310 (missile, purchased)
Czechoslovakia	No Indigenous Program	310 (missile, purchased)
Afghanistan	No Indigenous Program	175 (missile, purchased)
Hungary	No Indigenous Program	175 (missile, purchased)
Libya	No Indigenous Program	175 (missile, purchased)
Poland	No Indigenous Program	175 (missile, purchased)
Romania	No Indigenous Program	175 (missile, purchased)
Syria	No Indigenous Program	175 (missile, purchased)
Yemen	No Indigenous Program	175 (missile, purchased)
Cuba	No Indigenous Program	45 (missile, purchased)

* Under development

Notes: (1) All ranges are maximum, and assume missile is carrying standard payload for that missile system.
(2) The United Kingdom's and France's longest-range missiles are submarine based.
(3) Includes new Iranian and North Korean developments in 1998.

Source: Various sources as cited in Table in Lumpe, Lora, Gronlund, Lisbeth and Wright, David. C., "Third World Missiles Fall Short," *Bulletin of the Atomic Scientists* (March 1992), p. 32; also Erlanger, Steven, "Washington Casts Weary Eye at [Iranian] Missile Test," *New York Times* (July 24, 1998); and WuDunn, Sheryl, "North Korea Fires Missile Over Japanese Territory," *New York Times* (September 1, 1998).

THE PROBLEM AND THE SOLUTION

It was not until the twentieth century that our technological development finally put into our own hands the capacity to do catastrophic damage to ourselves and to the web of life on earth. The weapons of mass destruction which we so diligently created and still so assiduously work to refine have spread around the planet, exposing us to grave and perhaps terminal danger. On every major continent, nuclear power threatens us with the chronic problem of hazardous waste, the possibility of encouraging the proliferation of nuclear weapons and the catastrophic potential of accident.

The explosion of chemical technologies has injected a vast array of newly created chemicals into the world, many of them toxic. They came so quickly and on such a scale that the natural environment had no chance to find ways of breaking down and recycling them. Now more and more places all over the globe have become contaminated. Some have even become uninhabitable. And always there is the danger that a terrible accident will convert this chronic threat into an acute disaster. It has happened before. It will happen again.

We may be on the verge of generating a similar problem today through the rapid advance of biotechnologies. Introducing more and more novel, genetically engineered biological organisms into an unprepared natural world may pose an even greater threat. If we want to survive, let alone prosper, sooner or later we are going to have to face up to the limits that human imperfection unavoidably puts on our ability to control the products of our technological brilliance. Sooner would be much better than later.

Much of this book is an attempt to look at the many ways human fallibility and the characteristics of modern technology can combine to make things go not just wrong, but catastrophically wrong. To make intelligent decisions and realize the brighter future to which we all aspire, we must confront the downside of the technologies we choose—no matter how frightening—and not simply revel in the benefits they appear to provide. Otherwise, our natural tendency to overemphasize the advantages and undercount the costs, to "buy now and pay later," will get us into very deep trouble some day.

Because we humans are the most capable species on earth, we are also the most dangerous. Our confidence that we can indefinitely avoid catastrophe while continuing to develop and use powerful and dangerous technologies is a lethal arrogance. We cannot keep winning a technological game of "chicken" with nature. We are not perfect, and we are not perfectible. We are people, not gods. If we do not always remember the fallibility that is an inherent part of our nature, we will eventually do devastating and perhaps terminal harm to ourselves and to the other species that keep us company on this lovely, blue-green planet.

PART II

What Could Happen?

CHAPTER TWO

Terrorism and Dangerous Technologies

Eighteen minutes after noon on February 26, 1993, a blast shook the twin 110-story towers of New York's World Trade Center with the force of a small earthquake. A van loaded with some 1,200 pounds of powerful explosive had blown up in the underground garage. Walls and floors collapsed, fires began to burn and smoke poured into hallways and stairwells darkened by the loss of power. Dozens were trapped for hours in many of the Center's 250 elevators caught between floors. An estimated 40,000 people in the hundreds of offices and miles of corridors of Manhattan's largest building complex had to find their way out amidst the smoke, darkness and confusion. It took some of them most of the day to escape. When it was all over, six people were dead, more than a thousand were injured and property damage was estimated at half a billion dollars.[1]

The bombing of the World Trade Center was the worst terrorist incident on American soil—until the morning of April 19, 1995. A rented truck packed with more than 4,000 pounds of explosive made of widely available fertilizers, chemicals and fuel sat parked by the Alfred P. Murrah Federal Building in Oklahoma City, Oklahoma. Just after 9:00 A.M., when the building's workers were at their jobs and the second floor day care center was filled with young children at play, the truck blew up with a deafening roar. Walls, ceilings and much of the building's north face came down in an avalanche of concrete, steel and glass. The blast left a crater 20 feet wide and 8 feet deep, overturned cars, damaged 6 nearby buildings, and set dozens of fires. Nearly 170 people were killed, including many of the children in the day-care center, and hundreds more were injured. A building that was much smaller than the World Trade Center, in a city a fraction of the size of New York, had sustained a terrorist attack that took nearly 30 times as many lives.[2] Terrorism had come to America's heartland.

As terrible as these attacks were, they are dwarfed by the mayhem that could be caused by a successful terrorist assault on a nuclear power plant, toxic chemical manufacturing facility or radioactive waste storage site. Worse yet, imagine the magnitude of disaster that could be unleashed by terrorists armed with weapons of mass destruction. Even a crude, inefficient, homemade nuclear weapon would have turned the World Trade Center into rubble and taken the lives of more than 40,000 people. A more efficient weapon could have leveled much of Oklahoma City.

THE NATURE OF TERRORISM

Not every form of violent, destructive, antisocial activity is terrorism. Nor is terrorism defined by the ultimate goals terrorists seek to achieve. Calling violent groups "terrorists" when we don't like their objectives and "freedom fighters" when we do is a political game. It won't help us understand what terrorism is, judge how likely terrorists are to use dangerous technologies, or figure out what can be done about it. Instead, we need a working definition that is more than just propaganda or opinion.

Terrorism can be defined by its tactics and strategy: it is violence or the threat of violence carried out with the express purpose of creating fear and alarm. When an armed gang shoots bank guards in order to steal money, that is a violent crime, not an act of terrorism. The violence is perpetrated to stop the guards from interfering with the theft, not to frighten the wider population. But when a gang randomly plants bombs on city buses, they are not trying to stop the passengers from interfering with them, they are trying to frighten people. Their acts are intended to have effects that reach well beyond the immediate damage they are causing or threatening to cause. Whether their objective is to force the government to release political prisoners or to extort a ransom, they are terrorists because they are trying to terrorize.[3] Unlike other criminals, terrorists usually try to draw attention to themselves, often claiming "credit" for the acts they have committed. In many ways, terrorism is a perverse form of theater in which terrorists play to an audience whose actions—and perhaps, opinions—they are trying to influence. When terrorists hijack an aircraft, they may be playing to an audience of corporate managers who can assemble a ransom, government officials who can order their imprisoned comrades released, or whoever else has the power to meet their demands. But they are also playing to the public, whose mere presence as well as opinions and actions can put pressure on those in power to do what the terrorists want done. Those actually taken hostage on the plane cannot meet the terrorist demands, any more than can those maimed when a pub is bombed or killed by a murderous spray of gunfire in a hotel lobby. Nor are they in any position to apply pressure to the people who have that power. They just happened to be in the wrong place at the wrong time. Innocent victims, they have become unwitting players caught up in a real-life drama, the cannon fodder of terrorism.

Terrorists are trying to make the public feel vulnerable, unsafe, helpless. In some cases, choosing victims at random is the best way to accomplish this. If there is no clear pattern as to which particular bus is blown up, which airliner hijacked, which building bombed, there is no obvious way to avoid becoming a victim. That is very frightening. On the other hand, if the terrorist objective is more targeted, choosing victims randomly but within broadly defined categories may be more effective. The mercury poisoning of Israeli oranges in Europe in the late 1970s was targeted randomly but only at consumers of Israeli produce. It was intended to damage Israel's economy by creating fear that their agricultural exports were unsafe.[4] Economic damage was also the goal of the terrorist who poisoned some of the Johnson & Johnson Company's painkilling Tylenol capsules with lethal cyanide in the early 1980s.[5] The targets of the Oklahoma City bombing were also neither purely random nor very specific, but were chosen to intimidate federal employees and users of federal services, to express broad ideological antipathy to the government.

In sum, acts intended to instill fear in the public, committed against more or less randomly chosen victims not themselves able to meet the attackers' demands, define terrorism and set it apart from many other forms of violence. Bombing the barracks of an occupying military force is an act of war, violent and murderous, but it is not an act of terrorism. It attacks those who are directly involved in the activity the attackers are trying to oppose, not randomly chosen innocent victims. The act of a habitual sex offender in kidnapping, raping and murdering a more or less randomly chosen innocent victim is a vicious and brutal crime, but it is also not terrorism. Though it may well instill fear in the public, it is not done for that purpose, and it is not done to influence public opinion or behavior. Suicide bombing a city marketplace to precipitate a change in government policy is an act of terrorism. The more or less randomly chosen victims cannot directly change government behavior, but the indiscriminate slaughter is intended to shock and frighten people into demanding that the government change direction by convincing them that they will be in danger until those policies change. Whether or not the bombing achieves that objective, the act itself is still an act of terrorism.

It is important to emphasize that there is nothing in the definition of terrorism that prejudges the legitimacy or desirability of the terrorists' ultimate goals. Whether a group is trying to overthrow a legitimate democratic government and establish a rigid dictatorship, create a homeland for a long-disenfranchised people, trigger a race war, or get more food distributed to malnourished poor people, if the group uses terrorist means, it is a terrorist group.

Terrorism may be despicable, but it is not necessarily irrational. There are a variety of reasons why subnational groups with clearly political goals sometimes choose terrorist tactics to undermine support for the government and/or its policies. Domestic terrorist groups may believe that this is an effective way of convincing the public that the government does not deserve their support because it cannot keep them safe. Or they may believe that provoking widespread and repressive counterterrorist measures will turn the public against the government by exposing just how brutal and overbearing it can be. As paradoxical as it may seem, terrorist groups clearly believe that the end result of their terrible random acts of violence will be an uprising of the public against the government and increased support for the group's political agenda.

Since international terrorists attack only foreigners and their property and usually claim to be the avenging arm of their oppressed brothers and sisters, it is easier to understand why they might believe that their brutal actions will build public support for their cause at home. They may also see international terrorism as the only way to shock the world into paying attention to the plight of their people. For subnational groups, it is certainly true that terrorism is the "weapon of the weak." A powerful and influential group would not need to resort to such desperate and horrible tactics to make itself heard.

There is a tendency to think of terrorists as either small, disconnected groups of half-crazy extremists or expert paramilitary cadres bound tightly together in grand international conspiracies. In fact, the reality most often lies in between. It is true that terrorist groups do cooperate across ideological and political boundaries, sometimes even "subcontracting" with each other or carrying out joint attacks. But these coalitions are typically loose and transitory. Japanese Red Army terrorists, for example, carried out a grenade and rifle attack planned by the Popular Front for the Liberation of Palestine (PFLP) against a crowd of 250 people at Lod Airport near Tel Aviv in 1972, killing 27 and wounding 80 more.[6] Yet this was not a stable, tightly organized alliance. By the mid 1990s, white supremacist, antisemitic, paramilitary groups in the United States had established extensive networks of communication, cooperation and support, but they had by no means coalesced into a single, centrally controlled organization.[7]

Governments and Terrorism

Terrorism is not only a tactic of subnational groups. Governments can, and all too often do, carry out terrorist acts. In fact, the term "terrorist" appears to have first been applied to the activities of a government, the Jacobin government of France after the French Revolution.[8] The Nazi Gestapo, Iranian Savak and many other "secret police" organizations in many other countries have deliberately terrorized the population to suppress opposition and force the public to submit to the edicts of brutal governments. Because of the resources at their command, when governments engage in terrorism, their actions are often far more terrifying than the acts of subnational terrorists.

Governments have sometimes carried out official campaigns of terrorism. The Ethiopian government launched what it called the "red terror" in reaction to a revolutionary group's "white terror" campaign in the late 1970s. Within two months, more than a thousand people were killed, many of them teenagers. Their dead bodies were displayed in public squares with signs hung on them saying "The red terror must crush the white terror."[9] In 1998, testimony before South Africa's Truth and Reconciliation Commission revealed an apartheid-era campaign of chemical and biological attacks intended to murder political opponents.[10]

State-Sponsored Terrorism

The term "state-sponsored terrorism" has at times been used too loosely to brand the activities of governments with which we disagree. But there is a reality beyond the name-calling. Governments sometimes do directly aid subnational groups that stage attacks in the homelands or against the interests of opposing governments, groups that are "terrorist" by any reasonable definition of the term. Governments have provided safe havens, intelligence, weapons and even training. On the specious theory that "the enemy of my enemy is my friend," the United States, Saudi Arabia and others flooded Afghani revolutionary forces with weapons and money after the Soviet military intervened in support of the Afghan government in 1979. Much of it went to groups in Peshawar, the capital of Pakistan's Northwest Frontier province bordering Afghanistan. Long a violent area, the money and guns helped Peshawar descend deeper into lawlessness. According to a high-ranking officer, the Pakistani military was directly involved in training something like 25,000 foreign volunteers to fight with the Afghan guerillas. Most were Arabs, but there were also Europeans, Asians and some Americans. After the Soviets withdrew from Afghanistan in 1989, a large number of the foreign volunteers who survived the war stayed in and around Peshawar, working with organizations that have been accused of being fronts for international terrorist groups.[11]

In 1995, Pakistani police officials had the University of Dawat and Jihad in Peshawar under investigation. The officials claimed that the university may actually have been the training ground for terrorists responsible for attacks in the Philippines, Central Asia, the Middle East, North Africa and possibly North America. Ramzi Ahmed Yousef, convicted of both the 1993 bombing of New York's World Trade Center and a 1995 plot to blow up a dozen American airliners in East Asia, had used Peshawar as a base.[12] Several of the eight men arrested by the FBI later in 1993 and accused of plotting to blow up car bombs at UN headquarters, the Lincoln and Holland Tunnels and several other sites in New York City were involved in the Afghan War.[13] So too was Osama bin Laden, the Saudi businessman widely touted by the United States as a financier of anti-American terrorists in the 1990s. In the words of one senior Pakistani official, "Don't forget, the whole world opened its arms to these people. They were welcomed here as fighters for a noble cause, with no questions asked. . . . [N]obody thought to ask them: when the Afghan Jihad is over, are you going to get involved in terrorism in Pakistan? Are you going to bomb the World Trade Center?"[14]

Nuclear deterrence, a mainstay of the official security policy of the nuclear weapons states, is itself a form of international terrorism. Nuclear deterrence does not so much threaten to annihilate the leaders of opposing governments—those with the power to make decisions of war or peace—as it holds hostage and threatens to destroy the ordinary people of the opposing nation if their government decides to attack. Even in democracies, the general public under threat is not in a position to control the decision of their government to launch a nuclear attack. No referendum has ever been planned for "button pushing" day. That is even more true of the public in authoritarian countries.

Furthermore, the underlying objective of threatening nuclear attack is precisely to create such widespread fear that the opposing government will feel enormous pressure to avoid any behavior that would result in the threat being carried out. This is terrorism, plain and simple. In fact, during the Cold War, the threat of "mutually assured destruction" was officially called a "balance of terror." And a balance of terror is still terror.

Though it does not legitimize or excuse the use of terrorist tactics by subnational groups, the fact is even democratic governments have provided them something of a model for this type of behavior.

THE TERRORIST THREAT OF MASS DESTRUCTION

Terrorists have not yet committed violence on anywhere near the scale that would result from a successful attack on a nuclear power plant, toxic chemical manufacturing facility or hazardous waste dump. They have not yet used a homemade, store-bought or stolen nuclear weapon. They have not yet contaminated the water supply of a city or the air supply of a major building with deadly chemicals or virulent bacteria. Why not?

If it is because they do not have and cannot develop the capability to use dangerous technologies as a weapon or a target, we can relegate these frightening scenarios to the realm of science fiction and breathe a collective sigh of relief. But if instead the capability to do such nightmarish damage is within their reach, it is important to know what is holding them back. Is it just a matter of time before this modern-day horror becomes real?

Because their actions seem so immoral, abhorrent and repulsive, we usually assume that terrorists will do whatever harm they are capable of doing. But though their methods are similar, not all terrorists are alike. Some may actively seek the capability for committing mayhem that dangerous technologies provide, while others have no desire to do that much damage. It would be very useful to know which is which. At best, that might help us formulate more effective strategies for preventing a terrorist-induced catastrophe. At least, we would know what kinds of groups need to be most closely watched.

Then there is the biggest question of all: whatever the reason terrorists have not yet committed such atrocities, is there any reason to believe this restraint will continue?

Can Terrorists "Go Nuclear"?

Despite high priority and lavish government funding, it took years for a collection of the most brilliant scientific minds of the twentieth century to develop the first nuclear weapon. Potent nerve gas weapons emerged from technically advanced laboratories run by teams of highly trained chemists. A great deal of engineering and scientific effort has gone into designing nuclear power plants. They are protected by layers of backup and control systems intended to make catastrophic failure very unlikely, whether by accident or sabotage. Are terrorists really sophisticated enough to get their hands on and successfully use dangerous technologies as weapons or as targets of their attacks?

The image of the terrorist as a demented fanatic who stashes a suitcase full of dynamite sticks wired to a crude timing device in some forgotten corner of a building is out of step with the times. Such crude forms of terrorism can still be very effective. But terrorists and criminals using terrorist tactics have long since shown themselves capable of much greater tactical and technological sophistication. In August 1980, a box as big as a desk was delivered to the executive offices of Harvey's Casino in Stateline, Nevada. A three-page extortion note sent to the management warned that the box contained a bomb that would explode if any attempt was made to move it. Not knowing whether the threat was real, casino managers called in bomb experts from the FBI, the Army bomb disposal team and the U.S. Department of Energy (the agency in charge of nuclear weapons research and manufacture). They examined the plastic covered box carefully. X-rays revealed that it contained 1,100 pounds of explosives. The experts were struck by the highly sophisticated design of the device, but nevertheless believed they could safely disarm it. They were wrong. The bomb exploded, doing $12 million worth of damage to the casino.[15]

If there was any lingering doubt about the possibility that technologically advanced subnational terrorist groups could arise, it was dispelled in the mid1990s by the emergence of the Japanese doomsday cult Aum Shinrikyo. Nearly a dozen of the sect's top leaders were educated in science and engineering at top Japanese universities, as were some other members. When the police arrested members of the cult and accused them of using sarin nerve gas in a March 1995 attack on the Tokyo subways, they found hidden laboratories at the cult's compounds that could manufacture the gas. They charged that Aum also had facilities capable of producing biological warfare agents. Furthermore, Japanese police reportedly suspected that the purpose of the 1993 visit of a high cult official to Australia was to obtain uranium to be used in building nuclear weapons.[16]

As discussed in chapter one, thousands of people have been trained in designing nuclear weapons in the United States, the former Soviet Union, Britain, France and China over the past 50 years. These are people of widely differing political, ideological and religious views, personalities and life circumstances. Many of Russia's nuclear scientists are now living in such economic deprivation that the United States allocated $30 million in 1999 to help create nonmilitary jobs for them in the hope of discouraging them from selling their expertise to rogue nations or terrorists.[17] Can we really be sure that no terrorist political group or religious cult will ever be able to recruit, coerce or buy off any of these experienced weapons designers?

Unfortunately, the degree of technical sophistication required to acquire or use dangerous technologies as weapons is actually much lower than many people think. Poison gas can be made with the chemicals most of us have around the house. As long ago as 1977, a British military research laboratory was openly advertising the sale of infectious organisms at bargain-basement prices, including three strains of *Escherichia coli*.[18] That is the same bacterium that was responsible for a mysterious epidemic of food poisoning that shut down the entire Japanese school system during the summer of 1996. More than 9,400 people were sickened and 10 died.[19]

In March 1995, four members of the right-wing Minnesota Patriots Council were convicted in federal court of conspiracy to use ricin, a deadly biological toxin, to kill federal agents. They had manufactured enough to kill 1,400 people using information in a manual they bought from a mail order house.[20] Two months later, a member of the American white supremacist group Aryan Nations was arrested for (and subsequently pled guilty to) making another mail-order purchase—three vials of frozen bubonic plague bacteria—obtained using false credentials from the food-testing laboratory at which he worked.[21] What about nuclear weapons?

Designing a Terrorist Nuclear Weapon

The "secret" of designing nuclear weapons is out, accessible in the public literature to anyone moderately well trained in the physical sciences or engineering. More than 20 years ago, the Public Broadcasting System's *NOVA* science television series recruited a 20-year-old chemistry student at the Massachusetts Institute of Technology and gave him the assignment of designing a workable atomic bomb. He was required to work alone, without any expert assistance, and to use only publicly available information. He began by simply looking up references in the college science library. In the student's words,

> the hard data for how big the plutonium core should be and how much TNT I needed to use I got from Los Alamos reference books [purchased from the National Technical Information Service in Washington for about $5 each] and also other reference books I checked out of the library.
>
> . . . I was pretty surprised about how easy it is to design a bomb. When I was working on my design, I kept thinking there's got to be more to it than this, but actually there isn't.[22]

Only five weeks later, the student's fully documented, detailed report was given to a Swedish nuclear weapons expert for evaluation. The verdict: a fair chance that a bomb built to this design would go off, though the explosion would probably be no more than the equivalent of 1,000 tons of TNT, more likely less than 100.

The design was crude and unreliable, the yield unpredictable and small by nuclear standards—it would not be an acceptable military weapon. But none of these deficiencies is much of a problem for terrorists used to unreliable bombs with unpredictable yields. An explosion equivalent to even 50 tons of TNT would be gigantic by terrorist standards. That is 25 times as powerful as the explosives used in the Oklahoma City and World Trade Center bombings. Imagine what would have happened had those blasts been 25 times as powerful. Then add the death and destruction that would have been caused by the enormous release of heat and radiation from this crude nuclear weapon—designed by one undergraduate student in less than two months.

One year later, a senior at Princeton University duplicated this design feat, and then some. Working from publicly available sources and taking more time, he designed a 125-pound device about the size of a beach ball, which he estimated would explode with the force of about 5,000 tons of TNT. A specialist in nuclear explosives engineering reviewed the student's 34-page term paper, declaring that the bomb design was "pretty much guaranteed to work."[23]

In April 1979, FBI Director William Webster said that sufficient information was available in public libraries to design a nuclear weapon small enough to be carried on a terrorist's back.[24] A few months later, the *Progressive* magazine published an article called "The H-Bomb Secret."[25] The author, a journalist named Howard Morland, had unearthed enough detail about the design of the much more powerful hydrogen bomb that the U.S. government took Morland and the *Progressive* to court to prevent the article's publication on grounds of national security. After a long battle, the court refused to enjoin its publication because all of the information it contained was shown to be available in sources that had already been public for years. This included a report giving precise specifications for the key hydrogen bomb trigger mechanism and other important design details. It had been on the public shelves in Los Alamos since 1975. The Department of Energy argued that they should never have declassified the report and made it available. Why did they? It was simply a matter of human error.[26]

Building a "Homemade" Nuclear Bomb

Skeptics often argue that designing a weapon on paper may not be all that difficult, but actually building a nuclear bomb would require large teams of people with advanced skills and access to materials and equipment that is expensive and difficult to come by. Ted Taylor, a noted physicist and the Los Alamos nuclear weapons designer credited with the most efficient A-bomb ever designed, disagrees. In his view,

> Under conceivable circumstances, a few persons, possibly even one person working alone, who possessed about ten kilograms of plutonium oxide and a

substantial amount of chemical high explosive could, within several weeks, design and build a crude fission bomb . . . that would have an excellent chance of exploding, and would probably explode with the power of at least 100 tons of chemical high explosive. This could be done using materials and equipment that could be purchased at a hardware store and from commercial suppliers of scientific equipment for student laboratories.

The key person or persons would have to be reasonably inventive and adept at using laboratory equipment and tools of about the same complexity as those used by students in chemistry and physics laboratories and machine shops.[27]

The M.I.T. undergraduate who designed the workable A-bomb for *NOVA* estimated that if he had the plutonium, he could actually build the bomb from scratch in a year or less, with the help of three to four people and no more than $30,000 for supplies purchased from ordinary commercial sources. The finished product would be about as big as a desk, and weigh 550 to 1000 pounds.[28] In other words, it would be roughly the size and weight of the box terrorists delivered to Harvey's Casino in Nevada five years later. A 1980 article published in the British magazine *New Scientist* gave new meaning to the phrase "homemade bomb." It outlined a method for turning an ordinary two story house (with basement) into an atomic bomb! Aside from the 30 to 37 pounds of nuclear material required, the plan only called for about 20 feet of black iron pipe, 2 sticks of dynamite, 15 sacks of cement, 20 cubic yards of sand and gravel and a few easily obtainable miscellaneous bits and pieces. According to the author, such a bomb "detonated in New York City, ought to kill perhaps 250,000 people and injure another 400,000 . . . more than adequate for the average terrorist."[29]

Are the Necessary Nuclear Materials Available?

There are about a thousand metric tons of plutonium contained in stored spent fuel from nuclear power plants. That is four times as much plutonium as has been used in making all of the world's nuclear weapons.[30] For a long time now, the public has been told that nuclear weapons cannot be built from this "reactor grade" plutonium without an expensive and technically complex refinement process: it is much too heavily contaminated with plutonium-240 to be usable for a weapon. Not only is that untrue, it has definitely been known to be untrue for more than 30 years.

In 1962, the U.S. government assembled a nuclear bomb from the kind of low-grade, contaminated plutonium typically produced by civilian nuclear power plants, and brought it to the Nevada desert for secret testing. It blew up, producing "a nuclear yield."[31] Fourteen years later, a study done at the Lawrence Livermore nuclear weapons lab in California came to the conclusion "that the distinction between military and civilian plutonium was essentially false—that even relatively simple designs using any grade of plutonium could produce 'effective, highly powerful' weapons with an explosive yield equivalent to between 1,000 and 20,000 tons of TNT."[32]

Is it also possible to use uranium taken directly from nuclear power reactors? Natural uranium contains only 0.7 percent U-235, much too low a concentration to sustain a chain reaction in weapons. Although some nuclear power reactors use natural uranium, most use uranium enriched to higher levels of U-235. Uranium enriched to more than 90 percent is an excellent nuclear explosive, and is typical military weapons grade. But according to Ted Taylor, "It is probable that some kind of fission explosive with a yield equivalent to at least a few tens of tons of high explosive could be made with metallic uranium at *any enrichment level significantly above 10 percent*" (emphasis added).[33]

There are civilian nuclear power reactors fueled by 90 percent enriched uranium, but most use uranium less than 10 percent enriched. Military nuclear power reactors may use more enriched uranium. For example, most of the Russian Navy uses uranium fuel enriched to 20 to 45 percent.[34] Thus, the uranium used as fuel in many, but not all, nuclear power reactors could be used to make bombs without further enrichment. Lower enriched reactor fuel can still be used, but it would have to be processed to raise its concentration of U-235.

In 1996, *Time* reported that 17 scientists at the Los Alamos nuclear weapons laboratory were given the assignment of trying to design and build terrorist-type nuclear weapons using "technology found on the shelves of Radio Shack and the type of nuclear fuel sold on the black market." They successfully assembled more than a dozen "homemade" nuclear bombs.[35]

If terrorists were willing to settle for a device that dispersed deadly radiation without a nuclear blast, they would have a much wider variety of designs and nuclear materials from which to choose. The skills and equipment needed to build a dispersal device are also much simpler than those required for building a bomb.

Plutonium is so toxic that even a few grams would be enough for a radiological weapon. Dispersed as an aerosol in the ventilating system of a major office building, a few grams would pose a deadly threat to thousands of people. It wouldn't be an effective military weapon, but it would be nearly ideal for terrorists. Many other radioactive materials could also be used. Biological or chemical weapons dispersal devices would be even simpler. By one estimate, terrorist biological weapons might be developed at a cost of $100,000 or less, "require five biologists, and take just a few weeks, using equipment that is readily available almost anywhere in the world."[36]

Getting Access to Nuclear Explosives

The problems of record keeping and protection of inventories detailed in the next chapter raise the possibility that it may not be as difficult as one might hope for terrorists to get their hands on the "any grade of plutonium" or the "uranium at any enrichment level significantly above 10 percent" that they would need to build a crude nuclear weapon. Using a conservative estimate of error in best practice U.S. plutonium record keeping and a conservative estimate of the size of plutonium inventories worldwide, enough plutonium could have been diverted to make more than 200 crude nuclear weapons (or deliver millions of lethal doses in dispersal devices) without the record keeping system ever noticing that anything was missing.

More than 200 pounds of highly enriched uranium disappeared under suspicious circumstances from the privately owned Nuclear Materials and Equipment Corporation (NUMEC) plant in Apollo, Pennsylvania in the 1960s. Two decades later, during exercises run by the U.S. Department of Energy, "mock" terrorists successfully stole plutonium from the Pantex nuclear weapons plant in Texas and the Savannah River Plant in South Carolina. Both Savannah River and Pantex are facilities at which protection was presumably a very high priority, since they not only handle plutonium, but are involved in assembling and disassembling nuclear weapons. The U.S. Nuclear Regulatory Commission (NRC) also tested the ability of armed terrorists to penetrate commercial nuclear power facilities. Serious security problems were discovered at almost half of the nation's nuclear plants. In at least one mock attack, "terrorists" were able to sabotage enough equipment to cause a core meltdown. Yet in 1998, the security testing program was eliminated in the interests of cutting costs.[37]

Then there is the problem of safeguarding inventories of plutonium, enriched uranium and nuclear weapons in the former Soviet Union. The Soviet system of record keeping was so poor that even by 1996, Russia still did not have accurate records of the quantity, distribution and status of nuclear materials at many of the 40-50 nuclear locations and 1,500-2,000 specific nuclear areas throughout the former Soviet Union.[38] Economic and political turmoil in Kazakhstan, Ukraine and Russia have greatly increased the chances of nuclear theft. We know that enriched uranium was stolen from Russian facilities at Podolsk and Sevmorput in the 1990s. And persistent reports of "holes in the fences" and other poor security practices in Russia do not bode well for the future.

There is no doubt that conventional weapons, explosives and all manner of related goods stolen from Russian and American military arsenals have found their way into the hands of terrorists around the world. There is also little doubt that a black market in nuclear materials now exists, most likely fed in part by criminal sources in the former Soviet Union. Between 1991 and 1994, the German police recorded more than 700 cases of nuclear smuggling (though at least some of these were hoaxes). Speaking in April 1996, CIA Director John Deutch put it this way: "The chilling reality is that nuclear materials and technologies are more accessible now than at any other time in history."[39]

Once "nuclear materials and technologies" are acquired, they can be smuggled across international borders. In 1985, a West German businessman was convicted of illegally shipping an entire $6 million nuclear processing plant to Pakistan—along with a team of West German engineers to supervise its construction![40]

U.S. borders are so porous that it would be relatively easy to bring nuclear materials into the United States that had been stolen from poorly guarded nuclear facilities elsewhere. It wouldn't be all that difficult to bring stolen or homemade nuclear weapons or components into the United States either. In early 1996, *Time* reported that "U.S. intelligence officials admit that a terrorist would have no more difficulty slipping a nuclear device into the U.S. than a drug trafficker has in bringing in bulk loads of cocaine."[41] Decades earlier, special forces teams actually tried to smuggle simulated nuclear bombs into the United States dozens of times to see if it could be done. They carried the dummy weapons across the borders in trucks, small planes and boats. None of them were ever intercepted.[42]

Some Incidents

There is a long history of nuclear threats and related plots by terrorists and criminals.

- In 1978, the FBI arrested two men after they tried to recruit an undercover agent to take part in a plot to steal an American nuclear submarine. The men had showed him their plan to use a gang of 12 to murder the crew of the nuclear submarine USS *Trepang*, sail it from its dock in New London, Connecticut, to the mid Atlantic and turn it over to a buyer they did not identify. The plan included the option of using one of the ship's nuclear missiles to destroy an East Coast city as a diversion to help with the getaway.[43]
- Hours after Uganda's megalomaniacal dictator Idi Amin was overthrown in June 1979, top secret documents were discovered that allegedly revealed a plot Amin was preparing. Nuclear weapons small enough to fit into suitcases were to be built and carried into the nation's embassies around the world by teams of Ugandan diplomats, possibly for purposes

of nuclear blackmail. It was further reported that Amin was actively seeking terrorist help and expertise to carry out this grotesque plan.[44]

- On April 1, 1985, New York City officials received an anonymous letter threatening to contaminate the city's water supply with deadly plutonium trichloride unless all charges against Bernhard Goetz were dropped. Goetz was accused shooting four young black men in a subway confrontation. The charges were not dropped. Two and a half weeks later, tests of the city's water showed levels of plutonium 35 times greater than normal.[45]

- In December 1994, a Czech scientist named Jaroslav Vagner and two colleagues were arrested for nuclear smuggling when Prague police found nearly six pounds of weapons grade uranium in the back seat of his car. Eighteen months later, interviews with Czech police and newly released documents revealed that allies of the three conspirators had threatened to explode a nuclear weapon at an unspecified Prague hotel unless the prisoners were released. According to the Czech police detective handling the investigation, "It is possible they have the nuclear material to do it. We found out they were planning to bring out [of Russia] 40 kilos of uranium within several days and . . . one ton within several years."[46]

In an attempt to deal with the threat of nuclear terrorism, the United States put together the multiagency Nuclear Emergency Search Team (NEST) in 1975. NEST was set up to evaluate criminal or terrorist nuclear threats and if necessary, to conduct extensive high-tech bomb searches to find and disable nuclear devices. By early 1996, NEST's annual budget was up to $70 million. The team had evaluated 110 threats (including some nuclear emergencies not involving weapons) and mobilized to search or take other action 30 times. All but one of the threats were reportedly hoaxes.[47]

What are the chances that this specialized team could actually find a real terrorist nuclear device? In 1980, NEST's assistant director put it this way: "If you can cut it down to a few blocks, we have a chance. But if the message is to search Philadelphia, we might as well stay home."[48]

THREATENING DANGEROUS TECHNOLOGY FACILITIES

Terrorists would not have to steal or build weapons of mass destruction to wreak havoc. They could achieve much the same effect by sabotaging plants that produce or use large quantities of toxic chemicals, attacking nuclear or toxic chemical waste storage areas or triggering the catastrophic failure of nuclear power plants. Ordinary explosives or incendiary devices, placed in the right

TABLE 2-1

NUCLEAR SAFEGUARDS INCIDENTS BY CATEGORY, 1976-1994

CATEGORY	NUMBER OF EVENTS	PERCENT OF TOTAL
Bomb-Related	690	38.9%
Firearms-Related	540	30.4%
Tampering/Vandalism	120	6.8%
Intrusion	43	2.4%
Missing or Allegedly Stolen	29	1.6%
Arson	21	1.2%
Transport-Related	12	0.7%
Miscellaneous	321	18.1%
TOTAL	1776	100%

Notes: (1) The NRC's definitions for the categories are:
"Bomb-Related: Events concerned with explosives or incendiary devices"
"Firearms-Related: Events typically describe the discharge, discovery or loss of firearms"
"Tampering/Vandalism: Incidents of destruction or attempted destruction . . . which do not directly cause a radioactive release"
"Intrusion: Incidents of attempted or actual penetration of a facility's barriers or safeguards systems"
"Missing or Allegedly Stolen: Events in which safeguarded material was stolen, alleged to be stolen, discovered missing or found [includes material missing or stolen during transport]"
"Arson: Intentional acts involving incendiary materials resulting in damage to property"
"Transport-Related: Events . . . where safeguarded material was misrouted or involved in an accident"
(2). Even though there is some overlap in categories, each event was only included in a single category. There is thus no "double-counting."

Source: Operations Branch, Division of Fuel Safety and Safeguards, Office of Nuclear Material Safety and Safeguards, Safeguards Summary Event List (SSEL) (Washington D.C.: U.S. Nuclear Regulatory Commission, NUREG-0525, Vol. 2, Rev. 3; July 1995), pp. vii and A-7.

places and ignited at the right time, could cause more damage than that caused by the terrible accidents at Bhopal and Chernobyl. Many threats have been made against nuclear facilities over the years. NRC data on "safeguards events" involving nuclear materials, power plants and other facilities show nearly 1,800 threatening events during the 19 years from 1976 through 1994 (see Table 2-1).[49] Close to 700 events (almost 40 percent) were bomb related.

Like terrorist incidents in general, these events follow an uneven pattern over time, as shown in Table 2-2. It is good news that there were many fewer bomb-related incidents during the first half of the 1990s than in any earlier five-year period. But that is a slender reed on which to build up hope. Even in those five years there were 76 bomb-related incidents, an average of more than one a month. And the decline in bomb-related events was more than made up by a sharp rise in the number of other incidents. There does not seem to be any good reason to believe that credible threats against nuclear facilities will stop anytime soon.

TABLE 2-2

NUCLEAR SAFEGUARDS
INCIDENTS BY YEAR, 1976-1994

YEAR	TOTAL NUMBER OF INCIDENTS	REACTOR EVENTS ONLY	BOMB-RELATED INCIDENTS
1976	72	66	55
1977	34	29	28
1978	47	40	29
1979	118	111	94
1980	109	103	73
1981	74	70	48
1982	82	80	57
1983	56	54	39
1984	58	57	28
1985	67	58	23
1986	96	84	43
1987	95	91	35
1988	137	128	37
1989	148	142	25
1990	103	100	11
1991	152	150	32
1992	94	94	12
1993	116	116	12
1994	118	118	9
TOTAL	1776	1691	690

Source: Operations Branch, Division of Fuel Safety and Safeguards, Office of Nuclear Material Safety and Safeguards, Safeguards Summary Event List (SSEL) (Washington D.C.: U.S. Nuclear Regulatory Commission, NUREG-0525, Vol. 2, Rev. 3; July 1995), pp. A-1, A-2, and A-8.

Many of the safeguards events in these tables were only threats or hoaxes. Even so, they indicate the extent to which nuclear facilities are seen as vulnerable enough to make the threats and hoaxes credible. In any case, it is very clear from these data that nuclear facilities have been considered attractive targets by terrorists and criminals for some time now.

A Chernobyl by Design?

The same month that terrorists used a truck bomb to blow up the World Trade Center in Manhattan, a mental patient crashed his station wagon through a door at the infamous Three Mile Island nuclear power plant.[50] Coming so close together, these two events naturally raised the question of what would happen if a disturbed individual or terrorist gang attacked a nuclear plant with a truck bomb as powerful as the one that ripped apart six stories of the Trade Center. The devastating Oklahoma City bombing two years later and the deadly 1996 truck bombing that destroyed part of an American military housing complex in Saudi Arabia (despite barriers that kept the vehicle 35 yards away) further strengthened this concern.[51]

The reinforced containment structure that covers a nuclear power reactor is the last line of defense against the release of large amounts of radioactive material. Its presence at Three Mile Island and its absence at Chernobyl was one of the most important reasons why the Chernobyl accident did so much more damage. Early in a severe reactor accident, high pressure from the heat of fission products, gas generation and the like threaten the containment. At the same time, it is filled with many radioactive materials in vaporized and aerosol form, easily carried by the wind and easily inhaled. It is the worst possible time for the containment to fail. If a terrorist group were to trigger a major nuclear accident, perhaps by sabotaging the plant with the help of insiders, anything they could do to weaken the containment enough to make it fail would greatly increase the magnitude of the disaster. Coordinating sabotage of the plant with a vehicle bomb attack might just do the trick. Is such a scenario possible?

In 1984, Sandia National Laboratories completed a study of the truck bomb threat to nuclear facilities for the Nuclear Regulatory Commission. In April, NRC staff reported to the commissioners that "The results show that unacceptable damage to vital reactor systems could occur from a relatively small charge at close distances and also from larger but still reasonable size charges at large setback distances (greater than the protected area for most plants)."[52] The Sandia study concluded that nuclear facilities in the United States were vulnerable to terrorist truck bombs, and that putting a few barricades near the reactor building would not solve the problem. "Unacceptable damage to vital reactor systems" could be done by bombs detonated some distance away, in some cases possibly even off-site.[53]

Trucks are not the only vehicles that terrorists could use to attack a nuclear plant. In the early 1970s, Atomic Energy Commission Chairman James Schlessinger put it this way: "If one intends to crash a plane into a facility . . . there is, I suspect, little that can be done about the problem. . . . The nuclear plants that we are building today are designed carefully, to take the impact of, I believe, a 200,000 pound aircraft arriving at something on the order of 150 miles per hour. It will not take the impact of a larger aircraft."[54]

At least a few dozen nuclear plants still operating in the United States today have containments no stronger than that to which Schlessinger was referring. Quite a few jumbo jets have been hijacked by terrorists by now, and they are considerably heavier and faster than the aircraft whose impact those containments were designed to resist. A Boeing 747, for example, weighs more than 300,000 pounds and can travel at speeds over 500 miles per hour. But even a much smaller plane could seriously damage the containment if it were filled with explosives.

As to sabotaging a nuclear plant, documents written to help nuclear plant operators test their security have been publicly available since the 1970s. They would be very useful to a potential saboteur. One includes a computer program that could be used to determine the best route for a saboteur to follow. According to a former nuclear safeguards inspector, to use the program, "All you need is the values (dimensions) of the plants, some of which are available in the library."[55]

Then there is the *Barrier Penetration Database,* prepared by Brookhaven National Laboratory under contract to the NRC. It gives detailed information on the types of tools and explosives necessary to break through dozens of barriers, from chain-link fences to reinforced concrete walls. It includes other information important to planning an attack, such as the weight of the equipment required and how long it should take to break through each barrier.[56] A 1993 RAND study (sponsored by the Department of Energy) looked at 220 direct attacks by armed bands of guerrillas or terrorists that the researchers believed were analogous to the kinds of direct armed assault such groups might launch against American nuclear facilities. They found that the attackers were successful 74 percent of the time.[57]

Of course, sabotage would be easier with the help of insiders working at the facility. In a 1990 study, *Insider Crime: The Potential Threat to Nuclear Facilities,* RAND found that financial gain was the main motivation in the overwhelming majority of insider crimes they studied.[58] Terrorists might need to do little more than find a sufficiently money-motivated insider and pay him/her off. Guards themselves were responsible for more than 40 percent of the crimes against guarded targets. Their overall conclusion? "[N]o organization, no matter how ingeniously protected can operate without some trust in individuals on all levels. Beyond a certain point, security considerations in hiring, guarding, controlling, and checking people can become so cumbersome as to actually impede the operation of a facility. This creates a serious dilemma in the case of a nuclear facility. . . . total security can never be attained."[59]

It would be much simpler to attack a toxic chemical plant, radioactive or toxic chemical waste storage dump and the like than a nuclear power plant. If a successful terrorist assault on a nuclear plant is possible, terrorist assaults on other dangerous-technology facilities are even more likely to succeed.

Why Haven't Terrorists Yet Engaged in Acts of Mass Destruction?

Experts have warned about the dangers of nuclear terrorism and its equivalents at least since the 1960s. Dangerous technologies have already been used to do damage on a scale similar to conventional terrorism (such as in the 1995 Tokyo nerve gas attack). There have also been threats of mass destruction and hoaxes involving nuclear weapons. But as yet, there has been no publicly reported case of terrorists (or criminals) doing the kind of massive damage that could result from large-scale use of dangerous technologies as either a weapon or target. Why not?

It is clearly not because this kind of attack is beyond their capabilities. It cannot be because of a moral revulsion against taking innocent lives, since the taking of innocent lives is the terrorist's stock in trade. It is possible that terrorists might be inhibited by a belief that murder and destruction on a massive scale would invite ferocious retaliation. But decades of experience show that terrorists are willing to risk ferocious retaliation, and may even be trying to provoke it. Many who seek to retaliate against terrorists are already prepared to do them grievous, even deadly harm. Even in free societies, where the search for terrorists is complicated by constraints against hurting innocent people and forfeiting personal freedoms, terrorists are already pursued with dogged determination and severely punished (witness the death sentence meted out to Timothy McVeigh, convicted of the Oklahoma City bombing).

More likely, terrorists have simply not found acts of mass destruction necessary up to now. If acts of conventional terrorism still provoke enough fear to put pressure on decision makers to do what terrorists want done, there is no particular reason for them to go to the trouble, danger and expense of acquiring and using the means of mass destruction. If acts of conventional terrorism come to be more routine and so generate less shock and fear, terrorists may someday conclude that they must commit much greater violence to frighten people enough to achieve their objectives. They will find the tactics of mass destruction waiting in the wings.

Another possibility is that so far, terrorists have believed that committing acts of mass destruction would get in the way of achieving their objectives. The credibility of this explanation depends on what kind of group we are considering. Terrorist groups are not clones of each other. Understanding what makes them different is important to judging how likely it is that any particular group will use dangerous technologies. It is also the key to developing more effective countermeasures.

A Taxonomy of Terrorists

The first and perhaps most obvious distinction is between domestic groups and international terrorists. Germany's left-wing Red Army Faction, France's right-wing Federation for National European Action, America's white supremacist The Order, Spain's Basque separatist ETA and Peru's revolutionary Shining Path are examples of domestic terrorist groups active in the 1980s.[60] On the other hand, the Popular Front for the Liberation of Palestine, the Japanese Red Army, Hamas, the Jewish Defense League and the Armenian Secret Army are historical examples of internationally focused terrorist groups.[61] Some have straddled the boundary. The Irish Republican Army (IRA) carried out most of its attacks inside Northern Ireland (against both Irish and English targets), but it was also responsible for more than a few terrorist bombings in England.

Secondly, some terrorists have relatively well-defined, specific political goals, while the goals of others are much more vague, general, ideological and/or anarchic. The Palestine Liberation Organization (PLO) used terrorist tactics to raise public awareness of the plight of disenfranchised Palestinian Arabs and gain support for establishing an independent Palestinian state. Similarly, the goal of IRA terrorism was also clear, specific and political: to end British rule in Northern Ireland. By contrast, the Symbionese Liberation Army, made famous by its kidnapping of newspaper heiress Patricia Hearst, had only the most general, ideological, anticapitalist goals. The long terrorist career of the infamous Unabomber was aimed at promoting vague antitechnology goals. And the self-proclaimed goal of Shoko Asahara, the leader of the Japanese terrorist cult Aum Shinrikyo, was to "help souls on earth achieve 'ultimate freedom, ultimate happiness and ultimate joy.'"[62]

Asahara's statement leads naturally to a third, related characteristic that differentiates terrorists from each other: some are rational and some are not. There is a temptation and a desire to believe that anyone who would engage in brutal terrorist actions cannot be rational. But that is not true. Rationality is a matter of logic, not morality. It simply means that the tactics used are logically related to the goals being pursued. The German Nazis were vicious and their behavior despicable and profoundly immoral, but not illogical. At least in the short term, their tactics led them step by step toward their goals of conquest and control.

Nonrational behavior includes both irrationality (in the sense of craziness) and behavior that is not necessarily crazy, but is driven by something other than logic. Emotion, culture, tradition and religion are important nonrational drivers of human behavior. To refuse to do business with the lowest-cost supplier because that firm is run by someone who once caused you personal emotional pain may not be rational, but it is easy to understand and certainly not crazy. To sit for hours in the hot sun at a graduation ceremony wearing long, black robes appropriate to the much colder climate from which they came is not logical behavior, but neither is it a sign of mental imbalance. It is driven by culture and tradition. Similarly, belief in God and adherence to the ceremonies of a particular religion is not a matter of logic, it is a matter of faith. And while faith may not be logical, it is not irrational.

Whether or not terrorist tactics actually advanced the political goals of the PLO or the IRA, the decision to use them, though reprehensible, was still rational: terrorism might logically have led them where they wanted to go. It is difficult, however, to imagine by what logic the terrorist activities of Aum Shinrikyo can be related to their spiritual goal of helping people on Earth to attain "ultimate freedom, ultimate happiness and ultimate joy." Similarly, the brutal murders perpetrated in Los Angeles decades earlier by the "Manson Family" (see Chapter 8) were driven by Manson's psychotic fantasies, not some logical process of choosing among available tactics in the service of an achievable goal.

A fourth factor that differentiates terrorist groups is the degree of support for their underlying goals. Whether or not the public approves of their tactics, terrorist groups fighting for a popular cause behave differently than those whose *goals* are considered extremist and out of touch. Everything else being equal, when the cause is popular, it will be easier for the group to recruit members, raise funds and find hiding places for their weapons and for themselves. Those that are rational will cultivate this sponsorship, and take care when choosing their targets and tactics to avoid alienating their support-ers. They will try to avoid catching their potential allies in the web of innocent victims. Groups with little or no public support are less inhibited in their choice of victims. They may believe that acts of horrific violence are the only way to shock what they see as a complacent or submissive public into action. It is a safe bet that, despite their lack of support, most see themselves as the leading edge of a great movement, and think that once the public awakens, a mass of supporters will rise up and carry them to victory.

Finally, different terrorist groups face off against different kinds of opponents. Some see a very specific enemy, such as the top executives of a particular company (or industry) or the political leaders of a specific country. Others see a much larger enemy, such as all nonwhites or all non-Christians, or a vaguely defined group, such as the "international Jewish financial conspiracy." The nature of their perceived enemy affects the kinds of actions they take and the intensity of the violence they commit. The range of likely targets and the degree of violence tend to be greater when the enemy is "the federal government" or "the Jews" than when the target is the management of Exxon or the ruling party in Britain.

To summarize, the five distinguishing factors are:

- *geopolitical focus* (Are they domestic or international in focus and if international are they state sponsored?)
- *nature, specificity and achievability of goals* (Are their goals vague and ideological or specific and political?)
- *rationality* (Is their behavior driven by logic?)
- *public support for their goals* (How much public support is there for their goals as opposed to their tactics?)
- *size and character of their enemy* (Is their opponent a relatively small and specific group of decision makers or a much larger and more generalized class of people?)

Domestic terrorists with clearly defined and potentially achievable political objectives are most likely to see the terrorism of mass destruction as counterproductive. Because they see their terrorist acts as acts of resistance and rebellion that will eventually rally their silent, disempowered supporters to the cause, they must always balance the shock effect of the damage they do against the support they will lose if their violence becomes too extreme. Except in situations of complete desperation, groups of this type are almost certain to see in advance that acts of mass destruction would be disastrous tactical blunders.

International terrorist groups with clearly defined and potentially achievable political objectives are somewhat more likely to escalate the level of violence. If they are playing to a domestic audience and if most of the violence they do is outside the borders of their home country, they may feel that spectacular acts of destruction will not alienate and may even encourage those who support their cause at home. This might be especially true of a terrorist group attacking targets inside the borders of a nation whose military is occupying the terrorists' home country. However, if the terrorists are also trying to influence the wider international community to support their cause, acts of extraordinary violence are not likely to seem appealing.

Domestic or international terrorist groups whose objectives are much more general or ill formed and whose attitudes are much more nihilistic— such as doomsday religious cults, racist and ideological extremists, and nuts— are an entirely different story. They are less likely to be deterred by worries about alienating supporters and may find devastating of acts of mass destruction an appealing, unparalleled opportunity to exercise their power. Doomsday cults could even see such acts as a way of hastening the salvation they believe will follow the coming cataclysm. Groups of this sort are extremely dangerous.

With a doomsday philosophy and considerable scientific talent on board, Aum Shinrikyo is a good case in point. This well-financed cult was accused not only of committing an act of nerve gas terrorism, but of preparing for much more deadly and devastating uses of chemical, biological and even nuclear weapons. The powerful control exerted over Aum's members by its charismatic but apparently mentally disturbed founder combined with the group's nihilistic orientation and substantial resources is an almost perfect recipe for the terrorism of mass destruction.

By 1997, there were more than 200 right-wing, violence-oriented, white-supremacist militias in the United States, active in some 40 states.[63] The agendas of many of these groups are extremist by any reasonable definition of the word. Collectively calling themselves the "Patriot" movement, they picked up on the antigovernment rhetoric of mainstream political conservatives and distorted it to the extreme. Many Americans believe that some federal agencies are oversized, inefficient and sometimes abusive. But the rhetoric of the Patriots goes much farther. In the words of one analyst who has closely tracked the movement, "those in the Patriot movement are convinced that the government is evil. It is run by a secret regime ('The New World Order') that seeks to disarm American citizens and subjugate them to a totalitarian world government. Just about anything that can be described with the adjectives 'global,' 'international' or 'multicultural' is a Patriot menace."[64]

These groups are heavily armed, and not without financial resources. In 1995, 11 individuals associated with the Patriot group "We the People" were indicted on felony charges related to a scheme in which they allegedly collected almost $2 million from thousands of people they charged $300 each to be part of a phony class-action lawsuit against the federal government. A year before, the host of a popular Patriot radio program and founder of the Patriot group "For the People" took in more than $4 million.[65]

America's right-wing, extremist, paramilitary militias have many characteristics that make them prime candidates for the terrorist use of dangerous technologies. Some rank high on each of the five factors in the taxonomy of terrorist groups:

Factor 1: While they are domestic rather than international, they see themselves as fighters against a national government that has become the pawn of an international conspiracy ("The New World Order"). They believe the U.S. federal authorities constitute an "occupation government," and some think that no level of government higher than the county has any legitimacy. Thus, they see themselves more as fighters against a foreign government than as domestic revolutionaries.

Factor 2: Their goals are ideological, general and anarchic, rather than specific, limited and politically achievable.

Factor 3: They are not particularly rational. Living in a paranoid world of conspiracies, they are motivated by nonrational beliefs often clothed in the garb of some form of end-time religion. Some have been tied to the violence-oriented, postmillenial Christian Identity movement.

Factor 4: Though there is considerable cynicism about the federal government among many Americans, there is very little support for the nihilistic goals of the Patriot movement.

Factor 5: Rather than having a very specific and limited enemy, many of the Patriot militias consider all nonwhites, non-Christians and many mainstream Christians as well to be the enemy.

After studying a wide range of terrorist groups for the Department of Energy in 1986, RAND concluded, "Of the terrorist organizations active in this country, the right-wing extremists appear to pose the most serious threat to U.S. nuclear weapons facilities."[66]

To date, the lack of unity and coordination among the extremist militia has limited the scope of their terrorist efforts. There is, however, growing cross-fertilization among them. They count well-trained American military personnel (including at least some former Green Berets) among their members and are always trying to recruit more. They have not yet been able to attract many highly educated and technically skilled people to the movement, but the movement is relatively new and does seem to have significant financial resources. There is no reason to believe that the scope of their terrorist activities will stay limited indefinitely. These "Patriots" are prime candidates for becoming nuclear terrorists.

IS THE TERRORIST THREAT GROWING?

There are some useful data available on the frequency and severity of terrorist attacks and credible threats of attack, but they are limited and flawed. Because the "terrorist" label is so politically loaded, governments and other organizations that gather and publish such data often use different definitions of what is a terrorist incident. Beyond this, incidents that involve grave potential danger—such as threats against nuclear power plants, nuclear weapons facilities or radioactive waste

TABLE 2-3

INCIDENTS OF INTERNATIONAL TERRORISM*

YEAR	NUMBER OF INCIDENTS**	FATALITIES†	INJURIES	TOTAL CASUALTIES††
1968	120	20		
1969	200	10		
1970	300	80		
1971	280	20		
1972	550	140		
1973	350	100		
1974	400	250		
1975	345	190		
1976	457	200		
1977	419	150		
1978	530	250		
1979	434	120		
1980	499	150		
1981	489	380		
1982	487	221	840	1061
1983	497	720	963	1683
1984	565			1100
1985	635	825	1217	2042
1986	612	576	1708	2284
1987	665	633	2272	2905

storage areas—tend to be covered up, on the belief that making them public would spread undue fear and thus give the terrorists a partial victory. Clearly, such incidents are also kept secret to create a false sense of security so that the public will continue to support particular government institutions or policies.

It is easier to keep the attacks quiet when they are carried out by domestic terrorists than when international terrorists are involved. The incentives are also greater: the frequency and severity of domestic terrorist incidents are measures of internal opposition and political turmoil, something no government likes to

TABLE 2-3 CONTINUED

YEAR	NUMBER OF INCIDENTS**	FATALITIES[†]	INJURIES	TOTAL CASUALTIES[††]
1988	605	658	1131	1789
1989	375	407	427	834
1990	437	200	677	877
1991	565	102	242	344
1992	363	93	636	729
1993	431	109	1393	1502
1994	321	315	663	977
1995	440	165	6291	6456
1996	296	314	2912	3226
1997	304	221	693	914
TOTAL	12,971	7,619	22,065	28,723[††]
	(OVER 30 YRS)	(OVER 29 YRS)	(OVER 15 YRS)	(OVER 16 YRS)
AVERAGE	432/YR	263/YR	1471/YR	1795/YR[††]

* The U.S. State Department excluded intra-Palestinian violence beginning with 1984, apparently because it was considered to be domestic rather than international terrorism. Such terrorism had previously been included because the "statelessness" of the Palestinian people made the violence seem inherently international.

** Data for 1968-74 were extracted from a bar graph that was not specifically numbered, so they should be considered approximations, rounded to the nearest ten. The underlying data were not published.

† Data for 1968-81 were extracted from a line graph that was not specifically numbered, so they should be considered approximations, rounded to the nearest ten. The underlying data were not published.

†† Includes only years for which data on the sum of both fatalities and injuries were available. Therefore, the total of this column is less than the sum of the totals of the separate columns for fatalities and injuries.

Sources: Data on number of incidents for 1975-94 from the Office of the Coordinator for Counterterrorism, U.S. Department of State, Patterns of Global Terrorism, 1994, p. 65, 1995 data from the 1995 edition, p. 4, 1996 and 1997 data from the 1997 edition; data on number of incidents for 1968-74 from Office of Counterterrorism, U.S. Department of State (unpublished) as cited in Kegley, Charles W. Jr, International Terrorism: Characteristics, Causes, Controls (New York: St. Martin's Press, 1990), p. 15.
Data on fatalities for 1968-83 and injuries for 1982-83 from Cordes, Bonnie, et al., Trends in International Terrorism, 1982 and 1983 (Santa Monica, California: Rand Corporation, August 1984), pp. 6-7; these data were apparently based U.S. State Department data. Data for fatalities and injuries for 1985-97 from Patterns of Global Terrorism annual editions for 1985-97; approximate total casualties data for 1984 were from the 1992 edition, p. 60.

publicize. As a result, while all data on terrorism must be treated with caution, data on international terrorism are likely to be more reliable (see Table 2-3).

According to the U.S. Department of State, there were nearly 13,000 separate incidents of international terrorism from 1968 through 1997—an average of more than 430 per year for three decades. The average number of terrorist incidents per year rose from 160 in the late 1960s to more than 400 during the 1970s, peaking at more than 540 in the 1980s. With the end of Cold War rivalries and the beginning of serious ongoing peace negotiations in the

Middle East, the first eight years of the 1990s saw a significant drop in the average number of terrorist incidents per year to just under 400. Still, there were *almost two and a half times* as many incidents per year on average in the period 1990-97 as there had been in 1968-69.

International terrorism has also become a more deadly business over the years. The average number of people killed each year skyrocketed from 15 in 1968-69 to 150 during the 1970s, then grew rapidly to more than 500 in the 1980s. Fatalities dropped even more sharply than incidents in 1990-97, falling to about 190 per year. Even so, despite the Middle East peace process and the end of the Cold War, the average number of people killed each year by international terrorists was more than 25 percent higher in 1990-97 than it had been in the 1970s, and almost 13 times as high as it had been in the late 1960s.

These data are far from perfect, but they would have to be very far off the mark to overturn the basic picture they paint. International terrorism is, sad to say, still going strong. It shows no signs of fading away anytime soon.

In the United States, we have become accustomed to thinking of international rather than domestic terrorism as the gravest threat to life and limb. Yet fewer people were killed worldwide in 1995 in all 440 international terrorist incidents than died that year in the single domestic terrorist bombing in Oklahoma City.

Terrorists have not yet used dangerous technologies to do catastrophic damage, as weapons or as targets. But there is nothing inherent in the nature of terrorism that makes it self-limiting. Those who are ready, even eager, to die for their cause, who stand willing to abandon every constraint of civilized behavior and moral decency against the slaughter of innocents, cannot be expected to permanently observe some artificial restriction on the amount of havoc they wreak.

Many terrorists are political rebels, in the classic sense. They have chosen reprehensible methods to fight for rational, limited, clearly defined objectives. But others are striking out against a vast array of faceless enemies, out of touch with reality and trying to punish, even to destroy a world they cannot live in and do not understand.

CHAPTER THREE

Controlling Dangerous Inventories

The proliferation of dangerous technologies has spread vast stockpiles of nuclear weapons, radioactive materials, toxic chemicals and the like all over the globe. At the same time, there has been an enormous growth in the organization and sophistication of organized criminals and terrorists, the two groups that most find these inventories a tempting target. It is therefore urgent that strict control be exerted over dangerous inventories at all time and in all places. They must both be protected against theft or damage by outsiders and safeguarded against damage or unauthorized use by insiders. The fences, walls, vaults, locks and guards that impede the entry of outsiders won't stop insiders who know protection systems well enough to circumvent them. And there are always some insiders authorized to get past all the protective barriers. Preventing unauthorized use by authorized personnel requires record keeping systems designed to accurately track the location and condition of everything in the inventory, and warn of anomalies quickly enough to be useful.

"Protection" and "detection" are problems common to all inventories, whether they contain candy bars, shoes, diamonds, gold, toxic waste, nerve gas or nuclear weapons. The systems designed to solve these problems also have a lot in common, including, unfortunately, their inherent limitations. Because there are such strong incentives to track and protect stockpiles of fissile material and weapons of mass destruction as completely as possible, looking at the systems designed to control them should help us understand the theoretical and practical limits of our ability to control inventories of dangerous material in general.

KEEPING TRACK OF DANGEROUS MATERIALS

It is impossible to know constantly and with certainty the exact location and condition of every item in any inventory of size. There is always some degree of error, reflected in all inventory control systems by a special "margin of error" category. The U.S. Nuclear Materials Management and Safeguard System is the official government system for keeping track of fissionable materials, such as plutonium and highly enriched uranium. It too has such a category. Before 1978, it was called "material unaccounted for" (MUF). After 1978, in keeping with the trend toward euphemism, it was renamed "inventory difference" (ID). MUF or ID, it is defined as "the difference between the quantity of nuclear material held according to the accounting books and the quantity measured by a physical inventory."[1]

If it is known that some plutonium was destroyed in a fire, that missing material is considered "accounted for" and is not included in MUF/ID. Plutonium believed lost as a result of machining operations (such as cutting or grinding the metal) or trapped somewhere in the miles of piping in a facility is also not included in MUF/ID. Even if plutonium has been lost but permission has been granted to write it off for whatever reason, it is still not included in the MUF/ID. In other words, the MUF/ID reflects the difference between what the record keeping system shows should be in the inventory (after all these adjustments) and what a direct physical check of the inventory shows is actually there. That much material is, from a bookkeeping point of view, out of control.

Since some discrepancy between what the books say and what is actually in the physical inventory is inevitable, a decision must be made as to how large a difference is too large. The Nuclear Regulatory Commission (NRC) sets standards for how large an ID is too large. Their term for this is the "limit of error on inventory differences" (LEID). Concluding that the NRC's statistical tests on shipments of nuclear fuels had become so muddled that they were meaningless, staff statisticians decided to create a new LEID index in 1980. It was defined so that statistically there would be 95 percent confidence that no thefts or losses had occurred as long as the absolute value of the actual inventory difference (ID) was less than LEID. They applied this test to ID and LEID data on 803 nuclear materials inventories on which the NRC received reports between 1974 and 1978. Assuming no loss or diversion, the absolute value of the actual ID should have been greater than LEID in only about 40 cases (5 percent of 803). In fact, ID exceeded LEID for 375 inventories, nearly half the cases. The staff's understated conclusion: "Something is wrong!"[2]

The mere fact that there is an error in the inventory does not mean that anything has been stolen or otherwise "diverted" to unauthorized use. But it does mean that it *could have been* stolen or diverted, without the record keeping system noticing. In the words of the Department of Energy (DoE) itself, "It is not prudent to discount the fact that a small inventory difference could possibly be due to theft."[3]

How Good Can Detection Be?

After the Symbionese Liberation Army's kidnapping of Patricia Hearst in the mid 1970s made real the possibility of terrorism on American soil, the Atomic Energy Commission (AEC) decided to put together a special study group to look into the problem of safeguarding weapons-grade nuclear materials. The group included a former assistant director of the FBI, an MIT mathematician, a consultant on terrorism, and one representative each of the nuclear weapons laboratories and the nuclear industry. They addressed the question of just how good a record keeping system could be. With respect to both plutonium and uranium, the study group concluded:

> Because of the finite errors in the methods of analysis; because these errors cannot be reduced to a size which is much smaller than has been experienced in the past; . . . because of the human factors and the statistics of the system; there does not appear to be any way in which the measurement of the total inventory of an operating plant, or even a large segment of that inventory, can ever be known to better than one-tenth of one percent. At the present time, the real possibilities are much closer to one percent error.[4]

A decade and a half later, in June 1988, the European Parliament inquired into the inventory safeguards system of the International Atomic Energy Agency (IAEA), the agency charged with detecting the diversion of nuclear materials. Like the AEC study group, the IAEA concluded that a 1 percent uncertainty in measuring plutonium inventories was to be expected.

IAEA's standards require that it be able to detect the diversion of any "*significant* quantity" of nuclear material, defined as enough to make one crude bomb or several sophisticated weapons. For example, 8 kilograms (17.6 pounds) of plutonium is considered a "significant quantity." Yet in 1986, a leaked confidential IAEA safeguards report indicated that the agency's real goal (at 17 facilities it was responsible for inspecting) was much more modest: to detect the diversion of one to six significant quantities of nuclear material, a disturbingly large margin of error. The inquiry ultimately concluded that even if IAEA performance was up to its own standards, there would still be a 5 percent chance that a significant diversion of plutonium would not be detected.[5]

In February 1996, the DoE issued a landmark report, *Plutonium: The First 50 Years*, which focused on America's production, acquisition and use of plutonium from the dawn of the nuclear age (1944) through 1994. It gives 50-year MUF/IDs for each of the seven major DoE sites around the United States, as well as the cumulative total for the combined inventories of the Departments of Energy and Defense.[6] According to these figures, over the last half century, the MUF/ID averaged 2.5 percent of the inventory—much higher than any of the preceding analyses predicted.[7]

The DoE report claims that only 32 percent of the inventory difference occurred since the late 1960s, indicating that tracking accuracy improved. That is certainly heartening news. But even if the system had been that accurate throughout the entire 50-year period, the MUF/ID still would have been close to one percent.

If the IAEA detects anything unusual at the facilities it oversees, its governing board can only notify its member nations, the UN Security Council and the UN General Assembly. It has no direct enforcement powers. What happens if the MUF/ID is too high at a nuclear facility within the United States? According to the Department of Energy, "both DoE and contractors operating DoE facilities have always investigated and analyzed each and every inventory difference to assure that a theft or diversion has not occurred."[8] But a former NRC safeguards official paints a different picture:

> The NRC shows the flag and sends a task force to investigate, they have one or two reinventories, they find or lose 3 or 4 kilograms and arrive at a final figure. They then look at the whole operation and say the material unaccounted for could have disappeared into the river or hung up in the pipes or whatever, but say they find no direct evidence of theft or diversion. It's just a ritual with a lot of thrashing around. . . . The fact of the matter is they are playing a game and nobody really believes in the accounting system.[9]

Is Our Best Good Enough?

In 1995, the combined stocks of plutonium metal held by the U.S. Departments of Energy and Defense totaled 111.4 metric tons.[10] To be conservative, let's assume that the stocks of plutonium held by the former Soviet Union are no bigger than U.S. stocks, and ignore the stocks of plutonium held by all other nations.[11] Record keeping systems in the Soviet Union were fairly primitive, but to make the estimate even more conservative, assume that records for all plutonium inventories were always as accurate as the United States claims to have kept them since the late 1960s. This very understated estimate still leaves more than 1780 kilograms of plutonium (0.8 percent of 222.8 metric tons) in the margin of error. Even with 99.2 percent perfect control, enough plutonium could have been diverted to make more than 220 crude nuclear bombs (or perhaps twice that many well-designed weapons) without record keeping systems ever noticing that anything was missing. That is easily enough destructive power to cause devastating damage to any nation on earth.

The lowest MUF/ID believed theoretically possible—the AEC special safeguards study group's estimate of 0.1% (99.9 percent perfect control)—would still leave more than 220 kilograms of plutonium in the "uncontrolled fringe." That is enough for more than 25 city-destroying nuclear weapons, or 1 million lethal doses if dispersed as an aerosol terrorist weapon.[12]

Tracking Weapons and Related Products

There is no publicly available information on the structure, capabilities or performance of the systems used to keep track of weapons of mass destruction. We could use the MUF/ID from the plutonium accounts as an estimate of the weapons MUF/ID. But tracking the flow of materials passing through processing plants is different from keeping track of finished products like canisters of nerve gas or nuclear weapons. Since there is some information available on the control of military inventories of conventional weapons, ammunition and other supplies, it is worth investigating whether that experience might be a better guide.

A senior logistics officer in the Pentagon estimated that $900 million worth of weapons, spare parts, electronic equipment and other supplies disappeared in 1986.[13] Between 1979 and 1986, the Army reported recovering close to four million rounds of lost small-arms ammunition and explosives. That must have come as something of a surprise, since only two million rounds had been reported missing.[14] The deputy inspector general of the Department of Defense (DoD) confirmed that there had been many other cases in which even more powerful munitions (artillery shells, hand grenades and mines) had disappeared, without the military knowing they were lost until they were found. He said that 400,000 pounds of munitions were recovered in Europe in 1986 alone.[15]

From October 1984 to October 1985, the Army Audit Agency reported ammunition was found at unauthorized locations some 1,200 times.[16] According-ing to the Army inspector general, record keeping was so bad that the amount of ammunition and explosives lost by the Army each year *could not even be determined.*[17] The GAO concluded that "Controls are inadequate to detect diversion," then went on to point out, "The Army has known about many of its ammunition accountability deficiencies for years, but it has made no significant policy changes or efforts to eliminate systemic causes."[18]

According to the 1986 GAO report, the Air Force not only recognized that it too had problems with inventory control, it "comprehensively scrutinized" its methods, updated automatic data-processing equipment, implemented bar coding, introduced a new distribution system and automated its warehousing procedures in an attempt to rectify them. Yet "many longstanding problems in the Air Force supply system continue." "Inventory adjustments" (equivalent to MUF/ID) were growing, and the Air Force was often not able to identify the causes of these discrepancies. Inventory adjustments for the overall Air Force supply system *doubled* from 1982 to 1985, from 2.6 percent of inventory value to 5.2 percent.[19]

The GAO staff analyzed shipments from the San Antonio Air Logistics Center to and from three Air Force bases they visited. In 278 cases (more than 10 percent), they could not match records of shipments with records indicating that the shipments were received. Among these were 41 shipments from San Antonio that the bases had no record of ever receiving, and 19 shipments that San Antonio had records of receiving from the bases that none of the bases had any record of shipping![20]

At the same time, the Navy's judge advocate general was reporting that deficiencies in inventory control were systemic to the Navy's "afloat supply system." The chief of naval operations said "the deficiencies are long-standing" and that "they stemmed from a system that was not responsive."[21] The GAO found "problems at all levels of inventory management." Marine Corps inventory records were inaccurate for "relatively large percentages" of items they sampled. At one major supply management unit GAO studied "officials were unaware that their inventory contained classified items. . . . [They] were stored with general merchandise under conditions with less security than required."[22]

The military's inability to track inventories accurately persists despite the potential danger of loss of control over weapons and ammunition. GAO's "concurrent review of over 300 prior DoD [Department of Defense] and GAO reports showed that most of these systemic problems have existed for years."[23] The Pentagon and congressional officials admit that some of the lost arms and explosives wind up in the hands of criminals and terrorists.[24] According to the Bureau of Alcohol, Tobacco and Firearms, military explosives were used in 445 bombings in the United States from 1976 to 1985.[25] In the late 1970s, a supersonic Red Eye surface-to-air missile was reportedly found in operating condition in a New York City apartment believed to belong to a member of the FALN terrorist group. The heat-seeking missile could have easily been used to shoot down an airliner. It can destroy an aircraft two miles away at an altitude of 8,200 feet. The Army had no idea how the missile got into the apartment.[26] Two decades later, GAO looked into military record keeping for the Red Eye's modern descendents—portable, hand-held, nonnuclear, highly explosive and "extremely lethal" missiles like Stinger, Dragon and Javelin. In 1994, they reported that "the services did not know how many hand-held missiles they had in their possession because they did not have systems to track by serial numbers the missiles produced, fired, destroyed, sold and transferred. . . . [W]e could not determine the extent to which any missiles were missing from inventory." Three years later, GAO found that "the Army discovered it had not counted 3,949 missiles during the initial inventory" and pointed out that "Discrepancies still exist between records of the number of missiles and our physical count. . . . [M]issiles may be vulnerable to insider theft."[27]

Overall, "inventory adjustments" were nearly 5 percent of total inventory value for the American military as a whole in 1985, much higher than the MUF/ID rate for plutonium.[28] Surely weapons of mass destruction and other dangerous materials would be more tightly controlled. It is more conservative and more appropriate to use the much lower margin of error for plutonium to estimate the MUF/ID for nuclear weapons stockpiles. Since plutonium is very expensive, extremely toxic *and* a critical ingredient in nuclear weapons, it is reasonable to assume that the MUF/ID estimate for plutonium inventories is just about the best that record keeping systems can be expected to do. The lowest estimate of plutonium MUF/ID *capability* is 0.1 percent (from AEC's special study). That is 50 times lower than the U.S. military's overall inventory track record. As of the end of 1997, there were an estimated 36,000 nuclear warheads in the global stockpile (with an explosive yield equivalent to about 650,000 Hiroshima bombs).[29] Therefore, even if the records of *all* the nuclear nations were 99.9 percent accurate (an MUF/ID of 0.1 percent), there would still be more than 36 nuclear weapons unaccounted for at any point in time. That is enough to do catastrophic environmental damage and take tens of millions to hundreds of millions of lives.

There is good reason to believe that this still unacceptable theoretical level of record keeping accuracy is far beyond our practical capabilities. Apparently, we even have trouble keeping track of the track-keeping records. In late 1997, the *New York Times* reported that "The Energy Department has disclosed in private correspondence that it cannot locate the records proving that it dismantled and destroyed as many as 30,000 nuclear bombs at weapons plants across the nation between 1945 and 1975."[30]

By itself, none of this means that even a single nuclear weapon or a microgram of plutonium has actually been stolen or diverted. It also does not mean that the amount of weaponry or plutonium in the MUF/ID is lying around in vacant lots or unprotected basements. What it does mean is that—as long as there are large inventories—even the best record keeping systems cannot track products as dangerous as nuclear weapons or materials as dangerous as pluto-nium well enough to *assure* us that nothing has been stolen or otherwise found its way into unauthorized hands.

Why Is Accurate Detection Critical?

Despite the guards, locks, fences and alarms, imperfect record keeping matters. We must still protect against the threat of internal theft or blackmail. Internal theft is theft by those authorized to have access to the inventory, including the guards assigned to protect it. Accurate record keeping is the first line of defense against the internal thief for whom the critical problem is avoiding detection, not getting in.

Flaws in detection give internal thieves the room they need, especially if what they are trying to steal can be removed a little at a time. One of the most notorious losses of nuclear materials was the disappearance of nearly 100 kilograms (about 220 pounds) of highly enriched uranium over six to eight years from the Nuclear Materials and Equipment Corporation (NUMEC) plant in Apollo, Pennsylvania, in the 1960s.[31] Ten or more Hiroshima-strength bombs could be made with that much weapons-grade uranium, yet over six years it could have been stolen by removing as little as 1.5 ounces a day (about the weight of a medium-sized candy bar). Since uranium is heavier than lead, 1.5 ounces is very small in size.

Internal theft and malevolence is an age-old, worldwide problem. In 1986, two government studies found that most crimes against federal computer systems were committed by government employees. According to one report, officials responsible for security "are nearly unanimous in their view that the more significant security problem is abuse of . . . systems by those authorized to use them."[32] Half a year earlier, the Chinese government revealed that top officials on Hainan Island took advantage of economic reform policies to embezzle $1.5 billion by importing and reselling foreign consumer goods over 14 months. The amount of goods involved was staggering. It included 89,000 motor vehicles and 122,000 motorcycles, roughly equal to one-third of China's total annual domestic production.[33]

Inaccurate detection may also be enough to allow a blackmail threat to succeed. If an otherwise credible blackmailer threatens to attack a city with a stolen nerve gas bomb, extremely toxic chemicals, biological toxins or a nuclear weapon that he/she has built from stolen materials, the threat must be taken seriously unless the authorities can be sure it is bogus. Twenty-nine years ago, city officials in Orlando, Florida, got a real-life lesson in exactly what it was like to face such a threat.

On October 27, 1970, a letter arrived at City Hall, threatening to blow up Orlando with a nuclear weapon unless a ransom of $1 million was paid and safe conduct out of the country guaranteed. Two days later, city officials received another note, repeating the threat and the demands. This time, the blackmailer included a diagram of the bomb that he/she claimed to have built, adding that the nuclear material needed to make the bomb had been stolen from AEC shipments. City police immediately took the diagram to nearby McCoy Air Force Base and asked a weapons specialist there whether or not the bomb would work. It was not the best of designs, they were told, but yes, it would probably explode. The authorities then contacted the FBI, which asked the AEC's safeguards division if any nuclear material was missing. They were told that there was nothing known to be missing, but then again there was this thing called MUF. . . . So city officials put the ransom together and were about to pay it when the police caught the blackmailer. He was a 14-year-old high school honors student.[34]

Detection systems also play a critical role in highlighting unexplained losses of dangerous materials. With a sharp and timely warning, an intensive search can be launched to find out what happened and quickly recover any materials that have been diverted or simply misplaced. But if there is no warning, there is no search. For example, in January 1987, a Mexican lumber dealer got a bit more than he bargained for. He bought 274 wooden U.S. Army surplus ammunition crates from Fort Bliss in El Paso, Texas. When he unloaded the first truckload at his lumber yard in Mexico, he discovered that not all the crates were empty. He found 24 4.5-foot-long fully armed rockets with high-explosive warheads, the type fired by helicopter gunships against ground targets. Fortunately, the dealer immediately notified officials at Fort Bliss, and all the weapons were recovered. According to procedure, the crates should have been inspected twice before they were sold, with Army inspectors signing "several documents" to certify that they were empty. Not only had the inspectors done an inexcusably sloppy job, but nothing in the Army's detection system had indicated that anything was wrong. Fort Bliss officials apparently had no idea that the rockets were missing until the lumber dealer told them.[35]

PROTECTION: SAFEGUARDING DANGEROUS INVENTORIES

There is a basic conflict between protecting inventories from theft or loss and keeping their contents readily accessible. More fences, walls, gates, guards and checkpoints make it more difficult to divert materials, but these same controls also make it more difficult to get to the inventory for legitimate purposes. This is an especially vexing problem when rapid access is critical. For example, in crisis or high alert, militaries consider rapid access to battlefield weapons (even weapons of mass destruction) vital to their usefulness. Such "readiness" pressures guarantee that there will be less-than-perfect protection.

An ideal protection system would defeat any attempt to use force to gain access, provide quick and ready entry for authorized uses and permit no entry otherwise. It is not enough to be able to withstand a violent and determined assault. It is also necessary to rapidly distinguish between authorized and unauthorized personnel, and between legitimate and illegitimate uses by authorized personnel. Accomplishing this every time without error is simply not possible in the real world of imperfect human beings and the less-than-perfect systems we design and operate.

The full range of human and technical fallibility comes into play here. Guard duty is not a very pleasant job. Most of the time, it is excruciatingly boring, because most of the time nothing happens. And because nothing happens, it doesn't really matter if guards are alert or drowsy, clearheaded and ready to act or half asleep—most of the time. But at that crucial moment when thieves try to climb the wall and break into the warehouse or when terrorist commandos attack, everything can be lost if the guards are not fully alert, clearheaded, quick acting and forceful. The problem is that those crucial moments can come at completely unpredictable times. Guarding dangerous inventories therefore requires constant vigilance. Yet constant vigilance is less and less likely as hours turn into days, weeks, months and years in which nothing unusual ever happens.

Boredom dulls the senses. Extreme boredom is stressful, even maddening. It can lead to sluggishness, unpredictable mood swings, poor judgement and hallucinations (see Chapter 7). People will do almost anything to distract themselves when they are trapped in deeply monotonous routines—from reckless and ultimately self-destructive things such as abusing drugs and alcohol (see Chapter 6) to things that are just plain stupid. Consider, for example, the disturbing, but perhaps not-so-strange case of Tooele Army Depot.

Guarding Nerve Gas at Tooele Army Depot

Investigative reporter Dale Van Atta of the Salt Lake City *Deseret News* wrote a series of articles about security conditions at Tooele Army Depot (TAD) in 1978.[36] Located in the desert of western Utah, Tooele was the premier nerve gas storage area in the United States. An estimated 10 million gallons of deadly GB and VX nerve gases were stored there. Fatal within minutes of skin contact, even 1 percent of that much nerve gas is enough to kill everyone on earth.

The largely civilian guard force at Tooele does not appear to have been the world's most vigilant. According to Van Atta's sources, the "biggest problem in the force is constant sleeping on the job." One guard reportedly fell asleep on the transmitter button of his radio and snored undisturbed over the airwaves for an hour and a half. When a guard who was relatively new to the force argued with his partner about pulling their vehicle over and going to sleep, "He pulled a gun on me and held it for ten minutes, ordering me to sleep," an incident which shows a certain lack of emotional stability, clear thinking and good judgement apart from what it says about security practices. Not all that many people could go to sleep under orders at gunpoint. The supervisor to whom this incident was reported took no action, even though the guard who had threatened his partner admitted what he had done.[37]

When they were awake, the guards sometimes distracted themselves from the boring routine by racing their vehicles. For years, drag racing among patrols on isolated roads within the depot was common. "There was at least one every night, and it usually averaged about three races every night," one source said. According to one of the guards, "some supervisors will 'hot rod' their vehicles and try to elude guards chasing them. 'It's just . . . something for them to do.'"[38]

Another thing for them to do was play cards. There were "marathon card and cribbage games" among the guards; "a three hour game of solitaire" was not rare. Arsonists burned down an old railroad station at TAD while guards on the night shift played poker. Alcohol abuse was also a problem. According to one guard, "If they were to give a breathalyzer when we all line up for duty, a couple dozen men would be sent home throughout the shifts." An Army captain, a security shift supervisor, reportedly came to work so drunk one night that one of his subordinates sent him home. "He knows he's an alcoholic, and all the men know it."[39]

When not sleeping, drinking, drag racing or playing cards, some of the poorly paid guards were stealing. But one reported incident of malfeasance that came to light was far more serious than repeated petty theft. It involved an active conspiracy to cover up a serious discrepancy in the deadly nerve gas inventory.

In the summer of 1976, workers taking physical inventory at TAD discovered that there were 24 fewer 105 mm nerve gas shells than the records showed in storage. Instead of reporting the loss, the men created a dummy pallet of 24 empty shells painted to look like live shells that contained deadly GB nerve gas and then reported that the inventory matched the records. According to Van Atta's sources, at least eight men—including a lieutenant colonel—were involved in the cover-up. TAD used munitions workers and guards to count the nerve gas weapons, not independent inventory employees. Those responsible for handling and protecting the weapons were thus being asked to check on themselves.[40]

When the incident later came to light, the Army claimed that no nerve gas shells were actually missing. The problem, they said, had been an error in TAD's records, which showed 24 more shells received than the manufacturer had made.[41] Maybe so. Still, the willingness of TAD workers to cover up the apparent loss of two dozen deadly nerve gas shells is a horrifying breakdown of the inventory safeguards system. The workers who built the dummy shells had no way of knowing that the real shells were not in the hands of criminals or terrorists. There was enough GB gas in these shells to kill thousands of people. Yet they took deliberate action which, if undiscovered, would have stopped anyone from trying to track down the missing nerve gas. Suppose it had been taken by criminals or terrorists and used as a blackmail threat. Skeptical officials might have failed to take the threat seriously because the records showed nothing missing. Would the blackmailers, now angry and frustrated, have hesitated to carry out their deadly threat?

Unfortunately, this is not the end of the story. A few months later, TAD workers restacking pallets of nerve gas canisters found that one of the 24-shell pallets had been broken open, and one of the shells removed. The Pentagon reported five months later that its investigation of this incident was completed and "No conclusive accounting for the discrepancy was determined."[42] In plain English, they had no idea what happened to the missing nerve gas.

The most compelling test of any protection system comes when someone attempts to penetrate it. If guards can respond quickly and effectively at critical moments, it is not so important that they aren't all that impressive or even alert the rest of the time. Could the troubled guard force at Tooele pull itself together and repel any outside intruders that might have attacked? According to Van Atta's sources, the high security areas of Tooele were successfully infiltrated more than 30 times in the decade preceding his testimony. Fortunately, most were "friendly" assaults; that is, they were tests of the security system by special units of the Army and the Utah Army National Guard.

During test intrusions, one group of men entered the most sensitive chemical area at the depot and painted the date and time of their mission on top of two buildings they infiltrated; another group commandeered three vehicles and drove around a secured area for hours without being caught or even challenged. In another test, four prostitutes created a diversion that distracted six guards, while intruders entered a high-security area and left a message behind. Other test intruders let the air out of the tires of vehicles in which guards were sleeping and left dummy dynamite sticks on the door of a chemical storage igloo.[43]

Then there were the real infiltrations. Several times guards discovered the tracks of motorcycles that appeared to have been shoved under outer fences and ridden in unauthorized areas. Unidentified individuals systematically broke the windows and slashed the tires of so many military vehicles that it was estimated it would have taken more than four hours for several men to do that much damage. A number of times, unauthorized armed men discovered inside the depot told guards that they were rabbit hunters who had strayed—in every case the "rabbit hunter" was let go and the incident was never reported. In the words of an official letter of complaint signed by one frustrated TAD guard, "It is my opinion that the Tooele Army Depot Guard Force could not effectively counter an offensive by a troop of Girl Scouts."[44]

All this would have an air of comedy about it if it weren't for the fact that these people were at the core of the protection system at a facility that stored enough nerve gas to kill every man, woman and child on earth a hundred times over.

It was probably true that the majority of the guards at Tooele were conscientious, hard-working and competent people. But it is certainly true that the protection system there was a disaster waiting to happen. Why is all of this—which, after all, happened decades ago at a single facility—important to the broader issue of protecting dangerous inventories today?

If there were any real grounds for believing that highly peculiar circumstances had conspired to create a situation at Tooele unique in time and place, it would still be frightening to know that it could happen. What makes it important is that the most unusual thing about the situation at Tooele is that it became public knowledge. The conditions of work, human failings, attitudes and organizational problems that are so well revealed by Van Atta's reporting are much more common than most of us would like to believe. That is not to say that all, or even most, dangerous inventories face protection as disastrously flawed as at Tooele in the 1970s. If they did, it is likely that criminals and terrorists would already have done much more damage than they have. Yet there is plenty of evidence that the kinds of problems seen at Tooele are far from unique in time or place. In 1996, the vice president in charge of Westinghouse Hanford's tank program was quoted as describing the tank farm at DoE's Hanford nuclear reservation as "a Siberia for a lot of derelicts on the site."[45] These "derelicts" were tending 177 huge underground tanks (many the size of the U.S. Capitol dome) filled with enormous quantities of extremely radioactive high-level nuclear waste. Fifty-four of these tanks of lethal radioactive material are known to occasionally build up flammable gases that could lead to a chemical explosion.[46]

Protecting American Nuclear Weapons

Beginning in the mid 1970s, there was a great deal of congressional concern about the security of American nuclear weapons sites around the world. In September 1974, Rep. Clarence Long reported that at one location abroad "over 200 nuclear weapons are in storage. . . . Less than 250 feet from the facility is a host nation slum which has harbored dissidents for years. . . . [T]he United States deploys nuclear weapons at some foreign locations hundreds of miles from the nearest American installation."[47] Long also revealed a Defense Department consultant's report that found "[nuclear] weapons stored in the open, and [on] aircraft visible from public roads in the United States." He went on to specify that there were "vehicles and forklifts in restricted areas which could be used by an attacking force to capture or carry away nuclear weapons; . . . superficial checks of restricted areas; . . . [and] inspections which have failed to report deficiencies."[48]

On April 30, 1975, information was made public about two trips Senators John Pastore and Howard Baker had made a year earlier to a number of U.S. nuclear weapons sites overseas. At one unnamed site, they discovered "Nuclear weapons . . . stored at a barracks with minimum protection. . . . They are located in the basement of a building which contains administrative offices"; at another, the nuclear weapons were "located outside the base. . . . [N]uclear weapons must travel on a public road for 300 yards." According to Pastore, "We were told that in many of these places even a small group could actually invade one of these compounds and cause a lot of trouble." The senators concluded, "Nuclear weapons sites appear to be vulnerable to terrorist attack."[49] They were not alone in this assessment. In September 1974, General Michael Davison, commander of U.S. forces in Europe, had told a West German audience that his troops would have difficulty protecting nuclear weapons against a determined terrorist attack.[50]

Additional funding was provided to the Department of Defense for improving protection at American nuclear weapons facilities. Years later, a Cox Newspapers reporter named Joe Albright decided to investigate whether the deficiencies had been corrected. In February 1978, he testified before Congress:

> posing as a fencing contractor I talked my way past the security guards at two SAC [Strategic Air Command] nuclear weapons depots and was given a tour of the weak links in their defenses against terrorist attacks. . . . I also purchased government blueprints [by mail] showing the exact layout of the two weapons compounds and the nearby alert areas where [nuclear-armed] B-52s are ready to take off in case of war. . . . Blueprints E-2 and E-9 disclose a method of knocking out the alarm circuits. Blueprint G-5 shows two unguarded gates through the innermost security fence. . . .
>
> As an imposter at that SAC base [in December 1977] . . . I came within a stones throw of four metal tubes . . . [that the Air Force later acknowledged were] hydrogen bombs. At that moment, I was riding about 5 MPH in an Air Force pickup truck that was being driven by my only armed escort, an Air

Force lieutenant. Neither the lieutenant nor anyone else had searched me or inspected my bulky briefcase, which was on my lap.[51]

Albright wrote a series of newspaper articles about his exploits. *After* they were published,

An envelope containing a set of revised blueprints, disclosing . . . the wiring diagram for the solenoid locking system for the B-52 alert area, was mailed to me by the Corps of Engineers several days after Brig. Gen. William E. Brown, chief of Air Force security police, issued a worldwide directive to all Air Force major commands reemphasizing vigilance against intruders.[52]

Protecting American Inventories: The Problems Continue

Neither the public revelations of the 1970s nor giving the Pentagon more money to beef up protection has put an end to these problems. Quite the contrary, the record of dangerous inventories protection problems is long and disturbing:

- In late 1978, the officers responsible for protection at Offut Air Force Base near Omaha were relieved of duty after undercover Strategic Air Command (SAC) security investigators were able to penetrate the high-security area out of which the SAC airborne command post operates. This is the aircraft designed to be used, if necessary, to supervise the launch of American nuclear weapons in the event of nuclear war.[53]
- In September 1980, an 18-wheel tractor-trailer truck loaded with enriched uranium was found abandoned on a highway in Virginia. The uranium was packed in protective containers on the flatbed truck, two tires were missing, and the driver was nowhere in sight.[54]
- In the summer of 1982, a team of trained commandos hired to play "mock terrorists" walked into the center of the plutonium manufacturing complex at Savannah River, South Carolina. They also "visited" the nuclear bomb parts manufacturing operation at Rocky Flats in Colorado and the final assembly point for all American nuclear weapons at the Pantex plant in Texas.[55]
- In July 1983, federal authorities arrested eight men and charged them with conspiracy to sell $2 billion worth of advanced weapons to Iran and $15 million worth of machine guns to the underground Irish Republican Army. In a taped conversation, the key defendant also offered to sell a "stolen nuclear device" to government undercover agents.[56]

- In 1984, "mock terrorists" succeeded in three simulated infiltrations of the S site at Los Alamos National Laboratories. In two, they would have walked away with weapons-grade plutonium. In the third, they would have stolen an unlocked nuclear device constructed for the Nevada test site that could have been exploded within hours.[57]

- In August 1985, Navy officers found an armed band of Philippine Communist Party guerrillas camped *inside* the outer perimeter of the huge American naval base at Subic Bay, a mile and a half from the Navy ammunition storage area. At Clark Air Force Base, also in the Philippines, thieves had stripped fences, and perimeter guard towers had been abandoned. It is very likely American nuclear weapons were stored at both bases at the time. The Senate concluded U.S. military bases overseas were still extremely vulnerable to terrorist attack.[58]

- In October 1985, working with DoE investigators, an employee of the Pantex nuclear weapons plant smuggled a pistol, a silencer and explosives into the top-secret facility. A few days later, those same weapons were used to steal nuclear bomb components including plutonium from the plant. DoE also discovered that some of the guards at Pantex had found yet another way to alleviate the boredom of guard duty—they were bringing friends on-site and having sex with them in the guard towers. Although DoE officials said the security problems had been corrected, Texas Congressman John Bryant commented, "They made repeated assertions that everything was safe at these plants and . . . [yet] we had such lax procedures that an amateur could get out of there with the most dangerous weapons in the world." Rep. John Dingell, chair of the House oversight committee, added, "We found safeguards and security to be in a shambles at many of the nuclear sites."[59]

- During another exercise in 1985, "mock terrorists" escaped with plutonium after successfully "assaulting" the Savannah River Plant in South Carolina, where plutonium and tritium were being produced. The attackers succeeded even though the guards knew in advance that a test was coming—they had been given special guns that did not fire real bullets to be used during the test. One guard was unable to load his weapon, another's weapon jammed. The guards in a helicopter dispatched to chase the escaping "terrorists" could not shoot at them because they had forgotten their weapons. Twenty minutes after the attackers escaped, some guards at the facility were still shooting—at each other. Yet that same year the Department of Energy gave the company in charge of security at the plant a nearly $800,000 bonus for "excellent performance."[60]

- In 1986, a ring of international arms smugglers was caught tapping into the U.S. Navy's computerized supply system. They had stolen $25 million worth of "highly sensitive weapons and spare parts" for Iran. The regional commissioner of U.S. Customs in San Diego said they would have gone undetected if not for an anonymous letter. The agent supervising the investigation said, "It was like buying through a Sears catalog."[61]

- In 1986, it was revealed that the Navy was using at least two ordinary cargo ships with *civilian* crews to ferry long-range, submarine-launched nuclear missiles across the Atlantic.[62]

- In August 1987, the U.S. Army in Europe finally managed to locate 24 Stinger missiles that Army Missile Command had asked them to find *nearly a year earlier*. Some Stingers were not stored securely, kept in lightweight metal sheds with "Stinger" stenciled on the outside.[63] Yet these surface-to-air missiles can destroy aircraft flying three miles high.[64]

- In July 1989, authorities arrested two Air Force security police and charged them with stealing three F-16 jet fighter engines from Hill Air Force Base in Utah. The special agent in charge of the FBI's Utah office said, "The thefts at Hill appear to be part of a larger problem involving other states and other military establishments. Unfortunately, many of the participants are military policemen."[65]

- In October 1996, seven men, two of whom were civilian employees at Fort McCoy Army Base, were charged with stealing $13 million worth of military equipment, including 17 mobile TOW missile launchers and a Sheridan battle tank! All the equipment stolen was in operating condition.[66]

- In October 1997, federal agents arrested six marines, including a captain, in North Carolina. They were charged with offering to sell stolen machine guns, Claymore mines, hand grenades, antitank weapons and powerful explosives.[67]

Despite decades of investigations, hearings, newspaper accounts, special funding for improved security and protestations by the Departments of Defense and Energy that decisive action has been taken, American military stockpiles of weapons, explosives and other dangerous materiel are still less than completely protected.

Both local/state government and the private sector have also experienced some of the very same problems. For example, the New York State Power Authority and privately owned Consolidated Edison operated nuclear reactors at the Indian Point nuclear power plant (only about 24 miles from New York City). In October 1979, some of the more conscientious guards there made a series of allegations about what they considered to be serious deficiencies in the protection system at the facility. They alleged that warning alarms on fences were sometimes disconnected, access to keys for secure areas was not carefully controlled, and special codes were used to warn guards in advance of "surprise" federal inspections. They also said that guards were sometimes certified to have passed training courses that they never took—one guard cited an incident in which he and four others were sent to take a course on crowd control, shown an X-rated video, and then told to sign a document saying they had completed the course. In the words of one former guard, "It would take one commando or one well-trained wacko to go in, inflict heavy damage and leave without being detected."[68]

The following year, a cub reporter on his first assignment penetrated security at Metropolitan Edison's infamous Three Mile Island nuclear power plant. Posing as a guard, he was able to get a job at the plant that gave him wide access to the facility despite his complete lack of training and experience. For two weeks, he wandered around the plant, taking a camera into restricted areas in which photography was forbidden, and gathering information for a series of articles on plant security. The ease with which he infiltrated the facility is particularly shocking in light of the fact the Metropolitan Edison and Three Mile Island had been in the spotlight under intense public pressure since the notorious accident there only one year earlier.[69]

Safeguarding Inventories
in the Former Soviet Union

The military was not immune to the rapid rise in crime rates in the former Soviet Union during the late 1980s and early 1990s. The biggest problem was theft of small weapons, sold in a growing black market that helped arm the notorious organized crime gangs plaguing Russian society. But not all the weapons stolen were small.

In the spring of 1990, the Internal Affairs Ministry of the Republic of Armenia announced that 21 battle tanks had been stolen by two groups of armed men. The weapons were ultimately recovered. In two other separate incidents in Armenia, 40 armed men attacked one military base and as many as 300 armed men attacked another. The heavily armed Soviet state was breaking down. Fighting in January between Armenian and Azerbaijani groups near Baku posed a serious enough threat to major nuclear storage sites outside the city to cause elite Russian airborne units to be dispatched to the area. Six months later, the *Wall Street Journal* reported that "The Soviet Union is quietly moving nuclear warheads out of the Baltics and its volatile southern republics and into parts of the Russian Republic considered politically stable. The . . . shift of weapons is taking place . . . because of worries that a nuclear warhead could fall into the wrong hands."[70]

After the collapse of the Soviet Union, officials at the Russian nuclear regulatory agency (Gosatomnadzor) revealed that there were serious problems with the physical protection systems at both military and civilian nuclear facilities in Russia. They lacked operational devices to monitor doors and windows, and to detect unauthorized entry. Guards were poorly paid, not all that well qualified and not well protected. There were no vehicle barriers around the facilities, and communications between guards on-site and authorities off-site were primitive.[71]

Combined with the increasingly depressed state of the Russian economy, these protection problems predictably led to thefts, some of which have become public. In 1992, 1.5 kilograms of weapons-grade uranium was diverted from the Luch plant at Podolsk.[72] In 1993, 4.5 kilograms of enriched uranium was stolen from one of the Russian Navy's main nuclear fuel storage facilities, the Sevmorput shipyard near Murmansk. Six months later, three men were arrested and the stolen fuel recovered: the man charged with climbing through a hole in the fence, breaking a padlock on the door and stealing the uranium was the deputy chief engineer at Sevmorput; the man accused of hiding the stolen uranium was a former naval officer; and the alleged mastermind of the operation was the manager in charge of the refueling division at the shipyard![73]

Apparently, stealing uranium from Sevmorput was not that big a challenge. According to Mikhail Kulik, the chief investigator, "'On the side [of the shipyard] facing Kola Bay, there is no fence at all. You could take a dinghy, sail right in—especially at night—and do whatever you wanted. On the side facing the Murmansk industrial zone there are . . . holes in the fences everywhere. . . . [W]here there aren't holes, any child could knock over the half-rotten wooden fence boards'"[74] The thief made the amateurish mistake of leaving the door of the burglarized shed open. The guards noticed the open door within 12 hours of the theft. Had he not made that mistake, in Kulik's view the theft "could have been concealed for ten years or longer."[75]

At about the same time, a colleague of mine travelled to the former Soviet Union as part of an official group invited to visit military facilities there. Among other places, the group was taken to a nerve gas storage facility, a fairly simple wooden structure with windows in the woods. It had no apparent barriers or extensive guard force surrounding it, let alone any sophisticated electronic detectors. It did have what he described as "a Walt Disney padlock" on the front door. When the group asked how many active nerve gas shells were stored in this building, they were told that no accurate records had been kept.

In one of the more bizarre post-Soviet incidents, a local state electric power utility, annoyed that Russia's Northern Fleet was $4.5 million behind in paying its electric bills, cut off power to the fleet's nuclear submarine bases on the Kola Peninsula in April 1995. A fleet spokesperson said "switching off the power for even a few minutes can cause an emergency," though he refused to elaborate. The fleet commander called the power cutoff "an act of sabotage." But the power company would not relent until heavily armed sailors in bulletproof vests forced the engineers on duty to turn the power back on at gunpoint.[76]

This incident might have been more dramatic than most, but it was not unique. From 1992 to 1995, there were 16 cases of electricity being deliberately cut off to Russian military bases because of nonpayment of bills—including a cutoff of power to the central command of Russia's strategic land-based nuclear missile forces.[77] Aside from complicating inventory protection, such power stoppages could cause dangerous disruptions in military command and control. The turmoil extended into the civilian sector. In December 1996, more than a dozen employees took over the control room of the nuclear power plant that supplies most of the electricity to St. Petersburg. They threatened to cut off power to the city unless they received months of back pay. The next day 400 co-workers joined the protest.[78]

Without doubt, the most frightening news to emerge from Russia in recent times came from Alexander Lebed, the popular Russian general credited with ending the Chechnyin war. While national security advisor to President Boris Yeltsin, Lebed became concerned about the security of Russia's stockpile of some 250 one-kiloton nuclear "suitcase" bombs. According to Lebed, these nuclear weapons were built to look like suitcases and be carried by hand. They were designed to be detonated by one person with about 20 to 30 minutes preparation, and no secret arming codes were required. Lebed ordered an inventory to assure that they were all safe. In an interview on American television in September 1997, he was asked, "there are a significant number that are missing and unaccounted for?" He replied, "Yes. More than 100." He went on to say the suitcase bombs were "not under the control of the armed forces of Russia. I don't know their location. I don't know whether they have been destroyed or whether they are stored. I don't know whether they have been sold or stolen. I don't know." "Is it possible that everybody knows exactly where all these weapons are, and they just didn't want to tell you?" he was asked. "No." "You think they don't know where they are?" "Yes indeed."[79]

As thousands of nuclear warheads are disassembled in the United States and former Soviet Union in the 1990s, as much as 100-200 additional tons of plutonium and 500-1,000 tons of highly enriched uranium (HEU) will be removed from weapons and added to existing stockpiles of the metals.[80] The entire 1994 DoE/DoD inventory of plutonium not in nuclear warheads was 111 metric tons, so this constitutes a huge addition to the stockpile. Although plutonium and HEU is much, much less dangerous to us in the form of "scrap" metal than in the form of nuclear weapons, the accumulation of these weapons-grade metals still poses a very real and present danger.

Concerned about the possibility that economic pressures or failures of protection might lead to Russian weapons-grade material being stolen by or sold to terrorists, criminals or hostile nations, the United States concluded a history-making deal with Russia in 1993. For $12 billion, Russia agreed to sell the United States 500 metric tons of weapons-grade uranium extracted from the nuclear weapons it was scrapping. The uranium would be diluted to 4 percent U-235 by blending with natural uranium and used as reactor fuel. Unfortunately, implementation of the landmark agreement has been slowed by a variety of problems, not the least of which has been the reluctance of the first government-owned and later privatized U.S. Enrichment Corporation (USEC) to consummate the deal. With falling uranium prices and a plan to sell off its own stocks, the corporation no longer felt the arrangement would be profitable.[81] Since it is hardly news that a profit-driven company would put its own profits ahead of national security concerns, the wisdom of creating USEC to handle the U.S. end of the agreement is certainly questionable. In any case, by early 1998, only 36 of the 500 metric tons of highly enriched uranium contracted for had actually been delivered.[82]

Fabricating diluted HEU from weapons into low-enriched uranium reactor fuel and the parallel option of turning plutonium into mixed-oxide reactor fuel (chosen by DoE in 1997) are safer than leaving them as weapons-grade metals.[83] But nuclear power is also a dangerous technology. "Burning" uranium and mixed-oxide fuel in civilian nuclear power reactors does not eliminate the protection problem. It also generates more nuclear waste, which must then be securely stored. There are other ways of handling the plutonium and HEU extracted from weapons, but to date there is no "safe" option. This is one legacy of the nuclear arms race that we are going to have to live with, one way or another, for a long time to come.

SMUGGLING DANGEROUS MATERIALS

The inescapable limits imposed by both human fallibility and the inevitability of tradeoffs in technical systems prevent us from ever doing well enough at either record keeping or protection to completely safeguard inventories of materials that pose a potentially catastrophic threat to human life. But these limits do not inherently mean that that catastrophic potential will be realized. We know we have lost track of frighteningly large amounts of dangerous materials. We also know that at least some of that material has been stolen. But is there any real evidence that any of the material has actually made its way to the black market or been put to unauthorized use?

More than once, bomb-quantity shipments of nuclear materials were apparently diverted unintentionally—aboard planes hijacked to Cuba during the 1960s and early 1970s. Fortunately, neither the hijackers nor Cuban airport officials seemed to know about the nuclear cargoes. The planes were always returned with the nuclear materials still aboard.[84]

The circumstances surrounding the loss of 220 pounds of highly enriched uranium (enough for four to ten nuclear weapons) from the NUMEC plant in Pennsylvania in the mid 1960s were very suspicious. NUMEC had been repeatedly cited by federal inspectors for violating security regulations. More than a decade later, two declassified documents revealed that American intelligence agencies suspected Israel might have obtained the missing uranium by clandestine means and used it to manufacture nuclear weapons. The company's founder was a strong supporter of Israel, NUMEC had business ties to both the Israeli and French governments, and an Israeli was among several foreigners who worked at the plant. The Israeli government denied any involvement.[85]

Israel's search for enriched uranium for its secret nuclear projects allegedly began an illicit trade in nuclear materials centered in Khartoum, Sudan. According to a British television documentary aired in the late 1980s, the trade started in the 1960s and persisted for decades, with first Israel and later Argentina, Pakistan and South Africa allegedly buying nuclear materials secretly on Khartoum's black market. When Sudanese authorities seized four kilograms of enriched uranium smuggled into the country in 1987, Sudan's prime minister reportedly acknowledged that there was a black market in Khartoum. A few months later he reversed himself, publicly denying that Khartoum was a center of the black market.[86] A former Sudanese state security officer claimed that the Israelis were still buying highly enriched uranium on Khartoum's black market during the 1980s, while lower-quality nuclear materials had probably been sold to Iraq, Iran and Libya. Former CIA Director Stansfield Turner reportedly acknowledged that there was a black market in Khartoum for uranium, which he believed was usually stolen from nuclear industry in Western Europe.[87]

In the 1990s, a series of reports of smuggling nuclear and other dangerous materials out of the former Soviet Union emerged:

- On June 30, 1992, NBC-TV reported on the trial of four Russians arrested by German police three months earlier for trying to sell nearly

1.5 kilograms of enriched uranium allegedly smuggled out of the former Soviet Union. NBC's Moscow correspondent claimed that a clandestine tape shot in Moscow showed an agent of the Russian military posing as a representative of a Polish firm. The Russian allegedly offered to sell 30 Soviet fighter planes, 10 military cargo planes, 600 battle tanks and a million gas masks.[88]

- In May 1993, Lithuanian police raided a bank in Vilnius and found 2000 kilograms of beryllium metal, mixed with small amounts of enriched uranium, stashed in the bank's basement. Beryllium has a number of uses, but it is most highly prized for use in nuclear weapons. Because it is an excellent neutron reflector, it reduces the amount of uranium/plutonium required for a nuclear warhead. A Russian syndicate falsified a shipping order to get the beryllium—and 9 kilograms of deadly radioactive cesium—out of a restricted nuclear research facility in Obninsk, near Moscow. The beryllium was smuggled into Vilnius; the cesium simply disappeared.[89]

- On May 10, 1994, German police accidentally found 5.6 grams of nearly pure plutonium-239 in the garage of a businessman named Adolf Jäkle, while searching for other illicit material. Jäkle claimed he had gotten the sample of plutonium through a Swiss contact, and that as much as 150 kilograms of plutonium (enough for nearly 20 nuclear weapons) might have already been successfully moved out of Russia to storage in Switzerland.[90]

- On December 14, 1994, Czech police arrested three men and confiscated nearly three kilograms of HEU (87.7 percent uranium-235). Two were former nuclear workers (one from Belarus, the other from Ukraine), the third was a Czech nuclear physicist.[91]

- On April 13, 1995, suspicious Slovakian police followed a car with Hungarian license plates after it crossed into their country. When they finally stopped the car, they found it contained two Hungarians—and 17 kilograms of uranium. The car seemed headed for Hungary, but it was unclear where the uranium came from or where it was finally intended to go.[92]

- In January 1996, 7 kilograms of highly enriched uranium was reportedly stolen from the Sovietskaya Gavan base of Russia's Pacific Fleet. A third of it later appeared at a metals trading firm in Kaliningrad, 5,000 miles away.[93]

- In October 1996, the director of the international department at the Obninsk nuclear research institute in Russia (the same place from which the cesium and beryllium were stolen in 1993) reportedly claimed that three or four engineers and one perimeter guard working together could "remove significant quantities of uranium or weapons-grade plutonium from the institute's most secure experimental reactor site and conceal any evidence of the theft."[94]

- In February 1998, 14 members of the Italian Mafia were arrested by Italian police and charged with attempting to sell 190 grams of enriched uranium and what they said were eight Russian missiles. This was the first documented case of trade in nuclear materials involving organized crime.[95]
- In September 1998, Turkish agents arrested eight men and seized 5.5 kilograms of U-235 and 7 grams of plutonium powder the men had allegedly offered to sell them for $1 million. One of the suspects was a colonel in the army of Kazakhstan.[96]

Between 1991 and 1994, German police recorded more than 700 cases, some real, some involving con artists trying to "sting" unwary buyers. Furthermore, nuclear smuggling appeared to be growing: there were 41 cases in 1991, 158 in 1992, 241 in 1993, and 267 in 1994.[97] In early 1995, an intelligence report from western Europe, which excluded "fake" smuggling by con artists, claimed that there had been 233 such cases between 1992 and 1994.[98] International Atomic Energy Agency records listed 130 confirmed cases of nuclear smuggling from 1993 through 1996.[99] Variations in the numbers notwithstanding, there is no doubt that nuclear smuggling is no longer a theoretical possibility. It is a growth industry.

There is no question that the economic and political turbulence of the former Soviet Union has made inadequate control of dangerous inventories more of a problem. Yet, as decades of experience with dangerous U.S. inventories has shown, inadequate control is a problem even in the absence of such turbulence. The real lesson of this long, troubled history has nothing to do with the particulars of American or Soviet/Russian society. What we are seeing is not the result of some odd set of inexplicable flukes. It is instead a painfully clear demonstration that there are inherent, unavoidable limits to securing dangerous inventories anywhere.

In every specific case, there are particular personnel deficiencies, technical flaws or organizational problems to cite as the reason for the failure of control. Because there is always a proximal cause, there is always the temptation to think that some clever organizational or technical fix will eliminate the problem by eliminating that particular source of trouble. But it will not. There are two much more basic problems involved: the inherent, unavoidable limits to what technology can accomplish, and the inability of fallible human beings to control inventories perfectly any more than they can do anything else perfectly.

We can and certainly should do better than we have done. The technology, the design and organization of detection and protection systems, the training of guards can all be improved. But no matter how much we improve them, there is no question that in one form or another, the problems will continue.

CHAPTER FOUR

Accidents

Because all systems human beings design, build and operate are flawed and subject to error, accidents are not bizarre aberrations, they are a normal part of system life.[1] We call them accidents because they are not intentional, we don't want them to happen, we don't know how, where or when they will happen, and we hope they do not happen. But they are normal because, despite our best efforts to prevent them, they happen anyway.

Accidents and failures of dangerous technological systems differ from accidents and failures of other technological systems only in their consequences, not in their essence. Car radiators corrode and leak, so do toxic waste tanks. Textile factories explode and burn, so do nuclear power plants. Civilian airliners crash and burst into flame, so do nuclear-armed bombers. The reasons can be remarkably similar; the potential consequences could hardly be more different.

Although all accidents involving dangerous technologies are cause for concern, accidents involving weapons of mass destruction are the most worrying. Their ability to wreak havoc is not an unfortunate, unintended byproduct of their design, construction and operation, it is the very reason they were created. The frightful destructiveness of these weapons made them attractive to aggressors determined to terrorize opponents into submission. But it also made them attractive for defensive deterrence. We always knew just having these weapons was dangerous, but we reasoned that if our military forces were powerful enough, no nation—no matter how aggressive or antagonistic—would dare to attack us. We would be safe at home, and safe to pursue our national interest abroad without forceful interference. It seemed worth the risk.

In a world of nations that still have a lot to learn about getting along with each other, there is something to be said for having the capability to use force. Yet there are limits to what the threat or use of military power can achieve. The benefits do not continue to increase indefinitely as arsenals grow, but the economic costs do.[2] The likelihood of catastrophic human and technical failure also grows as they become larger, more sensitive and more complex (see Chapter 9). With limited benefits and growing costs and risks, it is inevitable that ever larger arsenals will eventually become a liability rather than an asset.

Since accidents involving weapons of mass destruction are the worst case, and since they differ from other accidents mainly in their consequences, we begin by focusing on them and assessing just what those consequences might be.

ACCIDENTS WITH WEAPONS OF MASS DESTRUCTION: POTENTIAL CONSEQUENCES

Nuclear Weapons

The accidental nuclear explosion of a nuclear weapon could be as devastating as its intentional use would be. The explosive power, radiation and heat released by the detonation of a modern strategic nuclear weapon would dwarf that of the bombs that destroyed Hiroshima and Nagasaki in 1945. In a matter of seconds, those primitive bombs flattened what were thriving, bustling cities. Within minutes, much of what was left of the cities caught fire, engulfing many of those who survived the initial blast. Within weeks, thousands more died a sickened and lingering death as a result of the lethal doses of radiation they had received. And the long-term effects of lower doses of radiation claimed still more victims— 10, 20, 30, even 40 years after the explosion.

Nuclear weapons use conventional (chemical) explosives to trigger the nuclear explosion that gives rise to the weapon's devastating blast, radiation and heat effects. Very precise conditions must be met for the conventional explosion to trigger the runaway fission/fusion reactions that produce a nuclear explosion.[3] If the weapon is not armed, there is very little chance that an accident could cause a full-scale nuclear detonation. But the fissile material contained in the core of every nuclear weapon is highly radioactive. An accident in which the core was scattered by the conventional explosive, burned up or otherwise released to the biosphere could do a great deal of damage, even in the absence of a nuclear explosion.

The nuclear core is made of either highly enriched uranium (HEU) or, more commonly, plutonium (Pu).[4] From the standpoint of radiological damage to biological organisms, plutonium is much more dangerous. It is classified in the very highest category of radiotoxicity in the International Labor Office's *Guidelines for the Radiation Protection of Workers in Industry.*[5] Another respected sourcebook, *Dangerous Properties of Industrial Materials,* also rates plutonium as "very dangerous," saying: "The permissible levels of plutonium are the lowest for any of the radioactive elements. . . . Any disaster which could cause quantities of Pu or Pu compounds to be scattered about the environment can cause great ecological stress and render areas of the land unfit for public occupancy."[6]

When radioactive substances decay, they emit alpha, beta and gamma radiation (among other things), all of which are harmful to living organisms. Gamma radiation is much more penetrating than alpha or beta. Shielding against gamma rays requires lead liners and thick concrete walls; alpha particles can be stopped by a piece of paper. Yet paradoxically, it is alpha radiation that is potentially the most biologically damaging—once a source of alpha particles is inhaled, ingested or otherwise absorbed into the body. Plutonium is a heavy alpha emitter.[7]

How could alpha particles be more biologically harmful? All particles, whether or not they have mass, transfer energy to nearby atoms as they pass into or through matter. But when particles that have mass lose energy, they slow down and so interact more strongly with atoms.[8] Because both alpha and beta particles have mass, they slow down when they lose energy, becoming more interactive with nearby atoms.[9] Gamma rays, which can be thought of as streams of massless particles (photons), do not slow down and therefore do not become more interactive. They always travel at the speed of light, losing energy uniformly along the path they travel.[10]

When gamma rays pass through living tissue, they tend mainly to rip electrons away from the periphery of atoms, and these are often quickly replaced by other nearby electrons. But when alpha particles pass through living tissue, they tend to transfer energy to the core (nucleus) of atoms, especially at or near the end of their path inside the tissue, when they are moving relatively slowly. If they transfer enough energy, they will break the atoms away from the molecules of which they are part, changing the very chemical nature of those substances. That is much more biologically damaging than stripping away electrons.[11] (Among the atoms abundant in biological organisms, hydrogen is the most vulnerable to being broken away from the other atoms to which it is bound. It has a mass closest to that of alpha particles, and so gains energy most efficiently when it is struck by an alpha particle.)

Because it is a heavy alpha emitter with high specific radioactivity and a long half-life, plutonium that enters the body is dangerous enough. But it is even more harmful because it tends to be deposited and retained in critical organs. Plutonium is stored in the lungs, lymph nodes, bones, liver and testes/ovaries.

It continues to irradiate surrounding cells as it sits in these vital organs, disrupting the cells and producing cancers and other forms of pathology. It is excreted from the body only very slowly, so the contamination is likely to be lifelong. The biological "half-life" of plutonium (the time it takes for the body to excrete half the plutonium contaminating it) is 40 years if it is deposited in the liver and 100-200 years if deposited in bone.[12]

As a result, a tiny amount of plutonium is extremely harmful. It is 30 times as potent as radium in producing tumors in animals.[13] It can be conservatively estimated from animal studies that inhaling no more than 200 milligrams (0.007 ounce) of plutonium will kill half the humans exposed within 30 days;[14] an inhaled dose of as little as 1-12 milligrams (0.00004-0.0004 ounce) will kill most humans (from pulmonary fibrosis) within 1-2 years.[15] Even a single microgram (0.000001 gram or 0.000000035 ounce) can cause lethal cancer of the liver, bone, lungs, and other organs after a latency period of years to decades. And a microgram of plutonium is a barely visible speck.[16]

It is reasonable to estimate that a typical plutonium-based nuclear weapon in the arsenal of one of today's nuclear-armed countries would contain about 9-11 pounds (4-5 kilograms) of plutonium.[17] Taking the smaller figure to be conservative, a single typical nuclear weapon would contain a theoretical maximum of 20,000 doses lethal enough to kill half of those exposed within 30 days, 330,000-3,300,000 doses lethal enough to kill at least half those exposed within 1-2 years, or 4 billion doses capable of causing cancer over the longer term. If an accident involving one such weapon resulted in people inhaling even 1 percent of its nuclear material, that would still be enough to kill more than 1,600-16,000 people within 2 years or to give up to 40 million people large enough doses to potentially cause them to eventually develop life-threatening cancer.

Chemical and Biological Weapons

Chemical and biological toxin weapons sicken and kill by disrupting the normal biochemical processes of the body. Probably the best-known class of chemical weapons of mass destruction are nerve gases, such as the World War II–vintage gas sarin, used in the 1995 terrorist attack in the Tokyo subways. They are neurotoxins that kill by interfering with the enzyme cholinesterase, which breaks down the chemical acetylcholine which triggers muscular contractions. Inhibiting cholinesterase leads to a buildup of acetylcholine in the muscles, causing convulsive contractions. Symptoms of nerve gas poisoning include abnormally high salivation, blurred vision and convulsions, usually leading to death by suffocation.[18]

In the years following the 1991 Persian Gulf War, thousands of American soldiers began to complain of a mysterious set of debilitating ailments, including memory loss, chronic fatigue, joint pain and digestive problems, that came to be called Gulf War Syndrome. For years the Pentagon denied that the syndrome could have been caused by exposure to chemical warfare agents. It was not until June 1996 that the military finally admitted that American troops might have been exposed to Iraqi nerve gas. Over the next half year, it slowly emerged that Czech and French troops had detected chemical weapons seven times in the first eight days of the war, that U.S. jets had repeatedly bombed an Iraqi arms depot storing chemical weapons during the war, and that Army engineers had blown up a captured Iraqi storage bunker at Kamisiyah only days after the war ended that the Pentagon knew (but accidentally failed to tell them) contained nerve gas.[19] The air strikes could have caused chemical agents to drift over thousands of American troops. But the demolition of the chemical weapons storage area may well have endangered many thousands more.[20] The Pentagon first estimated that only about 400 American troops might have been exposed to poison gas when Army engineers destroyed the Kamisiyah bunker. They later reluctantly upped their estimate to 5,000, then to 20,000, then to 100,000.[21]

Live biological pathogens can also be used as weapons of mass destruction. Rather than poisoning the body directly, these organisms infect their victims, causing sickness and death through the same biological pathways they would normally follow if the infection had occurred naturally. Unfortunately, biological warfare is no mere speculation by writers of science fiction. It has been a reality for a very long time. One of the worst epidemics in history began with an outbreak of the Black Death (a form of the plague) in Constantinople in 1334. In 1345-46, the Mongol army laid siege to Kaffa, an outpost of Genoa on the Crimean peninsula. During the siege, the plague swept through the Mongol forces. The Mongols used a powerful catapult to fling the diseased corpses of their own troops into the city.[22] This early use of biological warfare proved more devastating than even the Mongols imagined. It not only caused an outbreak of Black Death in Kaffa, but Genoese merchants later carried the plague to the Mediterranean ports of Europe.[23] Within 20 years after the epidemic began, it is estimated that the Black Death killed between one-third and three-quarters of the population of Europe and Asia.[24]

In the mid 1990s, the world began to learn about the horror of Japanese germ warfare activities during World War II. Japan had conducted a large-scale research program, experimenting with anthrax, cholera, the plague and a number of other highly infectious pathogens. The Japanese Army carried out forced tests on captive human subjects in China. At one testing area, victims were tied down while airplanes bombed or sprayed the area with deadly bacterial cultures or plague-infested fleas to see how lethal these biological weapons would be.[25]

The Japanese Army also regularly carried out field tests under more realistic conditions. They dropped infected fleas over cities in eastern and north-central China, and succeeded in starting outbreaks of plague, typhoid, dysentery and cholera. In addition to untold Chinese casualties, thousands of Japanese soldiers died from the diseases their own army had unleashed. As recently as 1997, China was still trying to get Japan to dispose of the remnants of up to two million chemical bombs (most filled with mustard gas, and many corroded and leaking) still in Manchuria, where the Japanese had manufactured both those weapons and the deadly germ warfare agents more than 50 years earlier. It took more than four decades for news of Japan's World War II germ warfare activities to surface because American officials had made a deal with the Japanese at the end of the war. They agreed not to reveal these activities or prosecute the perpetrators of this horror as war criminals in order to gain access to the data that the Japanese Army's germ warfare experimenters had gathered.[26]

During the Persian Gulf War, the specter of biological attack arose once again. In the aftermath of the war, UN inspectors searching Iraq discovered that some Iraqi Scud missile warheads had secretly been loaded with anthrax, a deadly and extremely persistent infectious agent.[27] Anthrax spores contaminating Gruinard Island (near Scotland), used as a germ warfare testing ground in the 1940s, were still viable 40 years later.[28]

Until relatively recently, biological weapons researchers were restricted to using naturally occurring pathogens or trying to breed more virulent strains. Now, remarkable advances in genetic research have opened the possibility of artificially creating still more deadly organisms that do not exist in nature. Some claim that the recent appearance of previously unknown or rare infectious organisms like HIV (the virus that causes AIDS), called "emerging viruses," may be the result of accidental releases from biological warfare research labs.[29] Others argue that they have surfaced because of the intrusion of human activity into what had been remote, inaccessible areas to which the viruses had long been endemic.[30] HIV's mechanism of transmission and long latency period make it a very poor choice for biological warfare. Some other emerging viruses, however, are an entirely different matter.

The Ebola virus (subject of the 1994 nonfiction best seller *The Hot Zone* and the fictional 1995 motion picture *Outbreak*[31]) was first identified in Sudan and Zaire in 1976. Ebola acts quickly, causing fever, headache, muscular pain, sore throat, diarrhea, vomiting, sudden debilitation and heavy internal bleeding. It kills 50 to 90 percent of those infected within a matter of days. People of all ages are vulnerable, the incubation period is only 2-21 days, and there is no known cure. However, to date the virus has only been shown to be transmitted by direct contact with body fluids (such as blood or semen) and secretions, although there is still some debate as to whether it can also be transmitted less directly.[32]

In September 1994, near Brisbane, Australia, 14 racehorses died quickly of a mysterious virus. Health officials and scientists became especially concerned when the virus infected two men working with the horses. One developed interstitial pneumonia and died within a week. The other recovered on his own, after suffering from a severe, influenza-like illness.[33] By the end of the month, researchers identified the pathogen as an entirely new and extremely virulent type of morbillivirus, which they called equine morbillivirus (EM). The same general group of viruses causes distemper in dogs and measles in humans. EM severely damages blood vessels by causing the cells lining them to clump. Fluids pour through the resulting holes into surrounding tissue. The infected horse or human is racked with fever and quickly debilitated. The rapid buildup of fluid in the lungs causes victims to drown in their own fluids.[34] As mysteriously and quickly as it had begun, the outbreak subsided. No new cases were discovered in either horses or people. Half a year later, researchers still had no idea where the virus came from.[35] If EM and/or Ebola prove to be transmittable through the air or water, they would be nearly ideal biological warfare agents (provided, of course, that the side using it had some means of protecting its own troops and population).

ACCIDENTS WITH WEAPONS
OF MASS DESTRUCTION: LIKELIHOOD

The best way of assessing the likelihood of any system's failure is by carefully analyzing its structure, operating characteristics and history. Because of the secrecy that surrounds the design and operation of systems involving weapons of mass destruction, it is very difficult to assess the likelihood of serious accidents this way. Furthermore, the historical record of accidents with these weapons tends to be incomplete and inaccessible. Until 1980, the U.S. military's official list of major nuclear weapons accidents, called "Broken Arrows," included a total of 13 accidents. Late that year, Reuters News Service made public a Defense Department document listing 27 Broken Arrows.[36] *After* Reuters' disclosure, the Pentagon officially admitted that even that list was incomplete—their records actually included 32 major U.S. nuclear weapons accidents since 1950, 31 of which ocurred between 1950 and 1968. Where did these 18 additional Broken Arrows come from? Why were they not part of the Pentagon's official list in the late 1960s? When Associated Press asked the Pentagon this, "Officials said they are unable to explain why the Pentagon at that time [the late 1960s] did not report the larger figure of 31 accidents occuring prior to 1968."[37] Then, as 1990 approached, a newly declassified report from the nuclear weapons laboratories revealed that there had actually been 272 nuclear weapons involved in serious accidents between 1950 and 1968—accidents in which there was an impact strong enough to possibly detonate the weapons' conventional high explosives.[38]

In June 1995, Alan Diehl, chief civilian safety official for the Air Force from 1987 to late 1994, publicly accused the military of covering up and playing down investigations of dozens of aircraft accidents. He documented 30 plane crashes that killed 184 people (pilots, crews and civilians) and destroyed billions of dollars worth of aircraft. Diehl charged they had been covered up to protect the careers of senior military officers. The *New York Times* reported, "Crashes of military planes are commonplace, Pentagon records show. The Air Force has experienced . . . about one every 10 days in recent months. Five of its 59 Stealth 117-A fighters . . . have crashed in recent years."[39] According to Diehl, that is normal; 10 to 20 percent of Air Force planes crash. A safety record like that would bankrupt any commercial airline.

Although none of the crashes Diehl documented seem to be Broken Arrows, his charges indicate an embedded pattern of secrecy and obfuscation. Taken together with the Pentagon's mysterious "discovery" in 1980 of 18 Broken Arrows they had earlier forgotten to mention—and the revelation of many more a decade later—it is possible, even likely, that the public record of serious nuclear weapons accidents remains incomplete. Still the public record is really all that we've got. It conveys at least a rough idea of how often such accidents occur.

The Public Record of
Nuclear Weapons–Related Accidents

Table 4-1 of the Appendix to this chapter (beginning on page 100) shows the date, weapons system, location and description of serious, publicly reported accidents involving U.S. nuclear weapons and related systems. Appendix Table 4-2 gives the same data for nuclear weapons–related accidents of other nuclear-armed countries.[40] Both tables include some accidents that involve major nuclear weapons capable delivery vehicles (missiles, aircraft, etc.) in which the presence of nuclear weapons is either unclear or was specifically denied. There are three reasons for including such accidents.

First, the track record shows that the involvement of weapons of mass destruction in an accident is downplayed or denied whenever possible. In fact, it is the U.S. military's longstanding official policy to neither confirm nor deny the presence of American nuclear weapons anywhere. Second, there is evidence that the distinction between major nuclear-capable delivery systems and those that actually carry nuclear weapons has often been more apparent than real.

Testifying before Congress, retired Admiral Gene La Rocque stated, "My experience . . . has been that any ship that is capable of carrying nuclear weapons, carries nuclear weapons. . . . [T]hey normally keep them aboard ship at all times except when the ship is in overhaul or in for major repairs."[41] For example, one of the reported accidents involved a series of conventional explosions and fires aboard the U.S. nuclear-powered aircraft carrier *Enterprise* on January 14, 1969. There was no mention of nuclear weapons aboard the carrier in accounts of the event. Yet an admiral who commanded such a ship during the 1960s assured me that there definitely were nuclear weapons aboard the Navy's carriers at that time.

But even if nuclear weapons were only present where their presence was specifically confirmed, accidents of this type would still be relevant. The absence of nuclear weapons at the time of an accident involving a system designed to carry them is fortunate, but their presence would obviously not have prevented the accident.

There is no clear evidence in the public record of the accidental nuclear explosion of a nuclear weapon. About the closest we have come was a major disaster in the southern Ural Mountains of the USSR in 1957-58, in which nuclear waste from the production of nuclear warheads exploded, apparently because of the buildup of heat and gas. The underground waste storage area erupted "like a volcano." Hundreds of people were killed outright, many thousands more were exposed to radiation and forced to relocate. A 375-square-mile area of Russia was heavily contaminated with radioactivity.[42] There was also an incident in early February 1970, in which an explosion rocked the main Soviet nuclear submarine yard at Gorki, killing an unspecified number of people. Exactly what exploded was never made public, but it is noted in reports that radioactive material was subsequently found contaminating the Volga River downstream of the shipyard.

On January 24, 1961, we came very close to an accidental nuclear explosion of an American weapon when a B-52, mainstay of the strategic nuclear bomber force, crashed near Goldsboro, North Carolina. The plane carried two 24-megaton nuclear weapons. There was no nuclear explosion. Part of one bomb was never recovered. The other fell into a field where it was found intact. According to Ralph Lapp, former head of the nuclear physics branch of the Office of Naval Research, five of the six interlocking safety mechanisms on the recovered bomb had been triggered by the fall. A single switch prevented the accidental explosion over North Carolina of a nuclear weapon more than a thousand times as powerful as the bomb that leveled Hiroshima.[43] Detonated at 11,000 feet, such a weapon would destroy all standard housing within a circle 25 miles and ignite everything burnable within a circle 70 miles in diameter.

We have occasionally had nuclear weapons accidents on the territory of other nations: the 1958 crash of a B-47 bomber carrying a nuclear weapon "in strike configuration" in Sidi, Slimane, Morocco; the 1968 crash of a nuclear-armed B-52 bomber near Toronto, Canada; and a major fire in late 1970 on board the U.S. submarine tender *Canopus,* loaded with nuclear weapons, in Holy Loch, Scotland, while two nuclear missile submarines were moored alongside. But perhaps most spectacular was the accidental bombing of the countryside near Palomares, Spain, on January 17, 1966.

A B-52 bomber, carrying four 20 to 25-megaton hydrogen bombs, was being refueled in the air by a KC-135 tanker plane when suddenly the planes lurched, and the fueling boom tore a hole in the side of the B-52. All four of the bombs fell out of the plane as it went down. One landed undamaged; the chemical explosives in two others detonated, scattering plutonium over a wide area of farmland.[44] The fourth bomb fell into the Mediterranean, triggering the largest underwater search ever undertaken. After three months, the bomb was found lying in the soft mud, dented but intact.

We have not been as lucky with all of the nuclear weapons accidentally lost at sea. On January 21, 1968, an American B-52 bomber with four megaton-class H-bombs on board crashed and burned in Greenland, melting through the seven-foot-thick ice covering Thule Bay. The conventional explosives in all four of the bombs went off, scattering plutonium. What was left sank and was never recovered. In 1989, the Pentagon disclosed that the nuclear warhead on board an A-4E Skywarrior attack jet (which sank after accidentally rolling off the U.S. aircraft carrier *Ticonderoga* 24 years earlier) was still on the bottom of the sea, only 80 miles from a Japanese island. They told the Japanese that the enormous water pressure at that depth had almost certainly burst the H-bomb, contaminating the ocean floor with plutonium.[45] In 1993, Russian scientists warned that plutonium from the nuclear warheads on board the nuclear submarine *Komsomolets,* which sank in Norwegian Sea in 1989, might soon begin leaking and poison important fishing grounds.[46]

In October 1986, a fire and explosion caused a Soviet nuclear ballistic missile submarine to sink in water three miles deep, 600 miles northeast of Bermuda. The sub went down with 16 ballistic missiles and 2 nuclear torpedoes aboard—a total of 34 nuclear warheads. Eight years later, Russian scientists informed American experts that the sub had broken up, that "the missiles and warheads were 'badly damaged and scattered on the sea floor,'" and that it was "certain that the warheads are badly corroded and leaking plutonium and uranium."[47] Russian and American scientists meeting at the Woods Hole Oceanographic Institute on Cape Cod issued a statement that the sub "contained the highest concentration of radioactivity of any of the many sources of high-level radioactivity dumped accidentally or purposefully onto the sea floor anywhere by any nation."[48] Strong ocean currents in the area raised the likelihood of wide environmental contamination.

According to a report by William Arkin and Joshua Handler issued in June 1989 and updated by Handler in 1994, there were more than 230 accidents involving nuclear-powered surface ships and submarines between 1954 and 1994. Although some of these were not very serious, many were. Arkin and Handler report that as of the late 1980s, there were "approximately forty-eight nuclear warheads and seven nuclear power reactors on the bottom of the oceans as a result of various accidents."[49]

Over all, for 1950-94, Appendix Tables 4-1 and 4-2 include 59 publicly reported nuclear weapons–related accidents involving American nuclear forces, and 30 more involving the forces of other nuclear-armed countries: 25 Soviet/Russian, 4 French and 1 British. (I do not claim that this is even a comprehensive list of publicly reported accidents, let alone a complete list of all the accidents that have occurred.) There is a curious pattern to these data. There are many fewer accidents reported publicly for the United States in the second half of this period than in the first, but there are many more reported for the Soviets/Russians in the second half than in the first. The American pattern could be the result of changes in policy. After 1968, the United States stopped routinely carrying fully capable nuclear weapons (with nuclear cores inserted) in American bombers, although they were still commonly deployed on land- and sea-based missiles. The Air Force also stopped insisting that part of the strategic nuclear bomber force had to be in the air constantly. The sometimes lengthy time delay in making reports of such accidents public, however, probably explains more. The increase in reported Russian accidents after 1970 might be due to greater openness, though only 8 of the 25 Russian accidents occurred after "openness" became official policy in 1985. Did the West become more determined or better at publicizing Russian mishaps after 1970? Or did the Russians simply become less careful?

In any case, it is reasonable to assume that there are nearly two and a half times as many U.S. as Soviet/Russian accidents listed because almost every kind of information is more available in American society. Given what has been learned about Soviet safety practices from events like the Chernobyl disaster, it is very unlikely that Americans are more accident prone. It is also hard to believe that the British have had only a quarter as many nuclear weapons–related accidents as the French, and impossible to believe the Chinese haven't had any.[50] Yet even assuming that every nuclear weapons–related accidents that ever occurred is included in the appendix tables, there were 89 from 1950 through 1994, *an average of 1 major nuclear weapons–related accident every 6 months for 45 years.* Fewer accidents (26) are listed between 1975 and 1994, the last 20 years of that period. But that is still an average of 1 major nuclear weapons–related accident every 9 months for 2 decades.

Other Accidents Related
to Weapons of Mass Destruction

There have been a variety of other potentially dangerous accidents involving systems related to nuclear and other weapons of mass destruction, associated equipment and facilities. Some of these are listed in Appendix Table 4-3, along with a number of cases of errant military missiles. Two of the most serious occurred years ago in the former Soviet Union. They are still shrouded in mystery and controversy.

In an accident that mimicked a biological warfare attack, a cloud of deadly anthrax spores was released into the air in April 1979 when a filter burst at Military Compound 19, a Soviet germ warfare facility in the Ural Mountains. As the population downwind inhaled the airborne spores, an epidemic of pulmonary anthrax broke out. Somewhere between 200 and 1000 people died. It could have been much worse. Prevailing winds had carried the organisms away from the largest part of the city of Sverdlovsk and over a sparsely populated area to the south-southeast. That, together with a late cold spell that sent temperatures below freezing, prevented the bacteria from claiming many more lives.[51]

The second incident occurred on September 12, 1990, when an explosion and fire at a nuclear fuel plant in Ust-Kamenogorsk in eastern Kazakhstan released a "poisonous cloud" of "extremely harmful" gas over most of the city's population of 307,000. It is unclear from reports whether the cloud was radioactive or some gaseous compound of the highly toxic element beryllium. There was no indication of casualties, but Kazakh officials did ask the central Soviet government to declare the area around the city an ecological disaster zone.[52]

The most serious American accident listed is the 1957 fire at the Rocky Flats nuclear weapons plant, only 17 miles from Denver. The fire may have burned up as much as 30 to 44 pounds of plutonium and resulted in the release of some plutonium and other radioactive materials to the atmosphere. Dr. Carl Johnson, the chief health official in the surrounding county in the 1970s, argued that there was a direct relationship between this and other accidents at Rocky Flats and abnormally high cancer rates in the area.[53]

The table also lists two examples of the military tendency to cover up questionable activities for a long time. It was not until 1975 that the U.S. Army revealed that 10 to 20 years earlier, three employees at their biological warfare research facility in Fort Dietrick, Maryland, had died of rare diseases being studied at the lab.[54] In 1975 and 1976, the U.S. Army also admitted that it had conducted 239 secret, "open air" germ warfare tests from 1945 through the 1960s in or near major American cities. In one test, live germs were released into the New York City subway tunnels.[55]

Design Flaws and Manufacturing Defects

Military systems—even those involving weapons of mass destruction—are not immune to such endemic problems of technical systems as flaws in design and manufacture (see Chapter 9). On May 23, 1990, nuclear weapons experts from DoE and Congress announced that more than 300 American tactical nuclear warheads deployed in Europe had been brought back to the United States and secretly repaired over the preceding year and a half. The massive recall was to correct design defects that could have led to accidental detonations of their conventional explosives powerful enough to disintegrate the core of the weapons and scatter plutonium into the environment.[56] Unfortunately, this does not seem to be all that rare an event. According to the *New York Times,* "Privately, Energy Department weapons specialists in California and former top officials of the Atomic Energy Commission . . . said that . . . [s]everal times in the 1950's, 1960's and 1970's, weapons found to lack proper safeguards have been returned to the Pantex [nuclear weapons manufacturing] plant for repairs."[57]

A few days later, it was reported that more than a dozen types of American nuclear warheads with different designs had been temporarily withdrawn for repairs since 1961.[58] On June 8, 1990, Secretary of Defense Cheney told the Air Force to withdraw hundreds of SRAM-A short-range nuclear-armed attack missiles from the bomber force until safety studies on their W-69 warheads were completed. Cheney did this two weeks after the directors of all three of the nation's nuclear weapons design laboratories (Los Alamos, Lawrence Livermore and Sandia) urged that the missiles be removed from bombers kept on 24-hour alert. The directors told the Senate Armed Services Committee about their concern that the warheads might explode, dispersing plutonium and other deadly radioactive materials, if they were subject to the extreme stress of a fire or crash.[59]

SPACE-BASED MILITARY NUCLEAR POWER ACCIDENTS

Nuclear-powered military satellites and spacecraft also pose a real danger. On April 21, 1964, an American SNAP-9A military navigation satellite carrying a nuclear generator fueled by plutonium-238 failed to attain orbit. It burned up in the atmosphere. Five years later, a group of Japanese scientists blamed the satellite for a near tripling in the measured fallout of plutonium over Japan between 1966 and 1967 (from 1,300 nanocuries to 3,300 nanocuries per square kilometer), along with a near tripling in the proportion of plutonium-238 (used in the satellite) as against plutonium-239 (used in nuclear weapons) in the fallout. The U.S. Atomic Energy Commission responded by saying that the radioactive fallout from the satellite posed no danger to fish or humans.[60]

Malfunctioning U.S. and Soviet moonshots have also created radioactive hazards. In January 1969, two separate Soviet nuclear-powered spacecraft headed for the moon were aborted in flight and released measurable radioactivity into the upper atmosphere. In April 1970, the ill-fated *Apollo 13* moon mission jettisoned its lunar lander into the Pacific Ocean. The lander was nuclear powered and its SNAP-27 plutonium power supply was not recovered; yet these facts were glossed over in reports of the mission.[61]

A launch malfunction on April 25, 1973, caused a Soviet nuclear-powered satellite to fall into the Pacific Ocean north of Japan.[62] Then, four years later, came the most spectacular and widely reported failure of a nuclear-powered satellite to date. In January 1978, after months of orbital decay, the crippled Soviet nuclear-powered military spacecraft *Cosmos 954* re-entered the atmosphere, scattering some 220 pounds of radioactive debris over an area of Canada the size of Austria. The satellite's reactor contained over 100 pounds of uranium-235. Thousands of radioactive fragments were recovered after a painstaking, months-long search, costing more than $12 million. Fortunately, the area over which the debris fell was only sparsely populated.[63] Five years later, after weeks haunted once again by the specter of radioactive debris raining down on populated areas, another nuclear-powered Soviet military satellite, *Cosmos 1402,* fell into the Indian Ocean.[64] These same fears were rekindled in the May 1988, when the Soviets disclosed they had lost contact with yet another nuclear-powered military satellite in serious trouble. Predicted to re-enter the earth's atmosphere sometime in the early fall, a last-minute maneuver by an automatic safety system on October 1 succeeded in separating the spacecraft's reactor and boosting it into a higher earth orbit.[65] In late 1996, an automated Russian Mars probe carrying 200 grams of plutonium failed to escape the earth, re-entered the atmosphere and plunged into the Pacific.[66]

The tragic explosion of the space shuttle *Challenger* during launch on January 28, 1986 took the lives of all of its crew. Yet the tragedy could easily have been much, much worse. The very next *Challenger* launch was scheduled to contain a liquid-fueled Centaur rocket carrying an unmanned spacecraft powered by 46.7 pounds of plutonium. Had *Challenger* exploded on that mission and triggered an explosion of the Centaur, the satellite's plutonium would have been dispersed into the air. According to John Goffman, former associate director of the Lawrence Livermore nuclear weapons laboratory, if that happened, "the amount of radioactivity released would be more than the combined plutonium radioactivity returned to earth in the fallout from all of the nuclear weapons testing of the United States, Soviet Union and the United Kingdom—which I have calculated has caused 950,000 lung cancer fatalities. If it gets dispersed over Florida, kiss Florida good-bye."[67]

Through 1988, the Soviets launched 33 military spy satellites powered by the fissioning of uranium-235, and at least 4 more spacecraft powered by the heat of naturally decaying radioactive materials aboard; the United States launched 1 satellite with a uranium-fueled reactor and 22 more spacecraft carrying a total of 37 heat generators powered by plutonium-238. Many of the 70-plus nuclear-powered systems in space are in near-earth orbit. Nine nuclear-powered American and Soviet spacecraft—more than 10 percent of the number in space—have failed to achieve orbit or otherwise re-entered the atmosphere. Yet in 1988, the Soviet Academy of Sciences revealed plans for a spaceship whose main engines would be nuclear driven; an expert panel of the American Physical Society reported (after studying secret federal documents) that some of the U.S. military's "Star Wars" plans for strategic missile defense called for 100 or more orbiting nuclear reactors to power space-based weapons, radars and sensors.[68]

Aside from the problems of failure to attain orbit and orbital decay, there is also the danger posed by "space junk." In 1997, it was estimated that about 8,000 objects, from used up satellites to lost tools, were circling the earth.[69] If a piece of this space debris smashed into an orbiting reactor, it could shatter it into "a million tiny pieces." Nicholas Johnson, a scientist at Teledyne Brown Engineering and an expert on space debris, estimated that some 1.6 tons of radioactive reactor fuel had been put into earth orbit by the late 1980s. As to the chances that this radioactive material would be involved in a space collision, Johnson said that if the number of satellites and space debris continued its pattern of increase, it was "a virtual certainty."[70]

ACCIDENTS WITH NUCLEAR WASTE

On July 23, 1990, more than 30 years after the spectacular explosion of Soviet nuclear weapons waste in the Ural Mountains (described earlier), the report of a U.S. government advisory panel warned that it could happen here. The panel was headed by John Ahearne, a physicist who was formerly a high official in the Departments of Defense and Energy and chair of the U.S. Nuclear Regulatory Commission. According to the authoritative study, there are millions of gallons of highly radioactive waste, accumulated during four decades of producing plutonium for nuclear weapons, stored in 177 tanks at the Department of Energy's Hanford nuclear reservation in Washington state. Heat generated inside the tanks or a shock from the outside could cause one or more of them to explode. Although the explosion would not itself be nuclear, it would throw a huge amount of radioactivity around. How imminent is the danger?

According to the panel, "Although the risk analyses are crude, each successive review of the Hanford tanks indicates that the situation is a little worse." According to the *New York Times*, "Experts outside the Department of Energy say the risk of explosion is now so high that the department has imposed a moratorium on all activity at one tank because of fear that a jolt or spark could detonate the hydrogen that has built up."[71]

In 1992, DoE released another report on Hanford, listing four decades of accidents at that nuclear weapons facility. It was the most thorough overview of human error, management mistakes and sloppiness at Hanford ever undertaken. It looked at fires, accidental explosions, failures of safety systems, incidents of fuel melting, and other events that exposed workers to radioactivity and toxic chemicals. The site was found to be extensively contaminated, and it was judged that the waste tanks were still in danger of exploding or otherwise dispersing radioactive material.[72]

On May 14, 1997, there was an explosion at Hanford's facility for recovering plutonium from nuclear waste. No one was in the chemical-mixing room where the explosion occurred, but liquid spilled out of the building and eight workers just outside reported a metallic taste in their mouths. The government claimed there was no evidence of toxic chemical or radiation exposure, but the workers were examined at a nearby hospital and subsequently sent home for the day.[73]

Only a few months earlier, another plutonium-recovery facility had been the site of the worst nuclear accident in Japan's history. At about 10:00 A.M. on March 11, 1997, in Tokai, a fire started in the remote-controlled chamber in which liquid waste from plutonium extraction is mixed with asphalt. Ten hours later, an explosion blew out most of the windows and some of the doors in the four-story concrete building. Containment systems failed, causing 37 workers to be exposed to radiation and allowing plutonium and other radioactive materials to escape into the air. Radiation was subsequently detected up to 23 miles away. Plant managers and the government said (as usual) that neither the workers nor the public had been exposed to harmful levels of radiation.[74]

ACCIDENTS WITH OTHER DANGEROUS TECHNOLOGIES

There have been many serious accidents involving dangerous technologies other than weapons of mass destruction, space-based reactors and nuclear waste—toxic chemicals and nuclear power among them. But, for a number of reasons, there is little point in exploring these here in detail. Our detailed exploration of the consequences and likelihood of weapons accidents clearly illustrates the disastrous potential of accidents involving dangerous technologies. There is no need to belabor the point. Beyond this, there is already an extensive and accessible literature on nuclear power that includes considerable discussion of the likelihood and consequences of accidents. And accidents at nuclear power plants and facilities handling dangerous chemicals also fit more naturally into and illustrate more strikingly our subsequent discussions of technical failure and human error (especially in Chapters 9 and 11).

Accidents involving these other dangerous technologies are a very serious problem. One of the largest accidental explosions in U.S. history, the Texas City, Texas, disaster of 1947, began with a fire in a French freighter carrying 2,300 tons of ammonium nitrate. The explosion caused by the fire blew out every window in this city of 15,000.[75] Shrapnel rained down for nearly a mile from the site of the explosion, triggering other fires and explosions in chemical tanks and ships. It was a week before all the fires were out. Nearly 600 people died and 3,500 more were injured.[76]

Accidents during shipment of hazardous chemicals are quite common. There were more than 200 railroad accidents in the U.S. involving hazardous chemicals in 1987 alone. Some came close to disaster. For example, on February 29, 1987, a 21-car train with 9 tank cars filled with butane derailed in Akron, Ohio. The butane in 2 of the tankers escaped, starting a fire that spread to a nearby chemical plant with large storage tanks on-site. It wouldn't have taken much more to turn what was a routine accident into a calamity.[77]

Although all accidents with dangerous technologies are cause for concern, those involving weapons of mass destruction have among the greatest catastrophic potential. As long as massive arsenals of these weapons continue to exist, accidents will continue to happen. The more of these weapons there are, the more actively they are moved around, the more nations possess them, the greater the probability of devastating accident. It is ironic. We have built these weapons to make ourselves secure. But in a post–Cold War world in which overwhelming nuclear deterrence adds little to the nation's security, the threat of accidental disaster these weapons pose has made them a source of insecurity instead.

The problem of individual weapons accidents is serious enough. But there is an even greater potential for disaster in the possibility that these weapons will again be used in war someday. And, as we shall see in the next chapter, that too can happen by accident.

APPENDIX

TABLE 4-1

MAJOR U.S. NUCLEAR
WEAPONS–RELATED ACCIDENTS: A CHRONOLOGY

DATE	WEAPON SYSTEM	LOCATION	DESCRIPTION
*Feb 13, 1950	B-36 Bomber	Pacific Ocean off Puget Sound, Washington	B-36 on simulated combat mission drops nuclear weapon from 8,000 feet then crashes. Chemical explosives detonate
*Apr 11, 1950	B-29 Bomber	Manzano Base, New Mexico	3 minutes after takeoff from Kirtland AFB, plane crashes into mountain, killing crew. Nuclear warhead case destroyed, high explosive burned
*Jul 13, 1950	B-50 Bomber	Lebanon, Ohio	B-50 on training mission from Biggs AFB crashes, killing 16 crew. Chemical explosive of nuclear weapon detonates on impact
Fall 1950	B-50 Bomber	Quebec, Canada	B-50 bomber, returning from U.S. military exercise, accidentally drops nuclear warhead into the St. Lawrence. 2.5 tons of conventional explosive detonates, dispersing HEU into river (*Christian Science Monitor Radio,* broadcast 2/18/97, KERA-FM Dallas)
*Aug 5, 1950	B-29 Bomber	Fairfield-Suisun AFB California	B-29 with nuclear weapon crashes on takeoff, killing 19. Chemical explosive detonates
*Mar 10, 1956	B-47 Bomber	Mediterranean Sea	B-47 with "two capsules of nuclear weapon material" disappears without a trace
*Jul 26, 1956	B-47 Bomber	U.S. Base at Lackenheath, U.K.	Unarmed B-47 crashes into "igloo" storing 3 nuclear bombs, each with 4 tons of chemical explosive. USAF general said, had blazing jet fuel set off chemical explosive, "part of Eastern England [might] have become a desert."
*May 22, 1957	B-36 Bomber	New Mexico	Mark 17, 10-megaton H-bomb accidentally drops on desert near Kirtland AFB, Albuquerque. There is a nonnuclear explosion
*Jul 28, 1957	C-124 Transport	Atlantic Ocean	En route from Dover AFB, C-124 loses power in two engines and jettisons two nuclear warheads over ocean. The weapons were never located.
*Oct 11, 1957	B-47 Bomber	Homestead AFB, Florida	B-47 with nuclear warhead and nuclear capsule aboard crashes shortly after takeoff. "Two low order detonations occurred during the burning"
Dec 12, 1957	B-52 Bomber	Fairchild AFB, Spokane, Washington	Crash on takeoff during "training mission"
*Jan 31, 1958	B-47 Bomber	U.S. Strategic Air Command Base at Sidi Slimane, Morocco	B-47 crashes during takeoff and burns for 7 hours with one nuclear warhead aboard "in strike configuration. . . . There was some contamination in the immediate area of crash"

DATE	WEAPON SYSTEM	LOCATION	DESCRIPTION
*Feb 5, 1958	B-47 Bomber	Hunter AFB Savannah, Georgia	B-47 on simulated combat mission out of Homestead AFB, Florida, collides midair with F-86. Nuclear weapon jettisoned by bomber after crash never found
Feb 12, 1958	B-47 Bomber	Off coast near Savannah, Georgia	No details available
Mar 5, 1958	B-47 Bomber	Off coast of Georgia	Atomic weapon jettisoned after midair collision
*Mar 11, 1958	B-47 Bomber	Florence, South Carolina	Malfunction of bomb-lock system causes Hunter AFB bomber to accidentally jettison A-bomb. High explosive detonates
Apr 10, 1958	B-47 Bomber	12 miles south of Buffalo, New York	Plane explodes in midair while approaching refueling tanker plane
*Nov 4, 1958	B-47 Bomber	Dyess AFB, Texas	High explosive in nuclear warhead goes off when bomber crashes because of fire shortly after takeoff
*Nov 26, 1958	B-47 Bomber	Chenault AFB, Louisiana	Catches fire and burns while on the flight line. Fire destroys one nuclear weapon aboard
*Jan 18, 1959	F-100 Fighter	Pacific Base	F-100 loaded with one unarmed nuclear warhead catches fire on ground
*Jul 6, 1959	C-124 Transport	Barksdale AFB, Louisiana	Plane crashes and burns, destroying one nuclear weapon being transported
*Sep 25, 1959	Navy P-5M Aircraft	Whidbey Island, Washington	Antisubmarine aircraft ditches in Puget Sound, with unarmed nuclear antisubmarine weapon. (*NY Times,* May 26, 1981)
*Oct 15, 1959	B-52 Bomber	Near Glen Bean, Kentucky	B-52 and KC-135 tanker collide in midair. Two nuclear weapons recovered undamaged
Before June 8, 1960	Nuclear Warhead	U.S. AFB near Tripoli, Libya	No details (*NY Times* report, June 8, 1960)
Before June 8, 1960	Nuclear Warhead	England	No details (*NY Times* report, June 8, 1960)
*June 7, 1960	Bomarc Surface-to-Air Missile	McGuire AFB, New Jersey	Antiaircraft missile damaged in explosion and fire; nuclear warhead is destroyed by the fire
Before 1961	Corporal Missile	Tennessee River	Missile carrying nuclear warhead rolls off truck and into Tennessee River
Jan 19, 1961	B-52 Bomber	Monticello, Utah	B-52 explodes in flight
*Jan 24, 1961	B-52 Bomber	Near Goldsboro, North Carolina	Bomber carrying two 24-megaton nuclear weapons crashes. Portion of one bomb never recovered; other fell into field without exploding, but 5 of 6 interlocking safety switches had been triggered by its fall
*Mar 14, 1961	B-52 Bomber	California	B-52 from Beale AFB carrying nuclear weapons crashes on training mission
Jun 4, 1962	Thor ICBM	Johnston Island, Pacific Ocean	A one-megaton nuclear warhead is destroyed in flight when Thor explodes at U.S. Pacific missile test range

DATE	WEAPON SYSTEM	LOCATION	DESCRIPTION
Jun 20, 1962	Thor ICBM	Johnston Island, Pacific Ocean	A second one-megaton warhead is destroyed in attempt to launch a high-altitude test shot
Apr 10, 1963	SSN *Thresher*	U.S. Atlantic Coastline	Nuclear attack submarine sinks on maiden voyage, presumed to be carrying SUBROC nuclear-armed missile
*Nov 13, 1963	Nuclear Weapons Storage Igloo	Medina Base, San Antonio, Texas	Explosion involving 123,000 pounds of high-explosive components of nuclear weapons; "little contamination" from nuclear components stored elsewhere in the same building
*Jan 13, 1964	B-52 Bomber	Cumberland, Maryland	B-52 from Turner AFB, Georgia, crashes with two nuclear weapons on board
*Dec 5, 1964	Minuteman I ICBM	Ellsworth AFB, South Dakota	LGM 30B Minuteman I missile on strategic alert seriously damaged when "retrorocket" fires accidentally during repairs
*Dec 8, 1964	B-58 Bomber	Bunker Hill AFB, Indiana	B-58 carrying at least one nuclear weapon catches fire and burns on the flight line; part of nuclear weapon burns
Aug 9, 1965	Titan II ICBM	Little Rock AFB, Arkansas	Explosion in missile silo, followed by fire
*Oct 11, 1965	C-124 Transport	Wright-Patterson AFB, Ohio	Plane carrying only nonexplosive components of nuclear weapons catches fire and burns on flight line during refueling stop.
*Dec 5, 1965	A-4E Skywarrier Attack Jet	200 miles East of Okinawa	Jet loaded with B43 nuclear warhead rolls off No.2 elevator of aircraft carrier *Ticonderoga* and sinks in 16,000 feet of ocean
*Jan 17, 1966	B-52 Bomber	Palomares, southern Spain	B-52 with four 20 to 25 megaton H-bombs collides with KC-135 tanker. One bomb lands undamaged; chemical explosives in two others go off, scattering plutonium widely over fields; one recovered intact from Mediterranean after intensive 3-month underwater search.
*Jan 21, 1968	B-52 Bomber	Thule Bay, Greenland	B-52 crashes,burns,melts through 7-foot-thick ice and sinks. Chemical explosives in all 4 H- bombs on board go off, scattering plutonium. (S.Sagan, *Limits of Safety,* Princeton U. Press, 1993, p.180)
Feb 12, 1968	B-52 Bomber	Near Toronto, Canada	Crash with nuclear weapons reportedly on board, some 20 miles north of Toronto
*May 27, 1968	SSN *Scorpion*	Mid-Atlantic Ocean	Mechanical problems cause sinking of nuclear attack submarine carrying two ASTOR nuclear-armed torpedoes
June 16-17, 1968	U.S. Cruiser *Boston*	Off Tonkin Gulf, Vietnam	The *Boston,* carrying Terrier nuclear-armed antiaircraft missiles, is accidentally attacked by U.S. fighter plane(s)
Jan 14, 1969	Nuclear-powered aircraft carrier *Enterprise*	75 miles south of Pearl Harbor, Hawaii	Series of conventional weapons explosions and fires aboard this nuclear weapons–capable aircraft carrier; 25 dead, 85 injured (*NY Times,* Jan.30,1969) The *Enterprise* was armed with nuclear weapons.

DATE	WEAPON SYSTEM	LOCATION	DESCRIPTION
Apr 3, 1970	B-52 Bomber	Ellsworth AFB, South Dakota	SAC bomber crashes on landing, coming to rest on fire, on top of 6 underground fuel storage tanks holding 25,000 gallons each (*Air Force Magazine,* Dec. 1970)
Nov 29, 1970	Submarine Tender *Canopus*	Holy Loch, Scotland	*Canopus* burns with Polaris nuclear missiles aboard and two Polaris missile submarines moored alongside (*Daily Telegraph,* Nov. 30, 1970)
Nov 1975	Guided Missile Cruiser *Belknap*	70 miles east of Sicily	After crashing with the aircraft carrier *J.F. Kennedy* during maneuvers in the Mediterranean, the *Belknap* suffered extensive fires and ordnance explosions. Both ships carrying nuclear weapons
Aug 24, 1978	Titan II ICBM	Rock, Kansas	Propellant leak causes toxic vapor cloud to spew from silo for 29 hours; two airmen die, several injured, hundreds evacuated (*Wichita Eagle-Beacon,* Aug.26&28, 1978). See parallel with Sep. 19, 1980 Titan II accident.
1980	FB-111 Fighter Bomber	Off Coast of New England	Nuclear-armed fighter bomber crashes
Sep 15, 1980	B-52 Bomber	GrandForks AFB, North Dakota	Nuclear-armed B-52 catches fire on runway
*Sep 19, 1980	Titan II ICBM	Damascus, Arkansas	Socket wrench dropped by Air Force repairman punctures Titan II's fuel tank, causing leak that leads to explosion demolishing silo door; 9-megaton nuclear warhead hurled from silo, but later recovered intact
Jan 11, 1985	Pershing 2 Ballistic Missile	Waldheide Army Base, near Heilbronn, Germany	First-stage engine of missile suddenly catches fire and burns during a "routine training exercise"; 3 killed, 7 injured; truck and maintenance tent also ignited (Assoc.Press in *West Palm Beach Post,* Jan.12, 1985)
*Sep 28, 1987	B-1B Bomber	Colorado	Pelican strikes unreinforced part of right wing near engine intake, causing fire. Plane rolls out of control and is destroyed (*NY Times,* Jan.21,1988)
Nov 8, 1988	B-1 Bomber	Dyess AFB Abilene, Texas	Fire breaks out in two of four engines, crew ejects, plane crashes and explodes (*NY Times,* Nov. 10, 1988)
Nov 17, 1988	B-1 Bomber	Ellsworth AFB, South Dakota	Bomber crashes trying to land in bad weather; explosion and fire. Crash occurred just 9 days after earlier crash causes grounding of all B-1s except those on alert and loaded with nuclear bombs (*NY Times,* Nov. 10 & 19, 1988)
Jul 24, 1989	B-52 Bomber	Kelly AFB, San Antonio, Texas	Bomber on ground catches fire and explodes; 1 killed, 11 injured (*NY Times,* July 26,1989)
Jun 24, 1994	B-52 Bomber	Fairchild AFB, Washington	Bomber crashes, killing all 4 crew; pilot reportedly avoided crashing into nuclear weapons storage area by pulling plane into fatal stalling turn (*NY Times,* Apr. 27, 1995)

* Officially acknowledged by the Pentagon

Sources: U.S. Department of Defense, "Nuclear Weapons Accidents, 1950-1980," as reprinted with comments by Center for Defense Information (Washington, DC), in *The Defense Monitor* (Vol. 9, No. 5, 1981). "Accidents of Nuclear Weapons Systems," in *World Armaments and Disarmament Yearbook* (Stockholm International Peace Research Institute, 1977), pp. 65-71. Arkin, William M. and Handler, Joshua M., "Nuclear Disasters at Sea, Then and Now," *Bulletin of the Atomic Scientists* (July-August 1989), pp. 20-24. Department of Defense, *Summary of Accidents Involving Nuclear Weapons, 1950-1980 (Interim),* as cited in Talbot, Stephen, "Nuclear Weapons Accidents: The H-Bombs Next Door," *The Nation* (February 7, 1981), pp. 145-147. "A Deadly White Blur," *Science* (Jan. 29, 1988), p. 453. Talbot, Stephen, "Broken Arrow: Can a Nuclear Weapons Accident Happen Here?," *KQED-TV* (San Francisco: Channel 9 Public Affairs, PBS, 1980). As cited in table.

TABLE 4-2

MAJOR NUCLEAR WEAPONS–RELATED
ACCIDENTS, OTHER THAN U.S.

DATE	WEAPON SYSTEM	LOCATION	DESCRIPTION
1957-58	**Soviet** Nuclear Waste (from nuclear warhead production)	Kyshtym, Southern Ural Mountains, USSR	Underground nuclear waste storage area explodes "like a volcano." 375 square mile area contaminated: hundreds of people die, thousands are affected
Jan 30, 1968	**British** Vulcan Bomber	Cottesmore, Rutland, United Kingdom	Strategic bomber crashes, burns; RAF claims no nuclear warheads aboard, but RAF firemen sent are equipped with radiation monitors (*Times,* Jan. 31, 1968)
Apr 11, 1968	**Soviet** *Golf* Class Nuclear Missile Submarine	Pacific Ocean, 750 miles west of Oahu	Submarine, carrying three ballistic missiles and probably two nuclear torpedoes, sinks after a series of explosions on board
Before 1970	**Soviet** Military Aircraft (unspecified)	Sea of Japan	American military reportedly recovers nuclear weapon from Soviet aircraft that crashed (*NBC Nightly News,* Mar. 19, 1975)
Early Feb 1970	**Soviet** Nuclear Submarine Construction Facility	Gorki, USSR	Large explosion rocks main Soviet nuclear submarine shipyard; several killed, radioactive wastes contaminate Volga River (*Daily Telegraph,* Feb. 21,1970 *Japan Times,* Feb.22, 1970)
Apr 12, 1970	**Soviet** *November* Class Nuclear Attack Submarine	300 Nautical Miles Northwest of Spain	While in heavy seas, submarine develops serious nuclear propulsion problem. Fails to rig towline to nearby merchant ship and apparently sinks. Submarine probably carrying 2 nuclear torpedoes
Feb 25, 1972	**Soviet** Nuclear Missile Submarine	North Atlantic Ocean, off Newfoundland	Submarine crippled by unknown causes, wallows in high seas (*Times,* Mar. 1, 1972; *International Herald Tribune,* Mar. 10,1972)
Dec 1972	**Soviet** Submarine	Off North America (East Coast)	Nuclear torpedo ruptures, leaking radiation (*SF Chronicle,* May1,1986)
Mar 30, 1973	**French** Mirage IV Stategic Bomber	Atlantic Ocean, near France	Bomber with landing gear problem told to ditch at sea; claim not nuclear-armed. Nuclear-armed Mirage carries bombs under wing, which could explain reluctance to land with landing gear problem (*LeMonde,* Apr. 1, 1973)
May 15, 1973	**French** Mirage IV Strategic Bomber	Luxueil, France	Strategic bomber crashes on takeoff (*Le Monde ,* Sep. 28, 1973)
June 18, 1973	**French** Mirage IV Strategic Bomber	Near Bellegard, France	Bomber crashes, allegedly on training mission (*Times,* Jun. 19, 1973)
Aug 31, 1973	**Soviet** Nuclear Missile Submarine	Atlantic Ocean	Sub, carrying 12 nuclear-armed SLBMs, has accident in missile tube
Sep 1973	**Soviet** Nuclear Missile Submarine	Carribean Sea	U.S. aircraft spot surfaced Soviet submarine with 8- foot gash on deck (*Daily Telegraph,* Sep. 6, 1973)
Sep 27, 1973	**French** Mirage IV Strategic Bomber	Off Corsica, in Mediterranean Sea	Bomber crashes when engine seizes due to oil loss (*Le Monde,* Sep. 28, 1973)
Sep 1974	**Soviet** *Kashin* Class Guided Missile Destroyer	Black Sea	Ship reportedly explodes and sinks (*New York Times,* Sep. 27, 1974)

DATE	WEAPON SYSTEM	LOCATION	DESCRIPTION
Sep 8, 1977	Soviet *Delta* Class Nuclear Missile Submarine	Off Coast of USSR, near Kamchatka	250-kiloton nuclear warhead accidentally thrown high into air when malfunction forces crew to open missile tube to correct dangerous pressure buildup and clear smoke filling missile compartment. Warhead falls into sea (later recovered)
1979–1980	Soviet "Echo" Class Submarine	Pacific Ocean	Sub, armed with nuclear torpedoes, is badly damaged in collision with another sub (rumored to be Chinese and to have sunk).
Sep 1981	Soviet Submarine	Baltic Sea	Series of powerful jolts rock sub; rupture in nuclear reactor; radiation leaks; some crew contaminated (*SF Chronicle,* May1,1986)
Oct 27, 1981	Soviet Submarine	Near Karlskrona Naval Base, Sweden	Diesel-powered sub believed to be nuclear armed runs aground in Swedish waters, unable to move for 10 days (*NY Times,* Nov. 6, 1987)
Jun 1983	Soviet Nuclear Powered Sub	North Pacific Ocean	Soviet submarine sinks, with about 90 people aboard (*NY Times,* Aug.11, 1983)
May 13, 1984	Soviet Northern Fleet Ammunition Depot	Severomorsk, Russia (on Barents Sea)	Huge explosion, then series of explosions and fires destroys major ammunition stocks, killing 200-300 people; over 100 nuclear-capable missiles destroyed; no evidence of nuclear explosion or radiation (*NY Times,* Jun. 23, 26 & Jul. 11, 1984)
Sep 20, 1984	Soviet *Golf 2-* Class Nuclear Submarine	Sea of Japan	Missile fuel catches fire, disabling nuclear-armed sub, setting it adrift (*Dallas Times-Herald,* Sep 21,1984)
Aug 10, 1985	Soviet *Echo-* Class Submarine	Chazma Bay (near Vladivostok)	One reactor explodes during refueling
Dec 1985	Soviet *Charlie/ Victor-*Class Submarine	Pacific Ocean	Human error causes reactor meltdown
1986	Soviet *Echo-* Class Submarine	Cam Ranh Bay, Vietnam	While on combat duty (with nuclear-armed torpedoes and cruise missiles) radiation detectors jump when crew adds wrong chemicals to reactor cooling system (possible meltdown)
Oct 6, 1986	Soviet *Yankee-* Class Nuclear Missile Submarine	600 miles Northeast of Bermuda	Liquid missile fuel catches fire, causes explosion. Submarine sinks, carrying 16 ballistic missiles and two nuclear torpedoes; 34 nuclear warheads aboard, (*NY Times,* Oct. 7, 1986)
Jun 26, 1987	Soviet *Echo 2-* Class Nuclear Submarine	Norwegian Sea	Pipes burst, crippling sub's nuclear reactor; sub carries nuclear warheads
Apr 17, 1989	Soviet *Mike-* Class Nuclear Attack Submarine	Norwegian Sea	Fire, reportedly caused by short circuit, causes submarine to sink with two nuclear torpedoes aboard
Dec 5, 1989	Soviet *Delta IV-* Class Submarine	White Sea	Control of missile lost during missile test launch accident
Mar 20, 1993	Russian Nuclear Missile Submarine	Barents Sea, Close to Kola Peninsula	Sub, carrying 16 nuclear-armed ballistic missiles, collides with U.S. nuclear-powered sub

Sources: Medvedev, Zhores A., "Two Decades of Dissidence," *New Scientist* (November 4, 1976); see also Medvedev, Z.A., *Nuclear Disaster in the Urals* (New York: Norton, 1979). Arkin, William M. and Handler, Joshua M., "Nuclear Disasters at Sea, Then and Now," *Bulletin of the Atomic Scientists* (July-August 1989), pp. 20-24. "Accidents of Nuclear Weapons Systems" in *World Armament and Disarmament Yearbook* (Stockholm International Peace Research Institute, 1977), pp.74-78. Handler, Joshua M., "Radioactive Waste Situation in the Russian Pacific Fleet, Nuclear Waste Disposal Problems, Submarine Decommissioning, Submarine Safety, and Security of Naval Fuel" (Washington, D.C.: Greenpeace, October 27, 1994; unpublished). As cited in table.

TABLE 4-3

STRAY MISSILES AND MISCELLANEOUS ACCIDENTS
RELATED TO WEAPONS OF MASS DESTRUCTION

DATE	SYSTEM	LOCATION	DESCRIPTION
1945-1969	U.S. Biological Warfare Tests	United States	Army report reveals it carried out 239 secret "open air" tests of "simulated" germ warfare around U.S. Includes dumping of bacteria into ocean near San Francisco (1950) resulting in 11 hospitalized cases of rare urinary infection; release of bacteria in NYC subways (1966) (*NY Times,* Dec. 23, 1976 & Mar. 9, 1977)
1951, '58, '64	U.S. Biological Warfare Laboratory	Fort Dietrick, Maryland	3 employees die of rare diseases (biologist & electrician of anthrax, caretaker of Bolivian hemorrhagic fever) being studied for possible combat use. Kept secret until 1975. (*NY Times,* Sep. 20, 1975)
Spring 1953	U.S. Nuclear Weapons Tests in Atmosphere	Lincoln County, Nevada a few miles to 120 miles east of Nevada Test Site	4,300 sheep grazing downwind of nuclear weapons tests die after absorbing 1,000 times maximum dose of radioactive iodine allowable for humans (*NY Times,* Feb. 15,1979)
Dec 5, 1956	U.S. Snark Guided Missile	Patrick AFB, Florida	Missile launched on "closed circuit" mission, fails to turn, flies 3,000 miles, crashes into jungle in Brazil (*NY Times,* Dec. 8, 1956)
1957	U.S. Nuclear Weapons Plant	Rocky Flats Plant, near Denver, Colorado	Fire results in escape of plutonium & other radioactive material. Plutonium-contaminated smoke escapes from plant for half a day. 30-44 pounds of plutonium may have burned up (*NY Times,* Sep. 9, 1981)
Early 1960	U.S. Matador Missile	Straits of Taiwan	Older version of U.S. cruise missile launched from Taiwan accidentally turns around and heads toward China; refused to self-destruct (*NY World Telegram,* Mar. 14, 1960)
1967	U.S. Mace Missile	Carribean Sea	Older version of cruise missile accidentally flies over Cuba after launch from Florida; refused to self-destruct (*Miami Herald,* Jan. 5, 1967)
Mar 1969	U.S. VX Nerve Gas	Skull and Rush Valleys Near Dugway Proving Grounds, Utah	6,000 sheep die in accident when wind carries nerve gas sprayed from Air Force jet well beyond test site
Apr 30, 1970	**French** Masurca Missile	Le Lavandou, France	Accidental missile launch from French ship. Missile lands near Riviera resort beach with many bathers. Windows and doors blown out but claim cause was missile shock wave, not explosion. (*Le Monde,* May 3, 1970)
Jul 11,1970	U.S. Athena Missile (used to flight test warheads and other ICBM components)	White Sands, New Mexico	50-foot, 8-ton Athena flies into Mexico by accident and crashes in remote area. Some release of radioactivity from small cobalt nose cone capsule. Third missile to crash into Mexico in 25 yrs.(*Japan Times* & UPI July 13, 1970; & *Daily Telegraph* Aug. 5, 1970)

DATE	SYSTEM	LOCATION	DESCRIPTION
Apr 1979	**Soviet** Biological Warfare Facility	Military Compound 19, near Sverdlovsk, Ural Mtns., USSR	200-1,000 near city die in anthrax epidemic resulting from explosion at secret germ warfare plant that released I-21 anthrax strain to air. (*NY Times,* Dec. 28, 1998; Jul 16, 1980); *Science News,* Aug. 2, 1980
Aug 7, 1979	**U.S.** Factory Producing Nuclear Fuel for U.S. Navy Submarines	Tennessee	300-3,000 grams of highly enriched uranium dust is accidentally released from plant's vent stack when a pipe clogs; up to 1000 people contaminated. Officials claim no health hazard (*NY Times,* Oct. 31, 1980)
Oct 18, 1982	**U.S.** GB Nerve Gas	Blue Grass Army Depot, Kentucky	Sensors register "small leak" of nerve gas; 200-300 evacuated; army claims no danger; two dead cows found on post; army says cows hit by truck (Assoc. Press, *West Palm Beach Post,* Oct. 20, 1982)
Dec 28, 1984	**Soviet** Cruise Missile	Barents Sea, off Northern Norway	Errant missile launched by Soviet Navy overflies Finland and Norway at high speed; Norway says it is cruise missile from submarine; Denmark says it probably is 1954-type missile used as target in USSR Navy exercises (*NY Times,* Jan. 4, 1985)
Jun 10, 1987	**US** NASA Rockets	Wallops Island, Virginia	Lightning causes accidental launch of three small NASA rockets; direction of flights not tracked because of unplanned firings (*Aviation Week & Space Technology,* Jun. 15, 1987)
Sep 12, 1990	**Soviet** Nuclear Fuel Plant	Ust-Kamenogorsk, Kazakhstan, USSR	Explosion releases "poisonous cloud" of "extremely harmful" (toxic? or radioactive?) gas over much of the city's 307,000 people; plant uses beryllium, which is toxic but not radioactive; Kazakh officials ask Moscow to declare city an ecological disaster zone (*NY Times,* Sep. 29, 1990)

Sources: Dugway nerve gas accident: Hearings on "Chemical and Biological Warfare: US Policies and International Effects," before Subcommittee on National Security Policy and Scientific Developments of the House Committee on Foreign Affairs, 91st Congress, 1st Session, December 2,9,18, and 19, 1969, p. 349; Hersh, Seymour, "Chemical and Biological Weapons: The Secret Arsenal," reprinted from *Chemical and Biological Warfare: America's Hidden Arsenal,* reprinted by Committee for a Sane Nuclear Policy, 1969. As cited in table.

CHAPTER FIVE

Holocaust by Accident: Inadvertent War with Weapons of Mass Destruction

In January 1987, the Indian Army prepared for a major military exercise near the bordering Pakistani province of Sind. Because Sind was a stronghold of secessionist sentiment, Pakistan concluded that India might be getting ready to attack and moved its own forces to the border. The two nations had already fought three wars with each other since 1947. Both of them were now nuclear-capable: India had successfully tested a nuclear explosive device more than a decade earlier; Pakistan was widely suspected of having clandestine nuclear weapons. The buildup of forces continued until nearly one million Indian and Pakistani troops tensely faced each other across the border. The threat of nuclear war hung in the air as they waited for the fighting to begin. Then, after intense diplomatic efforts, the confusion and miscommunication began to clear and the crisis was finally defused. India and Pakistan had almost blundered into a catastrophic war by accident.[1] More than a decade later, both nations have made great progress in developing their nuclear arsenals and no progress in resolving the tensions that brought them so close to disaster.

More than a few times, nations have found themselves fighting wars that began or escalated by accident or inadvertence. World War I is a spectacular example. By 1914, two alliances of nations (the Triple Alliance and the Triple Entente), locked in an arms race, faced off against each other in Europe. Both sides were armed to the teeth and convinced that peace could be and would be maintained by the balance of power they had achieved: in the presence of such devastating military force, surely no one would be crazy enough to start a war.

Then on June 28, 1914, Archduke Francis Ferdinand of Austria-Hungary and his wife were assassinated by a Serbian nationalist, an event that would have been of very limited significance in normal times. But these were not normal times. In the presence of enormous, ready-to-go military forces, the assassination set in motion a chain of events that rapidly ran out of the control of Europe's politicians and triggered a war that no one wanted.

Within two months, over a million men were dead. By the time it was over, 9 to 11 million people had lost their lives in a sickeningly pointless war, the worst the world had ever seen. The terms of surrender added to the damage by demanding heavy reparations from the losers. This helped to wreck the German economy, sowing the seeds of the rise to power of one of the most brutal, aggressive and genocidal regimes in all of human history. Within two decades, the Nazis and their allies once again plunged the world into war—a far more destructive war—that left some 40 to 50 million people dead (half of them civilians) and gave birth to the atomic bomb. The American bomb project was born primarily out of fear that the Nazi Germans might develop such a weapon first.

There may be many links in the chain, but there *is* in fact a connected chain of events leading from the accidental war we call World War I to the ovens of Auschwitz, the devastation of World War II, and the fear of nuclear holocaust that cast such a pall over the second half of the twentieth century. Yet the whole terrible chain of events might have been prevented, but for a simple failure of communications. The kaiser had sent the order that would have stopped the opening move of World War I (the German invasion of Luxembourg on August 3, 1914) before the attack was to begin. But the message arrived 30 minutes late. In a classic understatement, the messengers who finally delivered the belated order said, "a mistake has been made."[2]

NUCLEAR WAR STRATEGY AND ACCIDENTAL WAR

The American military's master plan for nuclear war, the Single Integrated Operating Plan (SIOP), is a complex and rigid set of integrated nuclear battle plans. SIOP bears a frightening resemblance to the mechanistic approach to war planning that led us to blunder into World War I. Reflecting on an interview with General Robert Herres, commander-in-chief of North American Aerospace Defense Command (NORAD), Daniel Ford writes, as General Herres described the precise time schedules . . . in the SIOP . . . he could just as well have been describing the famous Schlieffen plan . . . the meticulously crafted war plan . . . implemented at the outset of World War I. The Schlieffen plan was exceedingly rigid and mechanical, as were the plans of the opposing powers. . . . All it took was a catalyst—the assassination . . . to set the plans in motion. The various war machines began to work, and they stimulated each other in ways that no one could control. . . .

The present SIOP is such a machine, and once an alert began, it could quickly proceed to what . . . General Brent Scowcroft referred to as "the automatic phase of the war . . . the time at which the quick response systems are discharged against predetermined targets . . . and the battle plan unfolds more or less automatically."[3]

During the Cold War, the nuclear forces of the two superpowers were constantly watching each other, constantly ready to react. They were so tightly coupled that they were, in effect, one system. A serious glitch in either's forces could have led to a set of mutually reinforcing actions that escalated out of control and brought about a pointless war that neither wanted, just as it did in World War I. This time, the result would have been more than misery and carnage, it would have been mutual annihilation.

This problem has been with us for a long time. Late in the 1950s, at a UN Security Council debate, Soviet delegate Arkady A. Sobolev provided an early scenario for a catastrophic mutual interaction of the American and Soviet nuclear forces. At the time, it was standard practice for the U.S. Strategic Air Command to fly nuclear-armed bombers toward the USSR to the "fail-safe point" when there was an attack warning, at which point they would turn around and return to the United States unless they received special coded orders to proceed to target. Sobolev asked,

[W]hat would happen if the military Air Force of the Soviet Union began to act in the same way as the American Air Force is now acting? After all, Soviet radar screens also show from time to time blips which are caused by the flight of meteors, or electronic interference. If in such cases Soviet aircraft also flew out carrying atom and hydrogen bombs in the direction of the United States. . . . The air fleets of both sides having observed each other . . . would draw the conclusion . . . that a real enemy attack was taking place. Then the world would inevitably be plunged into the hurricane of atomic war.[4]

As the Cold War wore on, the problem became more severe. The growth of nuclear arsenals, along with increasing speed, range and accuracy of delivery systems, shrank decision time and required warning systems to become more sensitive and complex. The nuclear forces of the superpowers became more tightly coupled, while heightened sensitivity and complexity assured that they were still prone to error, despite great improvements in technology. Though many of the particulars changed, the essence of the Sobolev scenario remained a frightening possibility.

With the end of the Cold War and the disappearance of the "Soviet threat" in 1991, it was widely assumed that Russian and American missiles had been taken off hair-trigger alert and were no longer configured for launch on warning of attack. But that is simply not true. In April 1994, Admiral Henry Chiles, head of the U.S. Strategic Command (in charge of all American strategic nuclear weapons) told the Senate Armed Services Committee that the United States was still prepared to launch its missiles en masse after warning of attack but *before* enemy nuclear weapons reached their targets.[5] As of 1999—eight years after the demise of the Soviet Union—thousands of American nuclear warheads were still on launch-on-warning alert.[6]

Launch-on-warning is an extremely dangerous policy. It is based on the tenuous assumption that through careful design and redundancy, human/technical warning systems can be created that are not subject to catastrophic failure. Yet it is painfully obvious that that is not true (as we will see in the next five chapters). And once an attack is launched based on a warning that turns out to be false, there is a real attack underway that cannot be recalled or destroyed en route. Even if the other side's warning systems did not detect the launch, they will soon know for certain they have been attacked, as the incoming warheads explode over their targets. Given the horrendous damage that will result, a contrite admission that "a mistake has been made" (as the kaiser's messengers said in 1914) is unlikely to forestall retaliation. Some combination of human fallibility and technical failure will have brought about a disaster beyond history.

It is abundantly clear that the end of the Cold War has not put an end to the Cold War mentality that underlies the SIOP. As Bruce Blair of the Brookings Institution points out, "The planning apparatus of the U.S. command system has so far interpreted the political revolution in the Soviet bloc, as well as arms reductions, defense budget cuts and the curtailment of modernization programs for nuclear offensive weapons, mainly in terms of their implications for targeting. This orientation is deeply ingrained in the U.S. strategic culture."[7]

The Cold War may be dead and gone, but the Cold Warriors—and the weapons and institutions of the Cold War—are still very much with us. As long as that is true, accidental nuclear war remains a real possibility.

GENERATING WAR BY MISTAKE: TRIGGERING EVENTS

Accidents[8]

During a time of confrontation and crisis, a weapons accident that resulted in a nuclear explosion on the territory of a nuclear-armed country or its allies could trigger an accidental nuclear war. It is even possible that an accident involving the weapons of friendly forces might be surrounded in fog long enough to be misread as an enemy attack. But a weapons accident does not have to involve a nuclear explosion to trigger nuclear war.

There have been a number of publicly reported accidents in which the powerful conventional explosive in one or more nuclear weapons was detonated (see Appendix to Chapter 4). Suppose one of these bombs had fallen onto an oil refinery, tank farm or toxic chemical waste dump. Worse yet, if one fell into a nuclear waste storage area, the huge explosion and high levels of radioactivity that would result could easily be misinterpreted as an act of enemy sabotage or a deliberate attack—especially under the pressure and confusion of a crisis. The American military has made progress in developing conventional explosives that are less likely to explode accidentally. Still, problems with weapons design and manufacturing defects continue (see Chapter 4), and it is clear that such accidents remain possible.[9]

Scott Sagan has proposed a frightening accidental war scenario based on the January 21, 1968, crash of a nuclear-armed B-52 bomber into the ice at Thule Bay, Greenland.[10] During that crash, the chemical explosives in all of the four nuclear weapons on board detonated, scattering plutonium. Suppose, rather than crashing into the bay, the B-52 had crashed into the Thule communications center, an important part of the Ballistic Missile Early Warning System (BMEWS). In that case, a nuclear-armed B-52 would have been lost at the same time the Thule BMEWS station was blacked out. The information received by NORAD and SAC would then have been consistent with a Soviet attack against the warning system at Thule. The complete cutoff of communications from Thule would have caused bomb-alarm consoles at command posts in the United States to confirm the likelihood that an attack had taken place. If the crash had caused the *nuclear* explosion of at least one of the weapons carried by the bomber, the bomb-alarm system at Thule would have shown that a nuclear detonation had occurred. The loss of communications and the destruction of the B-52 would then have provided confirmation that the Soviets had launched a nuclear attack on Thule. This is precisely the sort of attack that the Pentagon expected to be followed by an immediate nuclear attack on the United States.[11]

An accidental explosion that scattered a great deal of radioactive material could also be misinterpreted without any weapons involved at all. Nuclear waste storage sites are attractive targets. Even a small, inaccurate attack would release huge amounts of radioactivity. During an international crisis, an explosion at a known nuclear waste site (especially near a populated area) might well be misinterpreted as the result of a deliberate enemy attack.

The crash of an aircraft or missile into a nuclear, chemical or biological weapons storage area could also produce a provocative combination of explosion and contamination. Such incidents appear in the public record: in the summer of 1956, a B-47 bomber crashed into a storage igloo containing three nuclear weapons in England; on June 24, 1994, a B-52 bomber crashed as the pilot pulled the plane into a fatal stalling turn in a successful last-minute attempt to avoid crashing into a nuclear weapons storage area. For years, the flight path of civilian airliners in Denver took them close to or right over a nearby nerve gas storage area at the Rocky Mountain Arsenal. In the mid 1990s, commercial airliners flew directly over the Pantex nuclear weapons plant on their way to and from nearby Amarillo airport. At the time, more than 9000 plutonium "pits" the size of bowling balls that had been removed from decommissioned nuclear weapons were being stored there, in World War II–vintage conventional weapons bunkers.[12]

In the right circumstances, an accident involving friendly forces might be misread as an enemy attack even in the absence of an explosion. Both the United States and the former Soviet Union had nuclear submarines sink at sea because of internal problems. If either side had lost a sub in the midst of a crisis that had escalated to the point of threat and counterthreat, it might have thought the sub had been sunk by enemy action. This would have been especially likely if the ship that sank carried nuclear missiles, such as the *Yankee*-class submarine the Soviets lost in the Atlantic in 1986. Since each of those ships carries enough nuclear firepower to destroy an entire nation, it is a prime target for early attack.

Nuclear-capable missiles have sometimes accidentally flown over, or crashed into or near, the territory of another country (see Appendix to Chapter 4, Table 4-3). One of these incidents, the launch of a U.S. Matador missile toward mainland China, occurred within a few months of the peak of the Quemoy-Matsu crisis, during which mainland China and Taiwan were actually shooting at each other. Relations between China and Taiwan remain tense to this day.

There is an elaborate set of controls in place intended to prevent accidental or unauthorized firing of American land-based ICBMs.[13] In addition to procedures that require a number of people to act together, higher command must send the proper enabling codes before on-site crews can physically launch the missiles. Of course, no system is ever failure proof, but it is very difficult for crews to launch inadvertently or without external authorization. However, more than a third of the U.S. strategic arsenal is carried on nuclear submarines, whose missiles are armed with thousands of city-destroying warheads. While submarine missile launch procedures require a number of people to act jointly, there is no external physical control against launch.

False Warnings

The equipment and people of the nuclear attack warning systems are expected to detect and evaluate any real attack quickly and accurately, whenever it might come, while at the same time never failing to recognize that false warnings are false. The time pressure is enormous.[14]

In the American attack warning system, satellites in geostationary orbit are supposed to detect the launch of a missile fired from land or sea within 30 seconds after liftoff. The satellites transmit data on the detection to a ground station for processing by computer. The result is evaluated by on-site personnel, who then have all of 15 seconds to decide whether to forward the information to the nation's missile warning centers at NORAD and elsewhere. The NORAD command director on duty then has about three minutes to judge whether the satellite warning is valid. Within 30 to 45 seconds, the director contacts the ground station operators by telephone and asks them to confirm that the warning is not the result of an equipment malfunction. The director also receives a report from the intelligence branch that keeps track of enemy forces that includes a statement of how probable they think it is that the missile warning is real. By the end of three minutes, the director is expected to verbally advise the Pentagon and the Strategic Air Command (SAC) as to how confident NORAD is that a missile attack is actually underway.

If NORAD reports "no confidence," the process is over. A report of "medium" or "high" confidence requires the Pentagon to notify the chair of the Joint Chiefs of Staff, who informs the secretary of defense. If warranted, a missile attack conference is convened, involving the president, secretary of defense and the chair of the Joint Chiefs of Staff. If possible, the senior commanders from all major nuclear commands also participate. The president is briefed on the SIOP options and given recommendations as to which types of targets to strike. Within at most ten minutes, the president has to decide whether and how to respond.

Asking fallible human beings dependent on fallible technical systems to make decisions of such enormous importance in so little time is asking for trouble. It is virtually inevitable that the system will fail catastrophically some day.

That day almost came only a few years ago, when Russian warning radars detected the launch of a rocket from the Norwegian Sea on January 25, 1995. About the size of a U.S. submarine-launched Trident missile, it seemed to be streaking toward Moscow, with projected impact in only about 15 minutes. The radar crew sent the warning of impending nuclear attack to a control center south of Moscow, where it was relayed up the chain of command. Soon the light flashed on the special briefcase assigned to President Yeltsin that contains the codes needed to launch a nuclear counterattack. The briefcase was opened, revealing an electronic map showing the missile location and a set of options for nuclear strikes against targets in the United States. Alarms sounded on military bases all over Russia, alerting the nuclear forces to prepare to attack. Yeltsin consulted with his top advisors by telephone. Tensions rose as the stages of the rocket separated, making it look as though several missiles might be headed for the Russian capital. About eight minutes after launch—only a few minutes before the deadline for response—senior military officers determined that the missile was headed far out to sea and was not a threat to the Russian homeland. The crisis was over.[15]

Despite the end of the Cold War in 1991 and all of the dramatic changes that have occurred since that time, we were once again at the very brink of nuclear war with Russia. In fact, this was the first time that the famous "nuclear briefcase" had ever been activated in an emergency. And the cause? The rocket the Russians detected, it turned out, was an American scientific probe designed to study the aurora borealis (northern lights). It was sent aloft from the offshore Norwegian Island of Andoya. Norway had notified the Russian embassy in Oslo in advance of the launch, but somehow "a mistake was made" and the message never reached the right people in the Russian military.[16]

False warnings of attack are probably the most likely trigger of accidental nuclear war. There have been many. Between 1977 and 1984 (the last year for which NORAD made these data available), there were nearly 20,000 routine false alarms, an average of more than 6 a day.[17] Most were quickly recognized as false. But there have also been much more serious false warnings of attack (see Table 5-1).

From 1977 through 1984, there were more than 1,150 serious false warnings of attack, an average of *one every two and a half days* for eight years. The rate of false alarms increased by nearly 500 percent between 1977 and 1983, before falling back to the average for the period in 1984. Since the data were no longer made public after 1984, it is impossible to say whether or not this trend continued. We do know that even as late as 1997, space junk that "might be construed as an ICBM warhead" continued to re-enter the atmosphere at the rate of about one "significant object" a week, according to the deputy director of the U.S. Space Control Center.[18]

Why do so many false warnings occur? All warning systems are subject to an inherent engineering tradeoff between sensitivity and false alarms. If a smoke alarm is designed to be so sensitive that it will quickly pick up the very first signs of a fire,

TABLE 5-1

SERIOUS FALSE WARNINGS OF ATTACK

YEAR	MAJOR MISSILE DISPLAY CONFERENCES	THREAT ASSESSMENT CONFERENCES
1977	43	0
1978	70	2
1979	78	2
1980	149	2
1981	186	0
1982	218	0
1983	255	0
1984	153	0
1985 NORAD stopped releasing data on false warnings		
TOTAL	1152	6

Source: Center for Defense Information, *Defense Monitor* (Vol. 15, No. 7, 1986), p. 6.

it is bound to go off from time to time because someone is smoking a cigarette or a little smoke from a backyard barbecue blows in the window. If the smoke alarm system is made less sensitive, it will not go off as often when it shouldn't. But then it also won't go off as quickly when there is a real fire. Modern nuclear weapons delivery systems are so quick and accurate that a military not prepared to ride out an attack requires very sensitive warning systems to get maximum warning time. Such systems are virtually certain to generate many false alerts.

The jump in false warnings after 1979 probably resulted from changes in NORAD's rules intended to increase the sensitivity of the warning system. Blair argues that this was done precisely to meet the time pressures imposed by a strategy of launch-on-warning:

Definitive confirmation of . . . surprise attack [previously] depended on actual nuclear detonations in North America. But . . . to meet the time constraints of launch on warning, NORAD dropped the requirement. High confidence [that a warning was real] *no longer depended at all on the presence of nuclear detonations.* . . . NORAD also revised its rules for threat evaluation during crises. Previously . . . high confidence depended on . . . indications of possible imminent hostilities . . . and positive attack information from at least two different . . . warning sensors. By the mid-1980s . . . NORAD . . . required a positive indication from only one . . . sensor system. . . . Moreover, NORAD procedures . . . allowed the loss of . . . [an attack warning] sensor . . . to be treated as a positive . . . indication of strategic attack.[19] (emphasis added)

These were very dangerous changes that dramatically increased the probability of accidental war. Multiple failures of the warning system, even in the absence of crisis, could have triggered a NORAD report of "high confidence" that an attack was actually under way when it was not. More frightening still, during a crisis the new rules meant that NORAD could report high confidence that an attack was under way *without any sensor warning*. If even one warning sensor had merely stopped working, and if the NORAD director believed (for whatever reason) that it had been destroyed by enemy action (including sabotage), NORAD could have reported high confidence to both the Pentagon and the Strategic Air Command that the United States was under nuclear attack. Under these rules, nothing more threatening than the sun's activity could have precipitated accidental nuclear war. Early-warning satellites have frequently been knocked out for hours by solar glare; ground radar warning sensors have often been severely degraded by adverse atmospheric conditions, some of which are caused by sunspots and related solar activity. In fact, it was typical for one attack warning sensor to be out for two hours or so on any given day; several times a year, two sensors were out at the same time for at least a few minutes.[20]

A number of other spectacular and threatening false warnings of attack have occurred:

- On October 5, 1960, the war room at NORAD received a top-priority warning from the Thule, Greenland, BMEWS station. Thousands of missiles were headed toward North America from the direction of the Soviet Union. While the deputy NORAD commander tried to verify the warning, something, most likely an iceberg, cut the undersea cable linking NORAD and Thule. Some 15 to 20 tense minutes passed before NORAD was able to determine that the warning was false. The radars had picked up the moon as it came over the horizon, interpreting its huge mass as an enormous number of missiles. Apparently, the BMEWS computers had never been programmed to disregard radar signals bouncing off the moon.[21]
- During the 1962 Cuban Missile Crisis, U.S. forces were on high alert. Near midnight, a guard at an Air Force facility in Duluth saw someone

climbing the base security fence. Fearing sabotage, he fired at the intruder and set off a sabotage alarm that was linked to alarm systems at other bases nearby. A flaw in the alarm system at one of those bases (Volk Field in Wisconsin) triggered the wrong warning. Instead of sounding the sabotage alarm, the klaxon that signalled the beginning of nuclear war went off. Fighter pilots rushed to their nuclear-armed aircraft, started their engines and began to roll down the runway, believing a nuclear attack was underway. The base commander, having determined that it was a false alarm, sent a car racing onto the tarmac, lights flashing, signalling the aircraft to stop. It turned out that the suspected saboteur shot climbing the fence was a bear.[22]

- During the fall of the Cuban Missile Crisis, Colonel Oleg Penkovsky, a Soviet intelligence officer spying for the United States and Britain, became an emergency warning system himself, code named DISTANT. Penkovsky was told to call one of two telephone numbers in Moscow if and when he had information that the Soviets were preparing to attack the United States. He was to blow three times into the telephone, wait one minute, then do it again. To avoid alerting anyone who was tapping his phone, he was not to speak. He was then to leave any additional information he had at a designated "drop" in Moscow, if he was able to do so.

  On November 2, 1962 (less than a week after the missile crisis ended), the DISTANT warning was given. The CIA immediately sent an agent to the drop. The agent was taken into custody by the KGB. Colonel Penkovsky had been arrested more than a week earlier, and apparently told the KGB how to give the signal. But did he tell them what the signal meant?[23] After all, he was facing severe punishment as a spy, and might have found it appealing to trick his captors into unintentionally triggering nuclear revenge on the KGB and the country he had already betrayed. It is hard to believe that KGB agents would have blithely given the signal, if they knew the chance they were taking of precipitating a massive nuclear attack against their country.[24]

- Because submarines cannot transmit long distance while underwater, they cannot notify command of problems they encounter that prevent them from surfacing. Soon after the loss of the nuclear attack submarine *Scorpion* in the Atlantic in May 1968, the U.S. Navy decided to fit its nuclear submarines with special Submarine Emergency Communications Transmitter (SECT) buoys. In case of trouble, these buoys would pop up to the surface and begin transmitting any of a variety of coded messages indicating what was wrong. In March 1971, near Soviet waters, a SECT buoy was accidentally released from an American nuclear missile submarine. It rose to the surface and began transmitting the proper coded message to indicate that the submarine was sunk by enemy action. Fortunately, nothing was actually wrong with the sub. As soon as the captain realized what had happened, he surfaced and countermanded the attack message. Then, within four days, it happened again. Had the submarine been unable to surface itself or an antenna because of internal problems, it would not have been able to transmit the cancel code.[25]

- At 10:50 A.M. on November 9, 1979, monitors at NORAD in Colorado, the National Military Command Center at the Pentagon, Pacific head-quarters in Honolulu and elsewhere simultaneously lit up. A nuclear missile attack was underway. The profile of the attack was realistic: a salvo of submarine-based missiles against just the kind of targets that the American military might expect the Soviets to strike. Ten jet interceptors were ordered into the air from bases in the United States and Canada. Six minutes later, it was determined that this was a false alarm. The attack profile seemed realistic to the U.S. military because they had created it. An operator had fed a test tape for a simulated attack into the computer as part of an exercise. According to Scott Sagan, the "software test information . . . was *inexplicably* transferred onto the regular warning display at Cheyenne Mountain and simultaneously sent to SAC, the Pentagon and Fort Richie."[26] Through some unknown combination of human and technical error, the realistic test attack data had gone out through the wrong channels and showed up as a "live" attack warning.[27]

- In the early morning hours of June 3, 1980, the displays at the Strategic Air Command suddenly showed a warning from NORAD that two submarine-launched ballistic missiles (SLBMs) had been fired at the United States. Eighteen seconds later, the warning indicated more SLBMs were on the way. SAC sent its bomber crews racing to their B-52s. The engines were fired up and 100 B-52s, loaded with more than enough nuclear warheads to devastate any nation on earth, were prepared to take off.[28] Launch crews in land-based missile silos across the United States were told to get ready for launch orders. Ballistic missile submarines at sea were alerted. Special battle-control aircraft, equipped to take over if normal command centers failed, were prepared for takeoff. One took off from Hawaii.[29] The American military got ready for Armageddon.

Just then, the warning disappeared. SAC contacted NORAD and was told none of the satellites or radars had detected any missiles at all. Bomber crews were told to shut down their engines, but stay in their planes. Then, suddenly, the displays at SAC again showed an attack warning from NORAD, this time indicating a barrage of Soviet land-based missiles launched against the United States. At the same time, the National Military Command Center at the Pentagon also received a warning from NORAD—but it was a different warning. Their displays showed Soviet submarine-launched missiles, not land-based ICBMs, speeding toward America.[30] Something was clearly wrong.

What turned out to be wrong was that a faulty computer chip (costing 46 cents) was randomly generating "2"s instead of "0"s in the transmissions from NORAD to each of the different command centers. Rather than saying that "000" sea- or land-based missiles were heading toward the United States, the computer was saying that there were "020" or "002" or "200". The faulty chip was the immediate problem, but the underlying cause was a basic design error. The NORAD computer was not programmed to check the warning messages it was sending against the data it was receiving from the attack warning sensors to make sure they matched—an elementary mistake. This kind of error-checking capability was routine in the commercial computer systems of the day. But the software that controlled a key part of the world's most dangerous technological system did not have this basic safety precaution.[31]

Three days after this false alarm, it happened again. SAC and the Pentagon received differing warnings of major missile attack, following an apparently random pattern. NORAD claims this second warning was the result of its trying to duplicate the problem while investigating the earlier event.[32] Maybe so. But the fact is, until this design flaw was discovered and repaired, it was bound to happen again. And if this basic a mistake could be made in such a critical system, is it unthinkable that there might be other hidden flaws that could surface at another unpredictable time? Software is more susceptible than electronic or mechanical hardware to hidden faults that do not surface until after years of seemingly flawless performance (see Chapter 10). Yet we retain enormous confidence in these systems. As General Hilsman, former director of the Defense Communications Agency, said about the June 1980 warning, "Everybody would have told you it was technically impossible."[33]

• In August 1984, not realizing his microphone was live, President Ronald Reagan joked about bombing the Soviet Union. Soviet Far Eastern military headquarters reportedly reacted by broadcasting a coded message reading, "We now embark on military action against the U.S. forces." Having intercepted and decoded this communication, American forces in the Pacific were put on alert. Within 30 minutes, the Soviet military cancelled their earlier message. It is not known why.[34]

There have also been incidents in which a number of false warnings occurred at the same time, reinforcing each other and making it easier to believe that a real attack was under way. Probably the most threatening of these (that became public) occurred in early November 1956. The Hungarian revolution was underway. The British and French militaries were attacking Suez. Moscow warned Britain and France that the Soviets might launch rocket attacks against London and Paris if the attacks on Suez did not stop. Meanwhile, the USSR was suggesting to Washington that joint Soviet-American military action should be taken in Suez.

On the evening of November 5, U.S. military command in Europe received a top-priority message: unidentified jet aircraft were flying over Turkey (a member of NATO). The Turkish Air Force went on alert. Command was also told that a British Canberra bomber had been shot down over Syria, where there were reports of 100 Soviet MiG-15 fighters in the air—the only enemy aircraft capable of shooting down the high-flying British bombers. And it was reported that the Russian fleet was moving out of the Black Sea, through the Turkish straits of Dardanelles into the Mediterranean.

The movement of the Soviet fleet was taken as important confirmation that hostilities were imminent. The Russian/Soviet fleet had been bottled up in the Black Sea during both world wars. They would surely move quickly to prevent that from happening again if they believed war was about to begin. U.S. General Andrew Goodpaster reportedly expressed fears that all this "might trigger off all the NATO operations plan." At that time, the NATO operations plan called for all-out nuclear war against the USSR.

The unidentified jets flying over Turkey turned out to be a flock of swans detected by the radars. The 100 Soviet MiGs over Syria turned out to be a much smaller number of planes escorting the aircraft carrying the Syrian president home after an official visit to Moscow. The British Canberra bomber had not been shot down, it had been forced down by mechanical problems. The Soviet fleet was indeed moving through the Dardanelles, but that was part of a long-planned routine military exercise. It had nothing to do with any of the other events. What appeared to U.S. military command as an ominous series of mutually reinforcing events portending war was in fact a remarkable set of coincidental misinterpretations.[35]

Soon after June 3, 1980, Pentagon officials commented that "the false alarm would not have been publicly disclosed except that hints of it had leaked to the press."[36] That attitude finally lead NORAD to stop releasing data on false warnings after 1984. There has been very little public information on major American false attack warnings since. There never was much public information about false alarms on the Soviet's attack warning system. Unlike weapons accidents, which sometimes happen in very public ways, failures involving false nuclear attack warnings are much easier to hide—that is, until the day the ultimate failure occurs.

As we have seen, warning systems can and do fail in many different ways. Sensors sometimes accurately detect real events, only to have computers or people misinterpret them (such as the moon rising in the 1960 BMEWS incident or the flock of swans over Turkey in 1956). Sometimes computers or people generate false warnings, without the sensors having detected anything (as in the June 3, 1980, NORAD incident or the Penkovsky warning). Sometimes the alarms themselves go off without sensors, computers or people generating any indication of a problem (as with the SECT buoys). And sometimes warning systems fail because they miss events they are supposed to detect.

Failures of Detection

Knowing that alarm systems have failed to detect what they are supposed to detect can lead to a state of hypervigilance in critical situations. Hypervigilance makes it easy for people to see danger where there is none. They are so determined not to miss the critical event that are primed to overreact. That overreaction can lead to tragedy, even to the tragedy of accidental war.

- On October 6, 1969, a fully armed MiG-17 jet fighter flew out of Cuba toward the United States. As it flew the 200 miles to Homestead Air Force Base in Florida, it was picked up on American radar only once, and then briefly. No American interceptor jets were sent aloft to check out the sighting because, in the words of an embarrassed Air Force official, "It might have been a whooping crane or a flight of geese." The plane, which had a Cuban flag painted on its tail, was not identified until it appeared over the tower at Homestead and requested permission to land. Permission was quickly granted, and the pilot brought the plane smoothly to rest not far from Air Force One, which was being refueled for the president's return trip to Washington.[37]
- During the 1970s, Soviet Bear bombers regularly flew as close as 150 miles from the U.S. coast. On April 14, 1977, American air defense radar detected one of these four-engine turboprops flying over the Atlantic, several hundred miles southeast of Charleston, South Carolina. The plane dropped closer to the water and disappeared from the radar. Two F-4 Phantom jets were sent up to intercept the bomber, but were unable to find it. The Bear was not seen again until it made a low pass directly over

a U.S. Navy task force only 60 miles off the coast, then simply continued on its way.[38]

- The frigate *Stark* was part of the U.S. Navy fleet patrolling the Persian Gulf during the Iran-Iraq War. On May 17, 1987, an Iraqi Air Force pilot flying a French Mirage fighter-bomber began to approach the *Stark.* Suddenly, the plane fired one Exocet missile, then another at the ship. The attack came as a total surprise. According to the Navy, the ship's electronic warning system detected that the Mirage had switched on a special radar used to pinpoint targets a minute before the first missile struck the ship—but no one told the captain. The *Stark's* state-of-the-art warning equipment did not pick up the missile firings. The only warning came from the oldest, lowest-technology warning system on board, a lookout watching from the deck. He spotted the 500-mile-per-hour missiles seconds before they slammed into the ship, setting it ablaze and killing 37 men. Iraq later said the attack was a mistake, and apologized.

  The *Stark* was equipped with a Phalanx defensive system, designed to automatically fill the air close to the ship with a deadly barrage of bullets. But Phalanx was not turned on. Since the system would shoot at anything approaching the ship, it was too risky to leave it on automatic. It could have been aimed manually, but there was not enough warning time. The *Stark's* radar couldn't detect the missiles fired from 12 miles away because they approached through a blind spot, where antimissile sensors and weapons were blocked by a part of the ship's own superstructure.[39]

- On May 28, 1987, a West German teenager named Matthias Rust filed a flight plan for Stockholm and took off from Helsinki, Finland, flying a single-engine Cessna 172. He had removed some seats and fitted extra fuel tanks to the plane to increase its range, but it was otherwise a perfectly ordinary small civilian aircraft. Rust diverted from his flight plan, turned south and flew into Soviet airspace over the coast of Estonia. About 7:00 P.M., he reached Moscow, circled low over Red Square, and landed near the Kremlin Wall. He had flown through sophisticated radar systems, across more than 400 miles of heavily defended territory to the heart of the Soviet Union, without ever triggering a single alarm.

  Rust climbed out of the plane, and was immediately surrounded—not by police or the military, but by a crowd of curious Russians and foreign tourists. Children climbed on the tail of the plane. He chatted with the crowd until the authorities arrived. Rust was arrested, imprisoned, then freed and expelled from the country a year later.

  The United States had spent many billions of dollars developing "stealth" fighters and bombers to sneak past Soviet radars and dense defenses such as those surrounding Moscow. But the whole system had been penetrated with ease by a boy flying an inexpensive, run-of-the-mill airplane he had rented from a flying club in his hometown. To add to the irony, he did this as the Soviets were celebrating Border Guards Day.[40]

Nuclear Terrorism, Catalytic War and Unauthorized Launch

Under the right circumstances, an intentional act of terrorism carried out by a group that had somehow managed to get hold of nuclear weapons might plausibly trigger an accidental nuclear war. The nuclear destruction of a city by a terrorist group might be misinterpreted as the result of a limited, purposeful attack by another nuclear-armed nation. In the confusion surrounding the event, a retaliatory strike might be launched against the presumed attacker, leading it to counterattack. A real nuclear war would then be underway.

It is more likely still that an accidental nuclear war could be triggered by an attack deliberately designed to start such a war. The attack would then be made to look like a limited, intentional attack by a rival nuclear-armed country. A renegade country would have an easier time accomplishing this than a terrorist group. Nuclear war planners have debated the possibility of "catalytic war" since the 1950s. They reasoned that a minor nuclear power might be tempted to try to fool rival major nuclear powers into destroying each other. When the dust, or rather the fallout, had settled, the country that started the war would still have its society and arsenal intact, allowing it to dominate the world.

The more we have learned about the devastating environmental effects of nuclear war, the more obvious it has become that this kind of catalytic war would ultimately destroy the nation that had triggered it as well.[41] Unfortunately, in a world in which very troubled human beings have more than once been in control of governments, the fact that catalytic nuclear war is both profoundly immoral and insane does not guarantee that it will never happen.

Most American nuclear warheads are fitted with safety devices called Permissive Action Links (PALs) to prevent them from being armed without proper authorization. There are also other devices built into the launch systems of land-based nuclear-armed missiles to prevent unauthorized launch. The protections are carefully designed, but no human/technical system is perfect and no safety system is without flaws (see Chapters 6-10). It is always possible that someone with sufficient know-how, equipment and access to PALs could work out a way to defeat them.[42]

There are no PALs on American nuclear weapons at sea. While other nuclear weapons cannot be armed and launched without "go-codes" transmitted by central command, the captain and crew of a missile submarine, acting together, can launch a nuclear attack on their own. They may not have the authority, but they do have the physical ability. Every American missile submarine has both the nuclear firepower and missile range to destroy any nation on earth.

The protection offered by go-codes is compromised when lower-ranking officers are given the codes in advance of the moment of need. Yet according to Bruce Blair, "Many duplicates of all these authorization and enabling codes were distributed throughout the nuclear command system. . . . [N]umerous U.S. military installations possessed all the codes needed to disseminate valid launch orders."[43]

The common belief that the president of the United States is the only one who has the authority to order the initial use of American nuclear weapons may be technically correct, but it could not be true in practice. If the system were set up so that it was not possible to use nuclear weapons without the direct consent of the president, a determined attacker could disable U.S. strategic nuclear forces by killing the president before striking the rest of the nation. Even highly centralized control of the nuclear arsenal not limited to the president would make it possible for an attacker to "decapitate" the military by attacking the central command center. A successful decapitation attack would render intact American nuclear forces unable to respond.

Alternate command centers are one protection against the decapitation of military command. One of these, SAC's computer- and communications-packed aircraft code-named "Looking Glass," went on constant airborne alert in February 1961. If SAC's underground command center were destroyed by enemy attack, Looking Glass would direct strategic strikes against the homeland of the attacker. Apparently the crew of Looking Glass can also fire the nuclear-armed missiles directly: "If enemy strikes also destroyed missile launch control centers, the battle staff [on board Looking Glass] could launch the entire force of *Minuteman* and *Peacekeeper* missiles."[44] To be sure, there are safeguards built into the missile launch centers to prevent the crew of Looking Glass from casually taking control.[45] But no safeguards are foolproof.

Military commanders have had the authority to order the use of American nuclear weapons for decades. Presidents Eisenhower, Kennedy and Johnson all predelegated the authority for emergency use of nuclear weapons to "six or seven three- and four-star generals."[46] Presidents Carter and Reagan empowered senior military commanders "to authorize a SIOP retaliatory strike and to select the SIOP option to execute."[47] In short, there have been a lot more "fingers on the button" than most of us realize. Given the many ways in which human beings can fail, this makes accidental war more likely. However compelling the military logic, this degree of decentralization is dangerous.

FAILURES OF VERIFICATION AND COMMUNICATION

No sane national leader would launch a retaliatory nuclear strike until an attempt has been made to verify that the triggering event really did mark the beginning of nuclear war. The most dangerous failure of verification is a false positive, in which the system confirms that the triggering event is real when it is not. This would sharply increase the likelihood that a nuclear strike would be launched in retaliation for an enemy attack that never actually happened. Fortunately, the kind of compound error that would be necessary to get false positive confirmation, though not impossible, is relatively unlikely. It is much more likely that verification will fail because there will not be enough time to find out if the triggering event is real. Those in authority will then have to decide whether or not to retaliate without being sure that the attack is actually underway. In such a situation, people tend to believe what they are predisposed to believe.

Only a year after the *Stark*'s failure to react cost the lives of three dozen men, the U.S. Navy cruiser *Vincennes* found itself in what appeared to be a similar situation. Part of the fleet in the Persian Gulf, the *Vincennes* was engaged in a firefight with several Iranian gunboats on the morning of July 3, 1988. Suddenly, the cruiser's combat information center detected an aircraft rising over Iran, headed toward the *Vincennes*. The combat center concluded that the oncoming plane was one of the 80 F-14 fighters that the United States itself had earlier sold Iran. The ship warned the plane to change course, but it did not respond and kept coming. The ship's crew then saw the aircraft descending toward the *Vincennes* from an altitude of 7,800 feet at a speed of 445 knots, as if to attack. Seven minutes after the plane was first detected, the cruiser fired two surface-to-air missiles, and blew it out of the sky.[48]

But the plane was not an F-14, and it was not attacking. The *Vincennes* had destroyed a European-built Airbus A-300 civilian airliner with nearly 300 people on board. Iran Air Flight 655 was a regularly scheduled flight on a well-known commercial route. Its transponder was automatically broadcasting a signal that identified it as a civilian aircraft. The A-300 has a very different shape than an F-14, it is nearly three times as long, and has more than twice the wingspan. And it was climbing, not descending.[49] How could such a thing happen?

The *Vincennes* was in combat in a war zone at the time the plane was detected. American commanders in the gulf had been warned that Iran might try to attack U.S. forces around the Fourth of July holiday. There were only 7 to 8 minutes after detection to react, only 3 minutes after the plane was declared hostile. There was also the specter of the tragic loss of life on the *Stark*, whose crew failed to fire at an attacking aircraft they could not believe was attacking. Under so much stress, with so little time and a predisposition to believe they were being attacked, the crew was primed for error. In the process of verifying the threat posed by the blip on the radar screens, they misinterpreted the electronic identification signal coming from the plane, misread the data that correctly showed that the approaching plane was gaining altitude (not descending to attack), and failed to find Flight 655, clearly listed in the civilian airline schedule they hastily consulted.

Five years before the destruction of Iran Air 655, at a time when the Cold War was still alive and well, the Soviets had made a similar mistake with similarly tragic results. On the night of August 31–September 1, after a long flight across the Pacific, Korean Airlines (KAL) Flight 007 drifted far off course and flew directly over Kamchatka Peninsula and Sakhalin Island, the site of a secret Soviet military complex. The area was well known by both military and civilian aviators to be supersensitive: the U.S. Federal Aviation Administration's standard area maps included a large-print warning over Sakhalin saying, "Aircraft infringing upon nonfree flying territory may be fired on without warning."[50]

Soviet radars detected the intruder and scrambled fighter planes to identify and intercept it. The radars tracked the plane for more than two hours, while the interceptors struggled to find it. They mistook the Boeing 747 civilian airliner with 269 people aboard for a U.S. Air Force RC-135 reconnaissance plane. A Soviet SU-15 fighter finally caught up with the plane, following it through the night sky. When the airliner did not respond to a warning burst of canon fire, the fighter pilot fired an air-to-air missile that blew the plane apart.

The Reagan administration's immediate reaction was to condemn the USSR as a brutal nation that had knowingly shot down a civilian airliner out of a combination of barbarism and blood lust. But within five weeks, it was reported that "United States intelligence experts say they have reviewed all available evidence and found no indication that Soviet air defense personnel knew it was a commercial plane before the attack."[51] There had been many previous instances of American spy planes intruding on Soviet airspace since 1950. In fact, there was an American RC-135 spy plane in the air at the time of the incident that had crossed the flight path of KAL 007. The Soviet military was preparing for a missile test in the area, which they knew the United States would want to monitor. As in the case of the *Vincennes*, all of this predisposed the Soviet armed forces to see an enemy military plane when they were actually looking at an ordinary civilian airliner.[52]

It is anything but reassuring that—with interceptor planes in the air and radars tracking KAL 007 for some two and a half hours—the Soviets still could not determine the nature of the intruder they had detected.[53] If they couldn't tell the difference between a harmless jumbo jet airliner and a much smaller military aircraft in two and a half hours, what reason is there to believe that their deteriorating military warning system today can verify that a missile attack warning is false in less than 30 minutes?

On April 14, 1994, years after the Persian Gulf War ended, two American Blackhawk helicopters flew over northern Iraq, carrying U.S. and other allied officials and representatives of the Iraqi Kurdish minority. They flew through the "no-flight" zone that allied forces had established to keep Iraqi military aircraft away from the region. Two U.S. Air Force F-15 fighters patrolling the area spotted the helicopters, mistaking them in broad daylight and good weather for Iraqi Hind military helicopters. An American Airborne Warning and Control System plane monitoring the no-flight zone failed to tell the F-15 pilots who queried it that the helicopters were American. The signalling device on the Blackhawks that identified them as friendly was not working. Without warning, the F-15s fired on the helicopters, killing all 26 people aboard.[54] "Friendly fire" incidents can lead to counterattacks against an enemy that has not taken any hostile action. In other words, *military forces might trigger themselves,* beginning a war by retaliating against enemy forces that they supposed were the source of what was really a friendly fire attack.

Despite modern equipment and well-trained military forces, it is amazing how easy it is to mistake completely harmless intruders for hostile forces when you are looking for and expecting to find an enemy. Between 1975 and 1980, Swedish government records showed an average of ten "certain" or "possible" intrusions per year of unidentified submarines into Sweden's territorial waters.[55] In October 1981, a nuclear-armed Soviet submarine ran aground near a Swedish military base. After that, the Swedes tracked dozens of intruders. After the Cold War ended, they continued to detect marauding foreign submarines, puzzled as to why the Russians would continue to provoke them.[56] Then, in early 1995, National Public Radio in the United States reported that the Swedish military might have solved the mystery: "the Swedes have discovered an error, and announced last week that for at least the last three years, their sensitive sonar equipment has been mistaking minks, the small furry animals, for submarines. . . . [T]heir chest cavities give off similar sound effects when they're under water."[57]

Problems with Command and Control

It is never easy to assure that military commanders, let alone political decision makers, have a clear and accurate picture of what is happening in critical situations. In the words of one senior military commander,

> Even in an ordinary war, you learn to discount the first report, only give partial credence to the second report, and only believe it when you hear it the third time. . . . The chances for misappreciation of information, for communications

foulups, for human errors are enormous. Even in tactical exercises—with people not under emotional stress . . . I've never seen a commander who was not surprised by what had actually happened versus what he thought had happened.[58]

If that is true in conventional war and war games, it is that much more likely to be true when decision time is as limited and stakes are as high as when nuclear war hangs in the balance.

During the nuclear power accident at Three Mile Island in March 1979, it took hours for operators in the control room to realize that a critical valve was stuck open. Because of a lack of telephones, federal officials on-site were unable to communicate quickly with officials in Washington who were trying to decide whether to evacuate the surrounding area. The chair of the Nuclear Regulatory Commission, preparing for a critical meeting with the governor of Pennsyvania, remarked, "His information is ambiguous, mine is nonexistent. . . . [I]t's like a couple of blind men staggering around making decisions."[59]

Trouble at the Top

High-level military communications do not always work smoothly even in normal times. When writer Daniel Ford was being taken around NORAD's command center by Paul Wagoner, the general in charge of NORAD's combat operations, he was shown five telephones next to the commander's battle station desk. The NORAD fact sheet Ford had been given said of their communications system, "When we pick up a telephone, we expect to talk to someone at the other end—right now." So he asked General Wagoner for a demonstration,

> starting with the special black phone that would be used to talk to the President. He picked up the phone briskly and punched the button next to it, which lit up right away. . . . About twenty seconds elapsed and nothing happened. The General . . . hung up, mumbled something about perhaps needing "a special operator", and started to talk about other features of the Command Post. I interrupted and pointed to another of the five telephones . . . a direct line to the Joint Chiefs of Staff. He picked it up. . . . Once again, nothing happened. . . . I looked at the other three phones, but General Wagoner decided to terminate his testing of the communications system."[60]

A similar but more official impromptu test of the nuclear command, control and communications system was staged by President Carter's national security advisor, Zbigniew Brzezinski, in 1977. He decided to check the effectiveness of procedures for evacuating the President from the White House in the event of a surprise nuclear attack. In Brzezinski's own words,

> I called in the person involved . . . [and] asked whether that [evacuation] schedule could be implemented; was assured that it could be. . . and I said, could you do it even now? He said yes. Whereupon I said I've been instructed by the president to test the system. I'm now acting as the president; you should evacuate me as if we were under attack. And the person . . . set the process in motion by calling the appropriate office. . . . We then waited for the helicopter to come. . . . It took much longer than it should have.[61]

A White House aide who participated in the exercise called it "a nightmare, just a complete disaster", adding that the Secret Service came close to shooting the helicopter down.[62]

Because the nuclear command-and-control system is no better than its key decision makers, the military high command occasionally holds war games designed to familiarize high civilian officials with the system. During the 1980s, Pentagon evaluation reports indicated that President Reagan did not seem to understand the system. According to a Pentagon aide, "I was disappointed in . . . his willingness to play it at the most superficial level." The president "acted like an automaton, like part of the set instead of the main actor . . . saying things like, 'What do I do now? Do I push this button?' . . . There were no questions from Reagan."[63]

Communications

The World-Wide Military Command and Control System (WWMCCS) is a system of computers at more than two dozen sites tied together by a network of sensors, coding and decoding devices, and the like. It was designed to warn the president and military high command of attack or developing security crises, provide information about the location and status of U.S. forces around the globe, and coordinate the dispatch of orders to the various military commands. A 1980 study of the system's performance by the U.S. General Accounting Office (GAO) found that design flaws made it prone to failure, especially under pressure.[64] The Pentagon defended the system, arguing that most of the time the "computers render effective support." But they admitted, "The principal exception occurs in crisis situations, when commanders need quick answers to a broad range of possible questions."[65] It is hard to have great confidence in a crisis-warning-and-response system that works fine as long as there isn't a crisis.

The results of most periodic tests of WWMCCS are not made public. One whose results did leak out was a 1977 exercise called Prime Target. The system's performance was appalling: 54 of the 124 messages transferred through the system by U.S. European Command failed to get through; the Atlantic Command experienced 132 failures out of 300 attempts; 44 of the Tactical Air Command's 63 attempts failed; and the Rapid Deployment Force's messages failed to get through more than 250 out of nearly 300 times.[66] Overall, WWMCCS worked properly less than 40 percent of the time.

Years later, in an exercise called Proud Spirit, overloaded WWMCCS computers shunted updating information into a buffer memory. The buffer then refused to feed the information back into the system, causing the Army's computers to fall silent for 6 hours while programmers tried desperately to figure out how to get the buffer to disgorge the information. By the time it was all over, WWMCCS had left military commanders without vital information about their units for 12 hours during the peak of the "crisis."[67]

Virtually every part of the military communications system has failed at one time or another:

- The famous "hot line"—the emergency communications link between Moscow and Washington set up to prevent accidental war—failed six times during the 1960s.[68]
- When NORAD's backup National Attack Warning System headquarters near Washington failed during a routine test in February 1984, it took over half an hour for officers to even notify the main Colorado headquarters that the system was down.[69]
- In April 1985, the Pentagon flew a group of reporters to Honduras to test its system for news coverage of emergency military operations. Their reports were delayed 21 hours because technicians could not make Navy telephones work.[70]
- The major earthquake that rocked San Francisco in October 1989 and disrupted the World Series also knocked out the U.S. Air Force center that manages America's early warning and spy satellites for 12 hours.[71]

Sometimes military communications go awry for reasons that have nothing to do with size, complexity or sophistication. For nine months, from the winter of 1974 through the summer of 1975, a Norwegian defense agency was mistakenly sending its military mail to the communist East German embassy, rather than the West German embassy! Reports strongly suggested that the agency involved was the Supreme Command, NATO's northern headquarters.[72]

Military communications have often been plagued with coordination and compatibility problems.[73] It was not until the American assault on Grenada in 1983 that the Army discovered its invasion forces were sometimes unable to communicate with the Navy's because their radios operated on different frequencies. Yet when the Army bought new communications systems specifically so that its battlefield commanders could coordinate battle plans with other units, it bought devices that were not compatible with those used by the Air Force. An additional $30 million had to be spent for adapters. What makes this story even more peculiar is that the Army had acted as purchasing agency for the Air Force equipment! An Army spokesman said that he could not explain why the two services wound up with incompatible systems.[74]

The Speed of Verification

In 1971, the House Armed Services Committee found that the global U.S. defense communication system took an average of 69 minutes to transmit and receive a "flash" (top priority) message. In a real nuclear attack, the missiles would have arrived before the message did. While there have been revolutionary improvements in the technology of communications since then, it is important to note that only about 5 of the 69 minutes was actual transmission time. The remainder was due to a variety of largely organizational and individual human foul-ups.[75] Looking at a few specific incidents makes it easier to understand the kinds of things that can and do go wrong.

- The 1967 war between Israel and Egypt began on June 5. At that time, the commander of the U.S. Sixth Fleet ordered all his ships to stay at least 100 miles off the coasts of both nations. But the U.S. intelligence-gathering ship *Liberty* was operating under different command, with orders that allowed it to get as close as 6.5 miles to the Israeli and 12.5 miles to the Egyptian coasts. On June 7, the Joint Chiefs of Staff (JCS) and the European Command decided to move the *Liberty* farther away from the coast. They sent a series of five messages, none of which reached the Liberty until more than half a day after the first message was released for transmission. In the words of a 1971 congressional report, "The circumstances surrounding the misrouting, loss and delays of those messages constitute one of the most incredible failures of communications in the history of the Department of Defense."[76]

  By the time the *Liberty* actually received the message directing it to observe a 100-mile distance, the ship had already been attacked by Israeli aircraft and patrol boats, and was slowly making its way back to port with

34 of the crew dead, 75 wounded and the ship so badly damaged that it later had to be scrapped.[77] According to Israel, the attack itself had been a human error, the result of mistaking the ship for an Egyptian supply vessel.[78]

- In January 1968, the U.S. spy ship *Pueblo* operating off the coast of North Korea was confronted by a North Korean patrol boat and ordered to stop or be fired upon. The Pueblo's commander then sent a message designated "Pinnacle" (meaning, "very important"), to be delivered immediately to National Military Command in Washington. It was given "flash" priority, the highest priority for rapid processing, transmission and delivery. This message was relayed to the commander of U.S. naval forces in Japan. Despite its Pinnacle and flash designations, it sat there for more than three-quarters of an hour before it was released for retransmission. The Joint Chiefs did not receive it until *two and an half hours* after it had been sent by the *Pueblo*.

  About 30 minutes after sending the first message, the captain of the *Pueblo* sent a second Pinnacle message with flash priority. It stated that the North Korean patrol craft had been joined by three others, and that the four ships were "escorting" the *Pueblo* out of the area. This message, reporting on the capture of a U.S. Navy vessel on the high seas at a time when quick action by other American forces might have freed the ship, took *an hour and 39 minutes* to reach the Joint Chiefs in Washington.[79]

- On May 12, 1975, the Cambodian Navy seized a U.S. merchant ship, the *Mayaguez*, in the Gulf of Thailand. The subsequent recovery of the ship and its crew by the American military has often been cited as an example of the effectiveness of improvements in communications and response made as a result of the *Liberty* and *Pueblo* incidents. Actually, it is quite the opposite. While this time U.S. military command could speak directly to the captains of Navy ships in the Gulf, U.S. forces in the area could not talk to each other. The Air Force had to fly in compatible equipment so that everybody in the area could communicate. The communications link with the Pentagon wasn't exactly trouble free either. A key military satellite failed to function properly, requiring some messages to be routed through the Philippines where, for some unknown reason, they were sent out to both the Air Force and the Navy uncoded over open channels. Because of this, the Cambodians knew the size of the U.S. Marine assault force coming to rescue the crew, and where they were going. When the marines landed by helicopter, they took nearly 100 casualties, 41 of them fatal, boarding a ship that had already been deserted.

Not only was that assault costly, it was also unnecessary. On May 14, *before* the marines attacked, Cambodian radio announced that the crew and the ship were being released. As the marines were paying the price for attempting to rescue the crew of the *Mayaguez,* the Cambodians were taking that same crew to the the the U.S. Navy destroyer *Wilson* for release. Either military command in the area had not been monitoring Cambodia broadcasts—an inexcusable and sophomoric error—or it had been unable to communicate with the marines quickly enough to cancel the assault. Upon release of the crew, the president ordered an immediate halt to the bombing of the Cambodian mainland. But, despite a direct presidential order, U.S. forces continued to bomb Cambodia *for at least another half hour.*[80] If nuclear war had been hanging in the balance, this slow a response would have been terminal.

THE CRISIS CONTEXT

Crises increase the chances of accidental war. During a crisis, forces armed with weapons of mass destruction are kept at a high level of readiness. If a nation under attack wants to react before the other side's warheads reach their targets and explode, they must be able to move quickly (within 5 to 30 minutes). Since normal peacetime controls will get in the way, security is tightened and peacetime controls are removed in layers, as forces are brought to progressively higher levels of readiness.

The American military operates under the DEFCON (for DEFense CONdition) alert system. There are five DEFCON levels: DEFCON 5 is the lowest level of alert, maintained under normal peacetime conditions; DEFCON 4 requires increased intelligence activity and continuous analysis of the developing political/military situation, along with some increased security and antisabotage measures; DEFCON 3 moves some forces to a higher level of readiness and reviews plans for even higher alert, since war is now considered possible; DEFCON 2 is a very high state of readiness, in which war is believed likely; and DEFCON 1 is the state of maximum readiness to execute war plans. War is imminent or actually underway.[81]

In addition to the heightened danger inherent in removing peacetime barriers to unauthorized use, crises increase the chance of accidental war because false warnings are easier to believe when real attack is more likely—especially when verification fails. Each side will be looking for any sign that the other is preparing to attack. They will be predisposed to see what they expect to see, and—as in the case of the *Vincennes*—that can lead to tragedy, this time on a catastrophic scale.

Bayesian statistics can be used to analyze how individuals might modify their original estimate of the probability of an event as they learn more about the situation. Blair uses a simple Bayesian statistical model to illustrate how individuals' initial expectations of the likelihood of attack would affect their willingness to believe what the attack warning system is saying. Blair assumes a warning system that is known to have only a 5 percent chance of generating a false alarm, and only a 5 percent chance of missing a real attack. Suppose prior to receiving any warning of attack, the estimate of the chance of a real attack was one in a thousand—as it might be in normal times. Because the original estimate of the chance of a real attack is so low, even two mutually corroborating warnings of attack would only result in upping that estimate to 27 percent. After a third positive warning, the estimated probability would still be less than 90 percent. But if a crisis raised the pre-warning expectation of a real attack to one in ten, two warnings would make the individual nearly certain (98 percent probability) that a real attack was underway. As the crisis deepened and the pre-warning estimate of the probability of a real attack rose to 30 percent, even one warning would result in near 90 percent confidence that the attack was real. If the chance of war was judged to be 50-50, a *single* warning would yield a 95 percent estimate that a real attack was underway; two warnings would make it seem virtually certain (99.7 percent).[82]

During the Cuban Missile Crisis, President Kennedy reportedly believed that the chances of war between the United States and the USSR were between 30 and 50 percent.[83] If the American attack warning system were as effective as Blair assumes, a single false warning would have led decision makers to believe that there was a 90 to 95 percent chance that a real attack was underway. Add to this the facts that (1) four of the six Soviet nuclear-tipped intermediate-range missile sites in Cuba had become operational in the middle of the crisis,[84] and (2) President Kennedy had publicly threatened to launch a "full retaliatory" nuclear strike against the USSR if even one nuclear missile was lauched from Cuba,[85] and you have a recipe for accidental nuclear war on a global scale.

To make matters worse, after a high-flying U-2 spy plane was shot down over Cuba in the midst of the crisis, Robert Kennedy told the Soviet ambassador that if even one more American plane was shot down, the United States would immediately attack all antiaircraft missile sites in Cuba and probably follow that with an invasion.[86] At the time, Cuban antiaircraft guns, not under Soviet control, were shooting at low-flying American spy planes entering Cuban airspace.[87] Without realizing it, the United States had given Fidel Castro, or possibly even an unreliable Cuban soldier operating an antiaircraft gun, the power to trigger an all out nuclear war.

Crises also make accidental war more likely because they put a great deal of stress on everyone involved, increasing the chances of mistakes, poor judgement and other forms of human error. High stress can cause serious deterioration in performance. Sleep deprivation, disruption of circadian rhythms, fatigue and even physical illness (common in national leaders) can magnify that deterioration (see Chapter 7). Crises also stimulate group pathology that can lead to very poor decision making (see Chapter 8). All of this gets worse as the crisis drags on.

The stress of a crisis can also multiply the dramatic and sometimes dangerous side effects of prescribed and over-the-counter medication, and increase the tendency to abuse drugs and alcohol (see Chapter 6). In the short term, chemical stimulants such as amphetamines may actually help decision makers to overcome stress and fatigue-induced impairment of performance. But there is no way of knowing early on how long the crisis will last, and prolonged use of stimulants can lead to impulsiveness, highly distorted judgement and even severe paranoia—a deadly combination in the midst of a confrontation between nuclear-armed adversaries.[88]

The Cuban Missile Crisis was undoubtedly the most dangerous of the Cold War crises. At its height, control over critical military forces was not nearly as precise and error free as is assumed by the pre-planned-options approach that underlies the SIOP. On October 22, President Kennedy publicly announced that the United States was beginning a naval blockade of Cuba and declared that any missile launched against the United States from Cuba would trigger a full-scale nuclear retaliation against the USSR. Hours before this speech, Air Defense Command had ordered nuclear-armed interceptors to immediately spread out to alternate bases. This was a violation of DEFCON safety regulations: the Air Defense Command was still operating at DEFCON 5 ("normal peacetime") and that action was not to be taken until DEFCON 1 ("war is imminent").[89] It was potentially a very deadly mistake, since Soviet detection of this activity might have led them to believe that we were at DEFCON 1 and getting ready to attack. Furthermore, the records of one fighter squadron show "Aircraft were dispersed . . . *fully armed and with all safety devices removed. Only the pilot stood between the complete weapon system and a full scale nuclear detonation.*"[90]

When DEFCON 3 was declared after Kennedy's October 22 speech, real nuclear warheads were placed on a number of rockets previously being readied to launch test warheads into the Pacific from Vandenberg Air Force Base in California. On October 26, while the United States was threatening to annihilate the Soviet Union if even a single nuclear missile were launched from Cuba, the Air Force launched a nuclear-capable Atlas missile from Vandenberg into the Pacific—an extremely provocative act. It was not done to be provocative. This test had been scheduled before the crisis began. It was launched because of bureacratic inertia (see Chapter 8), but it was nevertheless done without the consent or even the knowledge of the nation's top leadership in Washington.[91]

38   LETHAL ARROGANCE

On October 27, a crucial time during the crisis, an American U-2 spy plane based in Alaska accidentally flew well into Soviet airspace, causing the Soviets to scramble MiG interceptors to shoot it down. In reaction, U.S. air defense sent up F-102A interceptors to rescue the lost U-2 and prevent the MiGs from entering American airspace. The F-102As were carrying nuclear-armed Falcon air-to-air missiles, completely controlled by the pilot.[92] Had one of the American pilots accidentally (or intentionally) fired a Falcon, history's second nuclear war might have begun then and there.

This crisis also highlighted the importance of time in choosing an effective strategy for dealing with a very dangerous situation. In Robert Kennedy's words, "if we had had to make a decision in twenty-four hours, I believe the course that we ultimately would have taken would have been quite different and filled with far greater risks."[93] What if, as in the case of an attack warning today, they had to make a decision in 24 minutes?

The Cuban Missile Crisis was not the only Cold War crisis that could have served as a backdrop for accidental nuclear war. In 1954-55, fighting between mainland China and Taiwan over the islands of Quemoy and Matsu threatened to draw the United States into war with China; in 1969, a confrontation erupted over the border between the Soviet Union and China during which these two nuclear-armed adversaries were shooting at each other; and in 1973, the Middle East "Yom Kippur War" nearly brought the superpowers into direct conflict.

Now, despite the end of the Cold War, international crises continue to be fertile ground for accidental war with weapons of mass destruction. During the Persian Gulf War in 1991, Iraq, which had already used chemical weapons in war (against Iran and the Iraqi Kurds), threatened to launch a nerve gas missile attack against Israeli cities. Israel, which had already attacked an Iraqi nuclear reactor in the past and was believed to have an undeclared nuclear arsenal, could certainly have struck back. Never before had Israel sat out an Arab attack on its territory, military or terrorist, without retaliating. This time they made a commitment to show uncharacteristic restraint, and held back a counterattack even as Iraqi Scud missiles rained down on Israel. Had they instead adopted a launch-on-warning strategy like that of the United States, they might have launched a nuclear counterattack before the Scuds even reached Israel and exploded—before they knew that the missiles were carrying only conventional warheads.

In 1991, President Bush sent a message to Saddam Hussein implying that the United States would retaliate with nuclear weapons if Iraq used chemical or biological weapons in the war. This is often credited with having prevented Iraq from using these weapons. But the same letter reportedly attached the same veiled threat against burning the oil fields in Kuwait. And Iraq did that anyway.[94]

According to a London newspaper report, a senior British military officer with the U.K. forces sent to the gulf said of his brigade, "if they are attacked with chemical gas by Iraqi troops, they will retaliate with battlefield nuclear weapons." Prime Minister Thatcher later said there was "no authority" for that statement.[95] Whether or not British troops had the *authority* to respond with nuclear weapons, if they in fact carried such weapons with them to the battlefield, their mistaken, accidental or unauthorized use could have caused unpredictable escalation.

Nuclear weapons have never before been used in a war in which more than one side had deliverable weapons of mass destruction at the ready. It is impossible to predict what would have happened if that threshold had been crossed. In the Persian Gulf crisis, despite the end of the Cold War, we seem to have come very close.

There is no doubt that the dramatic lessening of tensions that followed the end of the Cold War reduced the chances of full-scale accidental war between nuclear-armed superpowers. Yet none of the critical elements that can combine to generate accidental nuclear war have disappeared: vast arsenals of weapons of mass destruction still stand at the ready; serious international crises and small-scale wars that threaten to escalate continue to occur; and human fallibility and technical failure still conspire to produce triggering events and failures of communication that might lead to nuclear war in the midst of a conventional war or crisis.

Beyond this, both the United States and Russia continue provocative Cold War practices in the post Cold War era. Thousands of strategic warheads are still on launch-on-warning alert. The 1994 U.S.-Russia agreement to aim strategic missiles at the open ocean rather than at each other was useful but made very little difference. It takes only seconds to retarget a missile.[96] During the 1990s, American strategic nuclear bombers still flew up to the North Pole on simulated nuclear attack missions against Russia. Even more frightening, Russia was still operating a "doomsday" system it had allegedly put in place some 20 years earlier to insure a devastating retaliation if America attacked first.

The almost completely automated system was designed to be switched on by the Russian general staff during a crisis, and to fire communications rockets in the event that nearby nuclear explosions were detected and communications with military command were lost. These rockets would overfly Russian land-based missiles and transmit launch orders causing the missiles to fire without any action by their local crews.[97] Given all the false warnings and other human and technical failures that have plagued the nuclear military, simply having such a system greatly increases the probability of accidental nuclear war. What justification can there possibly be for continuing these extraordinarily dangerous practices today?

After every major accident, false warning or communications disaster we are always assured that changes have been made that will reduce or eliminate the chances of it ever happening again. Yet, perhaps in a different form, it *always* seems to happen again.

We often fail to learn from even our most serious mistakes. Corrective measures that are taken are frequently too little and too late. We certainly can do better. But the problem goes much deeper. The fallibility inherent in human beings, our organizations and every system we design and build makes it impossible to do well enough.

As long as we continue to have arsenals of weapons of mass destruction, we must do everything possible to reduce the chance that we will blunder into cataclysmic war. Yet the reality is, we will never remove that threat unless and until we find a way to get rid of those arsenals. In human society, failure is an unavoidable part of life. And still we continue to bet our future survival on the proposition that the ultimate failure will never occur.

PART III

Why It Can Happen Here: The Many Sides of Human Fallibility

CHAPTER SIX

The Fallibility of Individuals: Alcohol, Drugs and Mental Illness

We all make mistakes every day, most of them so trivial and fleeting that we pay little attention to them. We misdial telephone numbers, misplace notes, misspell words. Our brains do not always stay focused tightly on the issue at hand. Attention drifts, and something that should be noticed goes unnoticed. Just as the complex technical systems we create and operate are subject to breakdown and less-than-perfect performance from time to time so is the complex biological system that creates and operates them—the human mind and body. We are very powerful, very capable, and yet we are prone to error, subject to malevolence and all the other limitations of being human. We are fallible in our actions and in our essence.

Our failures do not have to be continuous in order to be dangerous. There are many situations in which they are of little consequence much of the time, but can be catastrophic at critical moments. The pilot of a modern transatlantic airliner could be virtually asleep for most of the flight without causing any real problem. But if that pilot is not fully reliable at the critical moment when a problem develops and quick corrective action is required, hundreds of lives could be lost. Similarly, most of the time it is not a problem if a person guarding nuclear weapons is drug or alcohol impaired, because most of the time nothing happens. But if that guard is not alert and ready to act the moment that terrorist commandos try to break into the compound, a disaster of major proportions could result. The problem is that it is impossible to predict when those critical moments will occur. So when dangerous technologies are involved, there is little choice but to take even fleeting failures of reliability seriously.

Human error is a frequent contributor to commonplace technological mishaps. According to GAO, 30 percent of the 169 accidents involving major airlines investigated by the National Transportation Safety Board from 1983 to 1995 were caused in part by pilot performance.[1] Federal Aviation Administration (FAA) data show that 2,396 people died between 1987 and 1996 in what has to be one of the strangest categories of commercial aircraft accident, "controlled flight into terrain." These crashes, in which a properly operating aircraft is flown by its malfunctioning crew directly into the ground, accounted for 25 percent of commercial aircraft accidents around the world during that period.[2] Human error was a contributing factor in almost three-quarters of the most serious class of U.S. military aircraft accidents in 1994 and 1995.[3] In a 1998 study of ten nuclear power plants, representing a cross section of the U.S. civilian nuclear industry, the Union of Concerned Scientists concluded that nearly 80 percent of reported problems resulted from some form of flawed human activity (35 percent from "workers' mistakes" and 44 percent from poor procedures).[4]

The use and abuse of chemical substances is one of the most common ways that individuals render themselves less than fully reliable. Even in moderation, alcohol and drugs can at least temporarily interfere with full alertness, quickness of reaction, clarity of the senses and sound judgement. (When you think about it, that's part of the reason we use them in the first place.) When used to excess or at inappropriate times, they seriously threaten reliability. Mental illness and emotional trauma also can and do impair performance. All these problems are, unfortunately, all too common in the human population. And since every dangerous-technology design team and work force is recruited from the human population, the threat they pose can never be entirely avoided.

ALCOHOL

Alcohol is the most widely used drug. In 1996, the average American drank about 2.2 gallons of pure alcohol annually (roughly one and a quarter shots of hard liquor per day).[5] Since it is estimated that about one-third of the U.S. population does not drink at all, that implies the average *drinker* drank one and a half times that much. Even those in critical positions sometimes drink inappropriately despite the best efforts of the authorities to prevent it. The FAA and all commercial airlines have very strict rules concerning the use of alcohol by flight crews. That did not stop all three cockpit crew members of a March 1990 Northwest Airlines flight from Fargo, North Dakota to Minneapolis-St. Paul from flying the plane while under the influence of alcohol. The pilot of the plane reportedly "had so much alcohol in his blood . . . that he would have been too drunk to legally drive an automobile in most states."[6] The crew members were arrested in Minnesota, and their licenses were later revoked.

Drunken flying cost sixty-one commercial pilots their licenses between 1984 and 1990. About 1,200 airline pilots were treated for alcoholism and returned to flying from the mid 1970s to 1990 under a rehabilitation program run by the government. In the words of Dr. Joseph Pursch, a psychiatrist who helped begin that program, "Every pilot I treat, it's clear . . . should have been treated 10 or 15 years ago. . . . For every pilot I treat, there are five more who just sneak into early retirement."[7]

Alcohol in the American Military

The military is a natural and especially useful focal point for our exploration of substance abuse and mental illness in dangerous-technology systems. The armed forces are a large and important set of organizations intimately involved with a wide variety of dangerous technologies, from weapons of mass destruction to nuclear power reactors to highly toxic chemicals. Focusing on the military should also help avoid exaggerating the problem, since the military has much more power to regulate the behavior of the people who live and work within its confines than do most other organizations involved with dangerous technologies. That control should make it easier for the armed forces to minimize these human-reliability problems and take quick corrective action when they do occur. Furthermore, the military's public nature and prominence in society raises the chance that systematic data on alcoholism, drug abuse and mental illness would be available, especially in a democratic society. For all these reasons, the American military is worth special attention.

In 1980, the assistant secretary of defense for health affairs began a series of worldwide surveys in a systematic effort to collect data on substance abuse and related problems within the armed forces. Surveys were carried out in 1980, 1982, 1985, 1988, 1992 and 1995, looking at these problems in more or less comparable ways. Interestingly enough, after the 1988 survey was released, it was discovered that human fallibility had struck: a programming error had resulted in a substantial understatement of the "heavy drinkers" category for each of the military branches individually and all the branches combined in the 1982, 1985 and 1988 survey reports.[8] Data presented below are corrected for this error.[9]

The surveys classified alcohol consumption in five categories: "abstainer," "infrequent/light," "moderate," "moderate/heavy" and "heavy." Heavy drinking was defined as "consumption of five or more drinks per typical drinking occasion at least once a week." By this definition, 24 percent of all military personnel were heavy drinkers in 1982, and more than half (54 percent) were classified as either "heavy" or "moderate/heavy" drinkers. By 1995, there was a significant drop in reported alcohol use. Nevertheless, the consumption of substantial quantities of alcohol continued to be widespread, with more than 41 percent of military personnel still classified as either "heavy" or "moderate/heavy" drinkers, and more than 40 percent of those (17 percent of service personnel) in the "heavy" drinking category.[10]

Like most chemical substances, alcohol has a wide range of effects on the individuals who consume it. A typically "light" drinker who consumes an amount of alcohol normally classified as "moderate" may well become more drunk than a "heavy" drinker would. The survey therefore attempted to look separately at the problems respondents had experienced because of their use of alcohol. Negative consequences were reported by nearly 30 percent of all military personnel in 1995: 16 percent reported productivity loss (lost work time or ineffective performance), 6 percent reported that they were dependent on alcohol, and 8 percent reported other "serious consequences."[11]

These data may be significantly understated. It is difficult enough to collect accurate information on sensitive issues from members of the general public. People who live and work in an authoritarian "total institution"[12] like the military are much less likely to respond accurately. While the teams sent to collect the data "described the purpose of the study, assured the respondents of anonymity, and informed participants of the voluntary nature of participation,"[13] what incentive would anyone have to give accurate answers to questions that could potentially cause them a great deal of trouble, even ruin their careers, if anonymity broke down? What would serious abusers lose by understating their problem compared to what they could lose if they answered honestly and their answers were traced back to them somehow? Under the conditions of military life, verbal assurances of survey takers are not likely to be strong enough to overcome the powerful incentive to give answers that are more acceptable than accurate.[14]

There is no way of directly assessing the degree of such bias, if any, in the data. But since there is no reason to believe that respondents would *overstate* their abuse of substances or the consequences of that abuse, it is wisest to interpret these data as conservative estimates. Even so, it is quite clear that alcohol use is a widespread and serious problem in the U.S. military.

Alcohol in the Soviet/Russian Military

Russia has the only other military in the world that has arsenals of weapons of mass destruction comparable in size to those of the United States and is comparably involved with other dangerous technologies. While there are few systematic data available on alcohol use in the Soviet/Russian military specifically, the extent of alcohol abuse among the general population makes it clear that it is a real problem.

Vladimir Treml of Duke University tried to unravel the confusions and distortions of official data on alcoholism among the general population of the former USSR and in present-day Russia. He estimated that the average person (15 years of age or older) in the USSR consumed about 3.85 gallons of pure alcohol during 1984—45 percent more than the average American consumption that same year. Premier Gorbachev's nationwide antidrinking campaign began in May 1985. By 1989, Treml estimates that average Soviet alcohol consumption had fallen to about the same per capita level as in the United States.[15] But the unpopular campaign fizzled in the late 1980s, and by 1993, per capita alcohol consumption in Russia had climbed back to roughly the same peak level it had reached in the Soviet Union prior to Gorbachev's campaign (3.81 gallons).[16]

Treml's figures include both state-produced alcohol and *samogon* (a popular form of homemade alcohol), but not homemade wine or beer, or "stolen industrial alcohol and alcohol surrogates."[17] Since consumption of these forms of alcohol is substantial, these estimates should be regarded as conservative. Treml writes,

> The Soviet Union . . . [was] a country of heavy drinkers. . . . [C]linically defined alcoholics constitute[d] between 3 and 4 percent of the total adult population; heavy drinkers, . . . 9-11 percent; moderate drinkers, 70-75 percent. . . . On the eve of Gorbachev's [antidrinking] campaign, about 4.3 million people were officially classified as chronic alcoholics registered by health authorities. . . . [T]he country . . . [had] between 5 and 7 unregistered alcoholics for each registered one.[18]

Most of the available evidence on alcoholism in the Soviet/Russian military comes from comments made in various official military journals and interviews with émigrés who served in the armed forces before leaving the country. Combining analysis of Soviet military journals with a survey of several hundred former Russian soldiers, Richard Gabriel concluded in 1980 that

> Russian soldiers at all ranks have at least the same high rates of alcoholism and alcohol-related pathologies as the society at large—in some instances probably considerably higher. . . . The impression one gathers from reading Soviet periodicals is that heavy drinking and chronic alcohol abuse are common characteristics of Soviet military life. One soldier . . . noted . . . "even if you are not an alcoholic when you go into the Army, you are when you come out."[19]

Furthermore,

> as reported by official sources, . . . drunkenness seems to cut across all rank levels in the army. . . . Alcohol abuse appears to cut across all types of units as

well. Reports of alcoholism are just as common in the elite technical units, such as strategic rocket forces [the mainstay of the Soviet nuclear military] and the air defense corps, as they are in infantry and tank units."[20]

Nearly 65 percent of the former Soviet soldiers surveyed reported seeing or hearing of a noncommissioned officer drunk on duty. In the strategic rocket forces, it was more than 80 percent. Nearly 16 percent said ordinary soldiers were "often" or "very often" drunk on duty.[21]

Interviewed on the 1981 Boston PBS TV program "The Red Army," RAND national security analyst Enders Wimbush gave anecdotal evidence of what happened when the Soviet military tried to control drunkenness by restricting access to vodka:

> We know of one technique where shoe polish is spread upon bread, the bread is then set out in the sun, the alcohol content in the polish distills downward, and then the bread is consumed. Eau de cologne . . . is consumed in great quantities. Anti-freeze for trucks, brake fluid for all vehicles, and . . . the de-icer used by pilots [are also consumed]. The pilots will take off . . . radio back that they are encountering icing conditions and . . . activating their de-icers, when in fact they are not. They return to ground and pilfer the alcohol-based de-icing fluid, and then stage quite a party with it.[22]

For many civilians in Russia and most of the other Soviet successor states, economic life is harder than before, as they struggle to transform their economic and political systems at the same time they try to overcome the damage of the long Cold War. Russian military life has deteriorated even more dramatically. In the absence of any well-thought-out conversion process, drastic cutbacks in government weapons purchases have made life more and more uncertain for military-industry workers.[23] Both they and armed-services personnel have lost much of the status they once enjoyed. Many have been displaced. While those who have not lost their jobs may still have some privilege, they have fallen farther than the average citizen and that can be a great emotional shock.

DRUG USE

Beginning in 1971, the National Institute on Drug Abuse sponsored a series of national surveys on drug use in the United States, taken over in the mid 1990s by the Substance Abuse and Mental Health Services Administration. The

TABLE 6-1

NONMEDICAL DRUG USE IN THE UNITED STATES (1997)
(BY ESTIMATED PERCENTAGE OF POPULATION OVER 12 YEARS OF AGE)

	EVER USED	PAST YEAR	PAST MONTH
ANY ILLICIT DRUG*			
Total Population	35.6%	11.2%	6.4%
Males (Age 18-25)	50.3	30.8	19.6
MARIJUANA			
Total Population	32.9	9.0	5.1
Males (Age 18-25)	46.6	27.9	17.4
COCAINE (INCLUDING CRACK)			
Total Population	10.5	1.9	0.7
Males (Age 18-25)	12.1	5.7	1.9
HALLUCINOGENS			
Total Population	9.6	1.9	0.8
Males (Age 18-25)	19.1	9.4	3.8
STIMULANTS			
Total Population	4.5	0.8	0.3
Males (Age 18-25)	4.3	2.1	0.9
SEDATIVES			
Total Population	1.9	0.3	0.1
Males (Age 18-25)	2.2	0.9	0.2
TRANQUILIZERS			
Total Population	3.2	1.0	0.4
Males (Age 18-25)	6.2	3.3	1.3
ANALGESICS			
Total Population	4.9	1.9	0.7
Males (age 18-25)	8.3	4.8	1.7
INHALANTS			
Total Population	5.7	1.1	0.4
Males (Age 18-25)	13.8	4.7	1.6

* Includes marijuana, inhalants, cocaine, hallucinogens, heroin and nonmedical use of sedatives, tranquilizers, stimulants or analgesics.

Source: Substance Abuse and Mental Health Services Administration, National Household Survey on Drug Abuse: Population Estimates 1997 (Washington, DC: Department of Health and Human Services, 1998), pp. 17, 23, 29, 41, 47, 59, 65, 71 and 77.

population targeted included all civilian residents of households (including civilians living on military installations), and specifically excluded active-duty military personnel, persons living in institutional group quarters such as hospitals and jails, and children under 12. The data in the 1997 survey were collected through personal visits to each residence. An attempt was made to assure those responding of anonymity. Table 6-1 presents summary data from the survey conducted for 1997.

Excluding young children, it is estimated that more than a third of all Americans has used some illicit drug sometime in their lives. More than 1 in 9 surveyed admitted to using within the preceding 12 months; nearly 1 in 15 used some drug illegally within the past month. The most commonly used drug was marijuana, with 5 percent reporting recent use and almost twice that number reporting use within the past year. Two of the most troublesome categories of drug, from the human-reliability point of view, are cocaine and hallucinogens. For each of these, about 10 percent reported they had used the drug at some point, about 1.9 percent within the year and 0.7 percent within the preceding month. These look like small numbers, but when applied to the total 1997 U.S. population (about 216 million age 12 or older), they mean that an estimated 21 million Americans have used hallucinogens and 23 million have used cocaine: more than 4 million used one or both of these drugs within the last year and more than 1.5 million within the last month. Sedatives are the only category of drug in the table used by less than 0.3 percent over the previous 30 days, and 0.3 percent is more than 600,000 people.

The use of illicit drugs by males of the prime age group for entrance into the military (age 18-25) is far more common. More than half of those surveyed reported using such drugs some time in their young lives, nearly a third in the preceding year, and nearly a fifth in the preceding month. The rate of current drug use for young males is *two to five times* that for the population as a whole. That pattern holds across all drug categories in the table.

Sometimes the sheer size of the drug trade surprises even those who follow it closely. In the fall of 1984, a raid in northern Mexico yielded the largest recorded seizure of drugs ever—more than 10,000 tons of marijuana. That was eight times as much marijuana as authorities thought Mexico produced annually, and nearly as much as was believed consumed annually in the entire United States. Since the government used the same statistical techniques to measure the use of most other illicit drugs, the huge seizure sent Federal officials back to the drawing board, forcing them to rethink their estimates.[24]

Cocaine was very much at the forefront of public attention in the United States during the 1980s. The rate of cocaine use among high school seniors tripled between 1975 and 1985, from 2 percent to 6 percent.[25] By 1990, their rate of cocaine use still stood at more than 5 percent.[26] Cocaine was so pervasive in the America of the 1980s that it even managed to penetrate two of America's most venerable institutions, baseball and high finance. By 1985, scores of major league baseball players had been implicated as users and sometimes as sellers of cocaine in wide ranging criminal investigations. According to the *New York Times,* "One owner, who requested anonymity, said an agent had told him . . . that he could field an all-star team in each league with players who were using cocaine."[27] As to high finance, on April 16, 1987, "Sixteen brokers and a senior partner of Wall Street firms were arrested by federal agents . . . and charged with selling cocaine and trading the drug for stocks, information and lists of preferred customers." The special agent in charge of the federal Drug Enforcement Agency (DEA) in New York City said of the raid, "We don't believe this case is an aberration."[28]

Psychedelics

Although they are nowhere near the top of the drug abuse list, psychedelics (also known as hallucinogens), have special characteristics that make them particularly relevant to reliability. The premier psychedelic, d-lysergic acid diethylamide (LSD), was first synthesized in 1938,[29] but was not brought to widespread public attention in America until the 1960s, through the enthusiastic advocacy of such drug prophets of the era as Harvard psychologist Timothy Leary.

LSD is one of the most powerful psychoactive drugs known. Whereas the effective dose of most such drugs is measured in the tens or hundreds of milligrams, as little as 10 micrograms (0.01 milligrams) of LSD produces some euphoria, and 50-100 micrograms (0.05-0.1 milligrams) can give rise to the full range of its psychedelic effects.[30] And just what does the drug do?

LSD magnifies and intensifies sensory perception: "ordinary objects are seen . . . with a sense of fascination or entrancement, as though they had unimagined depths of significance . . . colors seem more intense . . . music more emotionally profound."[31] It can alter the drug takers' perceptions of the look and feel of body parts, cause the field of vision to quiver and swell, distort perspective. It can also produce "synesthesia," a kind of crossing of the senses in which colors are heard, sounds are seen, etc. The sense of time can become distorted. It may seem to pass much more slowly or even stop completely.

LSD has even more profound emotional effects: "feelings become magnified to a degree of intensity and purity almost never experienced in daily life; love, gratitude, joy, . . . anger, pain, terror, despair or loneliness may become overwhelming." These powerful emotions can produce an intense fear of losing control, even to the point of paranoia and panic. The drug takers may lose track of where their bodies end and the environment begins. The worst experience, the "bad trip" can be terrible, "a fixed intense emotion or distorted thought that can seem like an eternity of hell . . . [with] remorse, suspicion, delusions of persecution or of being irreversibly insane."[32]

Which of these effects will occur and at what level of intensity is extremely difficult to predict. "One person may feel only nervousness and vague physical discomfort from a dose that plunges another into paranoid delusions and a third into ecstasy.... [T]he drug ... is so unpredictable that even the best environment and the highest conscious expectations are no guarantee against a painful experience."[33] Furthermore, the same individual may experience radically different effects from the same dose taken at different times. Both the nature and unpredictability of the drug's effects render the drug takers completely unreliable while they are "under the influence."

LSD normally takes effect 45 to 60 minutes after the drug is taken orally. Its effects peak in 2 or 3 hours and can last as long as half a day, becoming more intense then less intense a number of times. The end of the 8 to 12 hour experience is not always the end of the drug's impact. Users are subject to spontaneous recurrences, called "flashbacks." Without having actually taken the drug again, the user may suddenly be on a complete or partial LSD "trip" at some unpredictable time in the future. A flashback "can last seconds or hours ... it can be blissful, interesting, annoying, or frightening. Most flashbacks are episodes of visual distortion, time distortion, physical symptoms, loss of ego boundaries, or relived intense emotion. ... [I]n a small minority of cases, they turn into repeated frightening images or thoughts."[34]

Flashbacks are common. Something on the order of a quarter of psychedelic drug users (perhaps more) experience them. They can occur up to several months after the last dose was taken, but rarely longer than that. Their timing is difficult if not impossible to predict, but flashbacks are known to be most likely at a time of high emotional stress, fatigue or drunkenness. Marijuana smoking is also a common trigger of LSD flashbacks.[35] All of which illustrates the importance of considering interactions among drugs, as well as the user's physical and emotional state, in evaluating how substance abuse affects reliability.

It is extremely difficult to detect the use of LSD through the kind of mass urine screening that is effective in detecting marijuana and heroin. Because LSD is so powerful, it is normally taken in extremely small doses. What little is taken is almost completely metabolized in the body—very little is excreted unchanged. Urine tests *can* detect LSD use, but the tests required are very sophisticated, time consuming, expensive, and often ambiguous. This, and the fact that LSD does not linger for days in the body, makes it "safer" (in terms of avoiding detection) for a determined drug user faced with an environment where random urine testing is common.

LSD is not the only psychedelic. Psilocyn, psilocybin, DMT (N,N-dimethyltryptamine), MDA (3,4-methylenedioxyamphetamine), MMDA (3-methoxy-4,5-methylenedioxyamphetamine), mescaline (peyote), DOM (2,5-dimethoxy-4-methylamphetamine), ketamine, a class of drugs called harmala alkaloids, DOET (2,5-dimethoxy-4-ethylamphetamine) and THC (the active ingredient in marijuana and hashish) among others, can all have effects that resemble those of LSD.[36]

Legal Drugs

Most people understand that the use of illegal drugs can affect mood, clarity of thought, judgement and reaction time in ways that compromise reliability. But it is much less widely understood that commonly prescribed legal drugs—even over-the-counter medications—can also seriously degrade performance and reliability. These legal drugs have interaction and side effects that range from the usually mild drowsiness that often accompanies use of antihistamines to such dramatic effects as depression, hallucinations, convulsions and rage.[37]

Among the 50 drugs most frequently prescribed in the U.S. in the mid 1980s were Dimetapp, Restoril, Inderal, Xanax and Valium. Dimetapp, used for the treatment of allergies such as hay fever, can produce giddiness, hypertension, dizziness, visual disturbances and irritability. Restoril, used to treat insomnia, may cause drowsiness, dizziness, mental confusion, lack of concentration, loss of equilibrium and even hallucinations. Inderal, used to treat migraine headaches and hypertension (among other things), can reduce alertness and cause visual disturbances, mental depression, hallucinations and congestive heart failure. Xanax, used in the treatment of anxiety, can cause irritability, difficulty in concentrating, loss of coordination, agitation, sleep disturbances and hallucinations. Valium, used to relieve anxiety and treat muscle spasms, has side effects ranging from drowsiness and fatigue to depression, blurred vision, hallucinations and such paradoxical reactions as heightened anxiety, insomnia and rage.[38]

By definition, over-the-counter medications are available without prescription, and therefore without having paid a visit to the doctor. The fact that they are presumed safe enough to be taken without medical oversight, along with their availability and relatively low price, leads to a more casual attitude in using them than in using prescription drugs. Yet they too have reported side effects that can seriously degrade performance and reliability. Among the 50 best-selling over-the-counter pharmaceuticals in the mid 1980s were Contac, Alka Seltzer Plus, Drixoral, Robitussin DM, Sudafed and Advil. The advertising that made these brand names household words also conveyed the impression that they are completely benign. Yet Contac, used for temporary relief of sinus and nasal congestion, has common side effects that include dizziness, trembling and insomnia. It also has been known to interact with a class of prescription drugs known as MAO (monoamine oxidase) inhibitors, used to treat depression, to produce a dangerous rise in blood pressure. Alka Seltzer Plus is used to relieve pain, fever and allergy symptoms. Its common side effects include nervousness, drowsiness, dizziness and insomnia; its rarer side effects include fatigue, changes in vision, agitation and nightmares. It can interact with alcohol, antidepressants and antihistamines to cause excessive sedation.[39]

Drixoral reduces the symptoms of allergies and can also induce sleep. Its common side effects include insomnia and agitation; its rarer side effects include nightmares, seizures and hallucinations. Drixoral has interaction effects similar to those of Alka Seltzer Plus. Robitussin DM, a very widely used cough suppressant, commonly produces nervousness, less frequently causes drowsiness and dizziness, and has been reported to interact with MAO inhibitors to cause high fever, disorientation and loss of consciousness. Sudafed, a sinus and nasal decongestant that has common side effects of agitation and insomnia, can on rare occasions cause hallucinations and seizures. Finally, Advil, a pain reliever and anti-inflammatory, commonly causes dizziness. Less frequently, Advil has been known to produce drowsiness and depression, and, in rare cases, blurred vision, confusion and convulsions.[40]

It is important to emphasize that many of the more striking and dangerous side effects of both prescription and nonprescription drugs are often temporary and occur as the result of interaction with alcohol or other drugs. They also tend to be rare. If that were not true, they would not be so commonly used. Nevertheless, rare events do occur.[41] And in anyone working with dangerous technologies in a job subject to sudden criticality, a temporary failure of reliability resulting from a relatively rare side effect or interaction can have permanent and catastrophic consequences.

Drug Use in the American Military

Table 6-2 presents summary data from worldwide surveys of self-reported substance abuse among American military personnel conducted under contract to the Department of Defense. Data are given for both 1988 and 1995 to convey some feeling for the change in the estimates over time. These are the same surveys that studied alcohol abuse, and hence suffer from the same potential biases related to the issue of anonymity discussed earlier.

In 1988, almost 9 percent of all U.S. military personnel worldwide reported using some sort of illicit drug within 12 months, nearly 5 percent reported doing so within 30 days before filling out the questionnaire. The most widely used drug was marijuana, with more than 6 percent reporting use within the year and 2.7 percent reporting use within the month. However, ignoring marijuana did not drop the percentage of military personnel reporting illegal drug use to insignificance: almost 6 percent reported using some other drug over the course of the preceding 12 months, and more than 3 percent within the past 30 days.

By 1995, a substantial drop in reported drug use had been recorded. Illicit drug use during the past year had fallen to 6.5 percent, and use within the month to 3 percent. The sharpest decrease had been in reported cocaine use, which fell

TABLE 6-2

NONMEDICAL DRUG USE IN THE U.S. MILITARY (BY ESTIMATED PERCENTAGE OF COMBINED MILITARY POPULATION)

	USED IN PAST YEAR	USED IN PAST MONTH
ANY ILLICIT DRUG*		
1988	8.9%	4.8%
1995	6.5	3.0
ANY DRUG EXCEPT MARIJUANA		
1988	5.9	3.1
1995	3.9	2.0
MARIJUANA		
1988	6.1	2.7
1995	4.6	1.7
COCAINE		
1988	2.5	0.9
1995	0.9	0.3
LSD/HALLUCINOGENS		
1988	1.3	0.4
1995	1.5	0.6
AMPHETAMINES/STIMULANTS		
1988	1.6	0.8
1995	0.9	0.5
BARBITURATES/SEDATIVES		
1988	0.6	0.3
1995	0.3	0.1
TRANQUILIZERS		
1988	0.7	0.4
1995	0.6	0.3
ANALGESICS		
1988	1.8	1.1
1995	1.0	0.6
INHALANTS		
1988	1.0	0.7
1995	0.7	0.4
PCP		
1988	0.1	0.1
1995	0.2	0.1
HEROIN/OTHER OPIATES		
1988	0.2	0.1
1995	0.2	0.1
"DESIGNER" DRUGS**		
1988	0.6	0.2
1995	0.5	0.2

*Nonmedical use of any drug or class of drugs listed in the table.
** Combinations of individual legal drugs created specifically for their psychoactive effects.

Source: Bray, R.M. et. al., *Worldwide Survey of Substance Abuse and Health Behaviors Among Military Personnel*, 1988 and 1995 editions (Research Triangle Institute), Table 5.1, p. 85 and Section 5.3, Table 12, respectively.

by nearly 65 percent for the 12-month period and slightly more than that for the preceding 30 days. The decline for drugs other than cocaine was less spectacular but still impressive. Not all reported drug use went down. Bucking the trend, LSD/hallucinogen use actually increased more than 15 percent within the year and 50 percent within the month, while PCP, heroin and "designer drug" use either rose or remained relatively stable.

Whether the striking decline in reported use of most drugs reflects a reduction in actual use or simply a higher degree of misreporting is hard to say. The military did intensify its antidrug campaign and the data could reflect the success of these efforts. But it could just as easily reflect greater underreporting of drug abuse triggered by increased fear of being caught and punished precisely because of the heightened antidrug effort. In any case, the military drug abuse problem has certainly not disappeared. There are no zero entries in Table 6-2.

These percentages are relatively low and may be substantially underestimated. But to be conservative, suppose we accept them as accurate. When converted to absolute numbers, they are much more impressive. In 1995, there were some 1,085,000 active-duty military personnel in the U.S. armed forces.[42] Applying the estimated drug use percentages to this overall force size implies that more than 70,500 U.S. military personnel used illicit drugs during the preceding year, and nearly 33,000 had used them during the preceding month. Excluding marijuana from the calculation still leaves more than 42,000 active-duty military personnel who used other illegal drugs within the year and almost 22,000 within the past 30 days. More than 16,000 active-duty military people had been tripping on LSD or another hallucinogen in the past 12 months and more than 6,500 within the past 30 days. Thus, even these conservative estimates project that thousands to tens of thousands of American military personnel abuse all sorts of illicit drugs, despite the military's ongoing antidrug efforts.

Over 40 percent of those surveyed in 1988 reported that they did not seek help for drug abuse because "seeking help for a drug problem will damage military career"; nearly 50 percent believe that they "can't get help for drug problem without commander finding out"; and more than 60 percent are kept from seeking help because they believe "disciplinary action will be taken against a person (with a drug problem)."[43] But distrust of the system is not the only thing the military's antidrug program is up against. For many in the military, especially the nuclear military, the conditions of life and work are a breeding ground for various forms of substance abuse (as we shall see in the next chapter). There are also both technical and human problems in implementing the antidrug efforts themselves.

Marijuana, heroin and amphetamine/methamphetamine use can be detected by simple, efficient and inexpensive mass screening of urine. But there is no cost-effective mass screening urine test for LSD.[44] Other psychedelics, such as MMDA, MDA, DOM and DOET, would probably produce a weak positive result on the sort of standard, inexpensive mass screening urine test done for amphetamine. But the gas chromatography follow-up test would fail to confirm the presence of amphetamine. The analysts could then plunge into the extremely complex chemistry required to ferret out the psychedelic involved, *if* they were sufficiently familiar with the highly sophisticated analysis required. Under normal circumstances, it seems much more likely that they would just report that no evidence of amphetamine was found and avoid the trouble. That is especially likely if very large numbers of urine specimens were being processed.[45]

All of this gives rise to the frightening possibility that at least some of the apparent drop in drug use detected by more comprehensive, mass random urine screening may simply represent the shifting of users from more easily to less easily detected drugs. Unfortunately, in some ways, drugs like LSD that are harder to detect are more likely to create serious reliability problems than the drugs they may be replacing. Given the potency of LSD, it would be relatively easy to smuggle some on board a nuclear missile submarine, for example, by soaking part of a T-shirt in LSD, then later cutting it up into little patches that could be swallowed whole.

Other limitations on the effectiveness of mass urine screening are implied by partial, anecdotal evidence. The substitution of "clean" urine borrowed from friends seems to have been controlled by direct observation of the taking of specimens. Still, those in the medical corps responsible for administering tests have been known to warn friends in advance about upcoming "random" drug tests. And commanders, anxious to show their superiors how effectively they have dealt with drug problems in their units, have been known to schedule random tests at times that are likely to show fewer positive results—for example, just before weekends or on the last Friday of the month, when most personnel have long since run out of money for replenishing depleted drug supplies.

There are no publicly available systematic data on drug and alcohol abuse in military industry or other weapons-related facilities. Yet periodic reports make it clear that these workplaces are not immune. For example, in 1988 a congressional oversight committee reported on an investigation of the Lawrence Livermore nuclear weapons lab that the Department of Energy undertook two years earlier. An undercover agent reportedly collected information suggesting that drugs were being bought, sold or used on the job by more than 100 Livermore employees. This included scientists and staff with high security clearances.[46]

Drug Use in the
Military of the Former Soviet Union

There is relatively little information available concerning the use and abuse of drugs in the former Soviet Union. For a long time, the problem was completely denied. As recently as the mid 1980s, the Soviets claimed that "'serious drug addiction does not exist' . . . and 'not a single case' of addiction to amphetamines, cocaine, heroin, or LSD has been recorded."[47] However, government attempts to address drug addiction through legislative and judicial means date back as far as 1926[48]— a historical record that gives the lie to such simple-minded propaganda.

While they no longer deny the problem outright, sources in Russia and other former Soviet states concede that they still do not have an accurate picture of the extent of the drug problem. It is generally admitted that official statistics on drug abuse had been substantially understated in the past. In 1988, for example, during informal meetings with representatives of the U.S. Drug Enforcement Agency, "Soviet officials estimated that in the USSR there were currently 150,000 - 200,000 drug 'addicts' (three to four times the number of officially registered 'addicts') and that the number of drug 'abusers' increased by 50 percent annually."[49]

Two surveys in the republic of Georgia, one in the mid 1970s and one in the mid 1980s confirm that drug use was high. "Over 80 percent of the individuals in each sample consumed drugs at least once a day and a majority of them did so at least twice a day."[50] There are vast areas in the former USSR where poppies (used to make opiates) and hemp (used to make hashish) are cultivated or grow wild. Many households, perhaps numbering in the hundreds of thousands, grow them illegally, presumably for their own use or for sale to supplement their income. According to a deputy minister of the Ministry of Agriculture in the former USSR, more than half of the hemp and poppy crop that was legally grown on collective farms was left in the fields after the harvest and became "a fertile source of drugs for local users and illegal drug producers."[51] Alternative sources of supply likely included theft from medical stores and smuggling from neighboring countries that produce drug-related crops (such as Afghanistan). Available data suggest that, for many individuals, monthly spending on illicit drugs exceeded the average monthly wage, creating the conditions for a great deal of drug-related crime.[52] The Russian Federation and other former Soviet states have been racked by rising crime rates as the rigid controls of their police states were relaxed in the presence of economic and political chaos. Drugs have played a part in this as well.

There is even less information available concerning drug abuse in the Soviet/ Russian military than there is for alcoholism. It seems to have been a relatively minor problem until the Soviets got involved in Afghanistan in the late 1970s.

Like America's war in Vietnam, the Soviet war in Afghanistan was unpopular, ill defined and fought against guerrilla forces in an area of the world that has long been a center of drug production. Though the evidence is anecdotal and very spotty, the Afghan War did apparently lead to increased drug abuse among the Soviet armed forces.[53] There is very little information available about the extent or nature of the drug problem today in the militaries of the Russian Federation and the other republics of the former Soviet Union. But there is little doubt that the armed forces are not drug free.

MENTAL ILLNESS

Mental illness is disturbingly common. In the mid 1980s, the National Institute of Mental Health (NIMH) estimated that almost one-fifth of the adult U.S. population suffered from some form of mental disorder.[54] That result was reinforced in the early 1990s, when NIMH analysis of the largest U.S. mental health survey to date revealed that an estimated 28 percent of American adults (nearly 45 million people) suffered from mental illness, drug abuse problems or a combination of the two. Only about half of those sought professional help.[55] Even though the disorders included in these statistics are not restricted to the most extreme behaviors, they are still relevant. Many subtler forms of mental illness can pose a serious threat to reliability—without being as easily noticed. No human life is wholly without trauma. From time to time, life events produce emotional shock waves in even the strongest and most solid among us. Sometimes the shock gives rise to prolonged anguish or triggers other mental or physical disorders. Sometimes it passes quickly and has little or no long-term effect on emotional stability. But even when trauma leaves no deep or permanent scars, it can still render us less than fully reliable during the time that the wave of troubling emotion is washing over us.

One common mental aberration that runs the gamut from the relatively harmless to the dangerously pathological is the dissociative experience. Dissociation occurs when any group of mental processes are split off from the rest of the mind's functions. Those processes operate almost independently of the rest of the psyche. A frequent and mild form of dissociative experience occurs when the mind "wanders"—for example, when a person is listening to someone else talk and suddenly realizes that a part of what was just said did not really register. At critical moments, even a wandering mind can cause major problems. But dissociation can also be a serious psychological disorder. It can take the form of amnesia, depersonalization (a feeling of loss of identity, of unreality, frequently accompanied by bewilderment, apathy or emotional emptiness), auditory hallucinations (such as "hearing voices inside your head") or even multiple personality.[56]

In a study of dissociation among the general population, Ross, Joshi and Currie found that one or another form of dissociative experience was very common. They administered the Dissociative Experiences Scale to a representative sample of more than a thousand adults in Winnipeg, Canada. The subjects were specifically instructed not to report any experiences that occurred when they were under the influence of drugs or alcohol. Nearly 13 percent of the people in the sample reported a substantial number of dissociative experiences. Some of the more extreme forms of dissociation were also unnervingly widespread: 26 percent of the people tested sometimes heard "voices inside their head which tell them to do things or comment on things they are doing," and over 7 percent heard such voices frequently (25 percent of the time or more). The results of that study, combined with the findings of previous clinical studies, led the researchers to tentatively predict that dissociative problems—including multiple personality disorder—may affect as much as 5 to 10 percent of the general population.[57]

Post-traumatic stress disorder (PTSD) has received considerable attention in the United States since the Vietnam War. In its most extreme forms, PTSD can be debilitating. Studies released in the mid 1990s suggest that severe emotional trauma may physically damage the brain as well as the psyche, shrinking the hippocampus, the brain structure vital to learning and short-term verbal memory.[58] The symptoms of PTSD range from emotional detachment and extreme suspicion to nightmares and intrusive thoughts about past traumatic events that can verge on flashbacks. During the same period, a national survey conducted at the University of Michigan estimated that PTSD has struck nearly 1 adult in 12 in the United States at some time in his or her life. In more than a third of the cases, symptoms have persisted for at least a decade.[59]

In view of all this, it is not surprising that mental disorders and emotional disturbances are a significant problem for the military. In a sample of 11,000 American naval enlistees, roughly 1 in 12 (8.7 percent) was discharged during first enlistment because of psychological problems.[60] During the decade of the 1980s, nearly 55,000 soldiers in the U.S. Army were diagnosed as having psychiatric disorders not specifically involving drugs or alcohol. Of these, more than 6,300 were schizophrenic, and nearly 6,400 exhibited other psychotic disorders.[61] Over the same period, 129,000 U.S. Navy personnel were diagnosed as having psychiatric problems.[62] Further, the data seem to suggest that the incidence of psychiatric disorders in the military increased during the last five years of that time span.[63] According to the 1995 worldwide survey of U.S. military personnel, more than 4 percent of those in the armed forces have considered hurting or killing themselves in reaction to life stress.[64]

ALCOHOL, DRUGS AND MENTAL ILLNESS
IN AMERICA'S NUCLEAR MILITARY

Across all branches of the military, it is the nuclear forces that consistently deal with the most dangerous technologies. The Department of Defense has a special Personnel Reliability Program (PRP) that covers anyone in the military who has direct access to nuclear weapons or components, control over the access of others to such weapons or components, or both. In practice, the PRP covers everyone who works with nuclear weapons, guards them, has access to the authenticator codes needed to fire them or is part of the chain of command that releases them. That is, anyone other than the president, a rather important exception.

There is an initial PRP screening intended to prevent anyone who is physically or mentally unreliable from ever being assigned to nuclear duty. It includes a security investigation, a medical evaluation, a review of personnel records and a personal interview. Psychological tests used to screen for mental and emotional fitness are notoriously unreliable. That is not the military's fault, it is a matter of the state of the art of psychological testing.

Once certified by the PRP and assigned to nuclear duty, military personnel are continually subject to temporary or permanent decertification and removal from nuclear duty at any time if their subsequent condition or behavior fails to meet PRP standards. This part of the program is the most effective, but, unfortunately, those who are decertified were *on* nuclear duty for some time *before* being removed. It is impossible to know how much of that time they were unreliable.

Every year from 1975 through 1990 (the latest year for which data are publicly available), permanent removals under the PRP have ranged from 2.6 to 5.0 percent of the personnel subject to the program, averaging nearly 4 percent (See Table 6-3). That means for a decade and a half, there were at least 1,900 people each year—and in some years as many as 5,800—judged unreliable enough to be *permanently* removed from nuclear weapons duties they were already performing.

Reasons for removal from nuclear duty range from poor attitude to substance abuse and mental breakdown. Drug abuse was by far the most common reason from 1975 through 1985; from 1986 through 1989, it was "poor attitude or lack of motivation." In 1990, the PRP category that includes mental disturbances and aberrant behavior became the largest, accounting for 27 percent of the removals that year. Finally, alcohol abuse grew in importance as a primary cause for removal, rising from 3 percent in 1975 to nearly 18 percent by 1990.

The number of people subject to the PRP fell by about 45 percent over this decade and a half. Since nuclear arsenals did not shrink by 45 percent, does this

TABLE 6-3

PERMANENT REMOVALS FROM MILITARY NUCLEAR DUTY
UNDER THE PERSONNEL RELIABILITY PROGRAM (PRP),
BY REASON (1975-90)

REASON	1975	1976	1977	1978	1979	1980	1981	1982
Alcohol abuse	169	184	256	378	459	600	662	645
Drug abuse	1970	1474	1365	1972	2043	1728	1702	1846
Negligence or delinquency	703	737	828	501	234	236	236	252
Court conviction or contemptuous behavior toward the law	1067	1333	1235	757	747	694	560	605
Mental or physical condition, character trait, or aberrant behavior, prejudicial to reliable performance	1219	1238	1289	1367	1233	941	1022	882
Poor attitude or lack of motivation	*	*	*	822	996	1128	1053	980
TOTAL REMOVALS (number)	5128	4966	4973	5797	5712	5327	5235	5210
(as % of people in PRP)	4.3%	4.3%	4.2%	5.0%	4.8%	4.7%	4.8%	4.9%
Number of people subject to PRP	119625	115855	118988	116253	119198	114028	109025	105288

mean that nuclear duty has somehow become more automated or that the coverage of the PRP has been narrowed? More automation may have some advantages, but, after all, technical systems also can and do fail spectacularly (see Chapters 9 and 10). If, on the other hand, this trend reflects a more restricted view of the PRP, some kinds of duty that were previously subject to PRP controls may no longer be covered.

Since the military has no reason to exaggerate failures of human reliability, assuming that the numbers in Table 6-3 are accurate will give a conservative estimate of the extent of this problem. According to these data, from 1975 through 1990, *more than 66,000 people were permanently removed from nuclear duty* they had already been performing because they were found to be unreliable—an average of more than 4,100 removals every year for a decade and a half! The problem of human reliability in the American nuclear military is serious and ongoing. There is no evidence to indicate, and no reason to believe, that human reliability is less of a problem in the militaries of the other nuclear weapons countries.

Some Incidents

Statistics can take us only so far in understanding this problem. Looking behind the numbers:

TABLE 6-3 (CONTINUED)

PERMANENT REMOVALS FROM MILITARY NUCLEAR DUTY UNDER THE PERSONNEL RELIABILITY PROGRAM (PRP), BY REASON (1975-90)

REASON	1983	1984	1985	1986	1987	1988	1989	1990
Alcohol abuse	621	545	500	395	415	388	365	337
Drug abuse	2029	1007	924	555	477	257	363	151
Negligence or delinquency	220	160	365	170	158	103	113	130
Court conviction or contemptuous behavior toward the law	607	580	327	447	473	486	437	340
Mental or physical condition, character trait, or aberrant behavior, prejudicial to reliable performance	704	646	550	408	437	486	481	510
Poor attitude or lack of motivation	904	828	627	556	564	609	633	432
TOTAL REMOVALS (number)	5085	3766	3293	2530	2524	2294	2392	1900
(as % of people in PRP)	4.9%	3.6%	3.2%	2.6%	2.7%	2.8%	3.1%	2.9%
Number of people subject to PRP	104772	103832	101598	97693	94321	82736	76588	66510

Source: Department of Defense, Office of the Secretary of Defense, Nuclear Weapon Personnel Reliability Program, Annual Disqualification Report, RCS: DD-COMP(A) 1403, Calendar Years 1975-77 (Washington, D.C.); and Nuclear Weapon Personnel Reliability Program, Annual Status Report, RCS: DD-POL(A)1403, Calendar Years 1978-90 (Washington, D.C.).

- In August 1969, an Air Force major was suspended after allowing three men with "dangerous psychiatric problems" to continue guarding nuclear weapons at a base near San Francisco. One of the guards was accused of going berserk with a loaded carbine at the base. His lawyer said that he had pleaded not to be assigned to a job in which he would handle explosives or weapons. Yet he was frequently on duty as senior officer of a two-man team guarding nuclear-armed missiles. The major testified that he had received unfavorable psychiatric reports on the three guards, but he had not removed them because he was short of staff and without them the "hippies" in San Francisco would try to steal the weapons.[65]
- In March 1971, three airmen with top-security clearance working with U.S. nuclear war plans were arrested for possession of marijuana and LSD. The men worked at the computer section that maintains the war plans for all U.S. nuclear armed forces, at the top secret underground SAC base near Omaha, Nebraska.[66]
- Army code specialist Donald Meyer said in a published interview that he had smoked hashish "two or three times every four hours nearly every day for 29 months" while stationed at a U.S. Army nuclear missile base

in Germany in the 1970s. "Meyer said he was under the influence . . . sometimes when he worked with secret material and that missile soldiers sometimes were high when they attached nuclear warheads to the missile. So were soldiers 'who connected the two pieces up' to make the missile operational."[67]

- On May 26, 1981, a Marine EZ-6B Prowler jet aircraft crashed and burned on the flight deck of the nuclear-powered aircraft carrier *Nimitz*. Fourteen people were killed, 44 injured and 20 other airplanes were damaged. The crew of the *Nimitz* and the Prowler's pilot were certified by the Personnel Reliability Program. Yet autopsies revealed that 6 of the deckhands who died in the crash had used marijuana, at least three of them either heavily or shortly before the crash. The pilot had 6 to 11 times the recommended level of antihistamine (brompheniramine) in his system, enough to cause such side effects as sedation, dizziness, double vision and tremors.[68]

- On November 7, 1986, Petty Officer John Walker, Jr. was convicted of espionage in one of the U.S. Navy's most serious spy cases. Walker, who had been a communications specialist on a Polaris nuclear missile submarine, was convicted of selling classified documents to the USSR from 1968 on. He not only had high-level security clearance, but was PRP certified reliable for nuclear weapons duty.[69]

Bad Year at Bangor

The Bangor Submarine Base in Washington State is home to some 1,700 nuclear weapons, about 1,500 of them aboard Trident nuclear missile submarines. There are additional nuclear weapons stored at Bangor's Strategic Weapons Facility, the only site on the West Coast for assembling and loading Trident missiles for the Pacific Fleet. More than a thousand of the five thousand military personnel at the submarine base are PRP certified. All of the following events took place in a single year:[70]

- On the evening of January 14, 1989, an 18-year-old marine, Lance Corporal Patrick Jelly, was on duty in the guard tower at the Bangor Strategic Weapons Facility. At 9:30 P.M., he shot himself in the head with his M-16 rifle. For weeks prior to committing suicide, "He had talked about killing himself, punctured his arms with a needle and thread and claimed to be the reincarnation of a soldier killed in Vietnam." Yet such obviously aberrant behavior was not reported or acted upon, and Jelly remained PRP certified until the night he died.

- Tommy Harold Metcalf was a fire-control technician on the Trident submarine *Alaska*. He had direct responsibilities in maintaining, targeting and firing the sub's 24 ballistic missiles, each of which carried a number of city-destroying nuclear warheads. On July 1, 1989, Metcalf went to the home of an elderly couple, responding to an ad for the sale of their motor home. He bound them and taped plastic bags over their heads, and murdered them both by suffocation. The keys to the motor home were found in his pocket when he was arrested. Metcalf was PRP certified as reliable at the time, and a subsequent review of his records reportedly found no problems with the screening process used to evaluate him.[71]

- In early August 1989, Commander William Pawlyk was arrested after stabbing a man and a woman to death. Pawlyk, who was subsequently sentenced to life in prison, had been commander of Submarine Group 9 at Bangor, and served aboard the nuclear submarine *James K. Polk* for five years. He was head of a reserve unit in Portland, Oregon, at the time of the murders.[72]

- Shyam Drizpaul, like Tommy Metcalf, was a PRP-certified fire-control technician, serving on the nuclear submarine *Michigan*. On January 15, 1990, he shot and killed one crew member in the lounge at his living quarters, then another in bed. Later, while attempting to buy another 9 mm pistol at a pawnshop, he grabbed the gun from the clerk, shot her to death and critically wounded her brother. Fleeing the scene of that crime as well, Drizpaul checked into a motel near Vancouver and used the same weapon to kill himself. A subsequent Navy investigation discovered that Drizpaul drank excessively, carried an unregistered handgun, "and claimed to have been a trained assassin." Apparently, none of this information had been conveyed to his superiors or acted upon.

It is important to note that all of these people had been PRP certified as reliable, and that relatively recent reviews had uncovered no behavioral or attitudinal problems that might have caused them to be removed from nuclear duty. The only reasonable conclusion that can be drawn from all of this is that there are serious limits to the effectiveness of the Personnel Reliability Program. It clearly cannot guarantee either the stability or reliability of those in America's nuclear forces.

Reliability in the Russian Military

Despite the spottiness of information, it is clear that the Russian nuclear military has parallel problems. In October 1998, the Chief Military Prosecutor's Office indicated that 20 soldiers in the strategic rocket forces were removed from duty during 1997-98 because of serious psychiatric problems. Some had been nuclear weapons guards. In a crime-ridden country, the strategic rocket forces had also achieved the distinction of having the highest increase in crime of any branch of the Russian armed forces between 1996 and 1997.[73]

The following illustrative incidents all reportedly occurred within a three-week period:[74]

- Novaya Zemlya is the only nuclear weapons test facility in Russia. On September 5, 1998, five soldiers from the Twelfth Main Directorate there murdered one guard at the site and took another hostage. They then tried to hijack an airplane and took some more hostages before they were finally captured and disarmed.

- On September 11, 1998, a young sailor aboard a nuclear attack submarine in Murmansk went berserk, killing seven people on the sub with a chisel and an AK-47 assault rifle. Then he barricaded himself in the torpedo bay, threatening "to blow up the submarine with its nuclear reactor." Sometime during the 20 hours he was in the torpedo bay, the distraught teenager reportedly killed himself.

- More than 30 tons of weapons-grade plutonium separated from civilian reactors is stored at Mayak. On September 20, 1998, a sergeant from the Ministry of Internal Affairs stationed there "shot two of his comrades and wounded another." He then managed to get away with an assault rifle and ammunition.

The alcohol and drug abuse, the mental and emotional disturbance that seriously compromise reliability often have their origins in the wide variety of traumas and frailties to which human beings are heir. Some of these lie deep within the past of the people involved. But the environment in which people live and work can also cause or bring to the surface these and other forms of unreliable behavior. That is certainly true of life in the military in general and the nuclear forces in particular. It is also true of those who work with other dangerous technologies. It is to these matters that we now turn.

CHAPTER SEVEN

The Fallibility of Individuals: The Nature and Conditions of Life and Work

In *Space/Aeronautics* magazine in the late 1960s, Charles Cornell wrote, "Seemingly inexplicable, inconsistent and unpredictable human 'goofs' account for 50-70 percent of all failures of major weapons systems and space vehicles. That puts human errors . . . ahead of mechanical, electrical and structural failures...as a source of system troubles. . . . The consequences range from minor delays to major disasters."[1] In the aftermath of the tragic midair destruction of off-course Korean Airlines Flight 007 by the Soviet military in 1983, a National Aeronautics and Space Administration (NASA) official testified before Congress that human error was responsible for more than two-thirds of the 950-plus incidents in which civilian airliners had strayed off course in the preceding five years.[2] According to an analysis by the Safety Studies and Analysis Division of the National Transportation Safety Board, pilot error caused or contributed to 54 of the 76 fatal crashes on major airlines in the United States from 1967 through 1981.[3]

In the preceding chapter, we considered how substance abuse and mental illness contribute to human error. But that is only part of the problem. Error is pervasive in human activity, even when lives and treasure hang in the balance. In the airline industry, in the space program, in the field of medicine, in the world of high finance, in the criminal justice system—nowhere are we completely insulated from its consequences. Witness:

- On a clear night in December 1995, the pilots of American Airlines Flight 965—both with thousands of hour of flying time and spotless records—made a series of "fatally careless mistakes" as they approached Cali, Colombia. After flying past the locational beacon 40 miles north of the airport, they programmed their navigational computers to fly toward it, and steered their plane into the side of a mountain. Flight recorders showed no evidence that they had discussed approach procedures, as is mandatory before every landing. It was the worst accident involving an American air carrier in seven years.[4]

- A Boeing 757 crashed off the coast of Peru in early October 1996, killing all 70 people on board. According to the National Transportation Safety Board, the plane went down "because maintenance workers forgot to remove tape and paper covers they had put over sensors while polishing the plane."[5]

- On July 17, 1997, one of three astronauts aboard the already troubled Russian *Mir* space station disconnected a critical electrical cable by mistake. That disabled the computer that controlled *Mir*'s position in space, causing the space station to drift. A backup system took over too late to correct the problem. With the solar panels no longer pointed at the sun, the power drained slowly out of the batteries. Data transmission to the ground was cut off, and the key lighting, temperature-control and oxygen-generation systems had to be shut down to save energy.[6]

- Six people attached to dialysis machines at the Albuquerque Kidney Center suddenly began to scream in excruciating, unexpected pain as a technician mistakenly threw a switch that sent a cleaning solution intended to rinse the machines into the patients' bloodstreams. Five of the patients recovered, one died.[7]

- In May 1995, the chief of neurosurgery at the world-renowned Memorial Sloan-Kettering cancer center in New York operated on the wrong side of a patient's brain. Six months later, the New York State Department of Health reported that "systemic deficiencies" at the hospital, such as failure to always follow medical practices as basic as reviewing diagnostic reports and medical records prior to surgery, had played a role in the incident.[8]

- Four minutes before the end of trading on March 25, 1992, the prestigious Wall Street firm Salomon Brothers unexpectedly sent huge "sell" orders for the shares of 400 companies to the floor of the New York Stock Exchange. The market was thrown into turmoil as the flood of sell orders alarmed traders and caused the Dow Jones industrial average to drop some 15 points. It was all a mistake. An investor had sent the firm an order to sell $11 million worth of stock in 400 companies, but a clerk put the figure in the wrong column, turning it into an order to sell 11 million shares. Even at an average price of $30 per share, that simple human error would have turned the $11 million sell order into an order to sell nearly one-third of a billion dollars worth of stock.[9]

- In April 1997, the inspector general of the Justice Department found that flawed scientific practices and sloppy performance were common at the FBI's world-famous crime laboratory. These "extremely serious and significant problems" jeopardized dozens of criminal cases, including the bombing of the Murrah Federal Building in Oklahoma. In that case, lab workers accused superiors of engaging in sloppy, improper or unscientific practices that so compromised bomb debris evidence that none of it could be tested.[10]

- At 4:00 A.M. on August 31, 1997, five bounty hunters wearing body armor and ski masks forced their way into a house in Phoenix, Arizona. They tied the hands and feet of a woman in the house, and held her, her daughters and her 11-year-old son at gunpoint. They then opened fire into another bedroom, killing a couple who had been sleeping there. The bail jumper the bounty hunters were tracking was not in the house, had never lived there, and was apparently unknown to the people whom they attacked. When asked why the bounty hunters had chosen that house, the Phoenix police said, "It's a mystery to us. There is nothing to indicate that the person they were looking for was at that house."[11]

It was not drug addiction, alcoholism or mental illness that caused these problems. People are prone to make mistakes even under the best of circumstances. But the physical, psychological and sociological circumstances in which we live and work are often not the best. They have powerful effects on our state of mind and thus our behavior on (and off) the job. Even emotionally, physically and mentally healthy people, not abusing drugs or alcohol, have real limits when subjected to the stress, boredom and isolation that are so often an inherent part of working with dangerous technologies. Spending endless hours interacting with electronic consoles, repeating essentially the same lengthy and detailed routine over and over, watching lighted panels and screens, flipping switches, checking and double-checking, sitting hour after hour in the control room of a nuclear power plant or missile silo, sailing for months in a submerged submarine isolated from most of humanity yet poised to destroy it—these are working conditions bound to aggravate the already strong human tendency to make mistakes.

The working environment in the nuclear military is among the most difficult of dangerous-technology work environments. It is isolating because it is enveloped in secrecy. No one on nuclear duty is permitted to talk about the details of their work with anyone lacking the proper security clearance. They cannot share what they do with friends and family. For much of the nuclear military, the work is also isolating because it requires long periods away from friends and loved ones.

Because of the constant repetition of routines and because it is so isolating, life in the nuclear forces is boring too. And it is stressful as well, for at least five reasons: (1) boredom itself is stressful; (2) isolation also creates stress (that's why solitary confinement is considered such a severe punishment for prisoners who behave badly); (3) for safety and security reasons, people on nuclear duty are always "on call," even when they are "off duty"; (4) Being highly trained to carry a task through, but never being able to take it to completion is frustrating and stressful; (5) many on nuclear duty are aware that if nuclear war comes, they will be part of the largest-scale mass murder in human history.

The nuclear military may be the maximum case of a stressful, boring and isolating dangerous-technology work environment, but it is far from the only case. One or more of these reliability-reducing characteristics are found in many other dangerous-technology workplaces. Nuclear power plant operators do work that is boring most of the time and stressful some of the time, though it isn't all that isolating. Working around highly toxic chemical or dangerous infectious biological agents has significant levels of background stress, punctuated by periods of intense stress when something goes really wrong. But because it combines most of the reliability-reducing elements relevant to dangerous-technology workplaces in general, the military work environment is worth special attention.

BOREDOM AND ROUTINE

In 1957, Woodburn Heron described laboratory research funded by the Canadian Defence Research Board in which subjects lived in what amounted to an exceedingly boring environment. There were sounds, but they were constant droning sounds, like the hum of a fan; there was light, but it was constant, diffuse light. There was no change in the pattern of sensory stimuli. The subjects became so eager for some kind of sensory stimulation that they would whistle, sing or talk to themselves. Some had a great deal of difficulty concentrating. Many lost perspective and found themselves on an emotional roller coaster, shifting suddenly and unpredictably from one emotion to another. Many also began to see or hear things that were not there.

People vary widely in their sensitivity to boredom. It is difficult to predict the threshold of monotony that will trigger these reactions in any particular individual in "real world" situations. Yet there is ample evidence that grinding boredom and dulling routine can and do produce such problems. For example, in 1987 it was reported that "Congressional committees, watchdog groups and the [Nuclear Regulatory] commission have repeatedly found operators of nuclear plants asleep or impaired by alcohol and drugs." Attempting to explain such behavior, a representative of the Atomic Industrial Forum (the industry lobbying group) said, "The problem is that it's an extremely boring job. It takes a great deal of training. Then you sit there for hours and hours and take an occasional meter reading."[12]

Boredom can be so painful that people feel compelled to try to escape, sometimes taking refuge in drugs and alcohol. In interviews of Vietnam veterans conducted by the Psychiatry Department of the Walter Reed Army Institute of Research during the Vietnam War (1971), soldiers often cited boredom as the main reason they used drugs. "Descriptions of work activities invariably included statements like, 'There was nothing to do, so we smoked dope. . . . We just sat around. . . . You had to smoke dope, or drink, or go crazy doing nothing. . . . It was boring until I started smoking skag [heroin]; then I just couldn't believe how fast the time went.'"[13]

A sailor who served as helmsman on the nuclear aircraft carrier USS *Independence* claimed that he regularly used LSD *on duty* during the late 1970s and early 1980s. It was the only way, he explained, to get through eight hours of extremely boring work. He said that there was almost never a day in his whole tour of duty that he was not on either LSD or marijuana, most often LSD.[14]

The dulling effects of routine can create great danger in systems subject to sudden criticality. According to Marrianne Frankenhaeuser of the Karolinska Institutet Department of Psychiatry and Psychology in Stockholm,

> An early sign of understimulation is difficulty in concentrating . . . accompanied by feelings of boredom, distress and loss of initiative. One becomes passive and apathetic. . . . [Then] when a monotonous situation all of a sudden becomes critical . . . the person on duty must switch instantaneously from passive, routine monitoring to active problem solving. . . . The sudden switch . . . combined with the emotional pressure, may cause a temporary mental paralysis. During a brief but possibly critical interval, the person in charge may be incapable of making use of the information available. The consequences of such mental paralysis—however brief—may be disastrous.[15]

Eastern Airlines Flight 401 was on a routine final approach to Miami International Airport on a dark night in December 1972. The crew included three able-bodied and experienced pilots. Weather conditions were fine. As they prepared to land, the crew noticed that the nose landing gear light wasn't working, so they couldn't tell whether the gear was extended and locked. An emergency landing with a possible nose gear problem is neither very risky nor particularly rare. But because of the otherwise routine conditions, the crew was so removed from the primary job of flying the plane that they became fixated on the light bulb. They didn't notice that the autopilot had disengaged and the plane was slowly descending. They ignored the altimeter, and didn't even react when the altitude alert sounded. By the time they realized what was happening, they were "mentally paralyzed," unable to react fast enough to prevent disaster. The jumbo jet crashed into the Florida Everglades, killing 99 people.[16]

The pilot, first officer, flight engineer and a guest crew member aboard Pacific Southwest Airlines Flight 182 were talking intensely about retirement benefits as the plane made a routine visual approach to San Diego in clear weather on the morning of September 25, 1978. The San Diego air traffic controller gave standard landing instructions, twice advising of a light aircraft in the area. Each time, the flight crew acknowledged the information. The second time, the captain said he had the Cessna in sight and would maintain visual separation. The approach seemed so routine that the conversation about retirement benefits continued unabated. The crew paid no attention to the fact that they had lost sight of the light plane. Shortly after 9:00 A.M., the jet slammed into the Cessna, and 144 people died.[17]

In 1904, Sigmund Freud published a very popular book, *The Psychopathology of Everyday Life*. He had collected hundreds of examples of inconsequential everyday errors: misreadings, misquotes, slips of the tongue, etc. Freud interpreted these not as meaningless accidents, but as unintended revelations of the unconscious mind, the famous "Freudian slip." His interpretation aside, the fact is that such seemingly trivial errors—trying to open the house door with the office key, calling one child by another child's name, misdialing the telephone—are exceedingly common.

Having analyzed many slips reported by some one hundred subjects over a number of years, psychologist James Reason of Manchester University reported that nearly half the absent-minded errors involved deeply ingrained habits: "The erroneous actions took the form of coherent, complete sequences of behavior that would have been perfectly appropriate in another context. In each case the inappropriate activity, more familiar to the subject than the appropriate one, had been carried out recently and frequently, and almost invariably its locations, movements and objects were similar to those of the appropriate action."[18]

Under normal circumstances these errors are easily corrected and of little consequence. Under abnormal circumstances, following familiar routines can lead to disaster. It is important to understand that *the difference between a trivial and a catastrophic error is situational, not psychological.* What creates the disaster is the context within which the error occurs, not the mental process that caused it. During special NATO training exercises over West Germany in the 1980s, a Royal Air Force Phantom jet pilot followed the same routine he had followed in the more common training missions he had been flying for eight years. Completely forgetting that this time he was carrying live Sidewinder missiles, he fired one and destroyed a multimillion-dollar Royal Air Force Jaguar aircraft.[19] In 1977, the experienced Dutch pilot of a Boeing 747 jumbo jet departing from Tenerife in the Canary Islands failed to wait for takeoff clearance, roared off and crashed into another 747 that was still taxiing on the runway. How could a well-trained, experienced pilot who had actually been head of KLM Airlines' flight training department for years make such an elementary error? He had spent some fifteen hundred hours in flight simulators over the preceding six years and had not flown a real aircraft for three months. To save costs, simulator pilots are never required to hold position while waiting for takeoff clearance. Apparently, the pilot simply reverted to the routine of the flight simulator with which he was so familiar. What in a different context might have been a trivial error, instead cost 577 lives.[20]

STRESS

Though we often think of stress as an unalloyed problem, a little stress "gets the juices flowing," increasing alertness, effectiveness and reliability. But as the pressure continues to mount, performance tends to level off, then decline, sometimes very sharply. Excessive stress can create all sorts of physical, mental and emotional problems that affect reliability, ranging from irritability to high blood pressure to complete mental breakdown. On the physical side, there is evidence that high levels of mental stress can adversely affect the body's immune system. In research reported in the early 1990s, hundreds of healthy British adults were exposed to cold viruses under controlled conditions. Those in the highest-stress group were found five times as likely to become infected and twice as likely to develop full-fledged colds as those in the lowest-stress group.[21] In late 1991, a team of cardiologists showed that stress can cause abnormal constriction of blood vessels in patients whose coronary arteries are already clogged with atherosclerotic plaque. The stress-linked narrowing of the arteries further impedes blood flow to the heart, raising the chances of heart attack.[22]

Psychiatrists widely accept the notion that stress can play a significant role in triggering episodes of severe depression. In the late 1980s, National Institute of Mental Health psychiatrist Philip Gold suggested how this linkage might work: When confronted by a threat, we naturally experience a "fight or flight" response—a complex biochemical and behavioral mobilization of the mind and body that includes increased respiration rate, a general sense of alertness, and a feeling of released energy. In the short run, this is a healthy, normal reaction that is vital to our survival individually and as a species. But sustaining this level of arousal for long periods produces serious, even dangerous effects.

Gold's work focused on melancholic depression, a relatively common form of severe depression with a clear and consistent set of symptoms (including low self-esteem, hopelessness and intense anxiety about the future). He suggests that depression results when the mechanisms that normally regulate the stress response go awry and a free-running state of constant stress develops. The "fight or flight" stress response may work well as a reaction to an acute, short-lived physical threat, the kind of threat for which it evolved. But the emotional stress so common to modern life tends to be ongoing, longer term and cumulative. Extended periods of high stress (particularly emotional stress) can overload the system, leading to maladaptive reactions like severe depression.[23]

The effects of chronic stress may be temporary, subsiding when the sources of stress are removed, or they may have a very long reach, possibly even affecting the physical structure of the brain.[24] Acute stress from emotional traumas such as job loss, divorce and the death of a loved one can have both powerful short-term effects and long-term impacts that can last from a few years to a lifetime. The most extreme, longest-term reactions to stress seem to occur where the level of stress is both high and prolonged.

In recent years, one of the more celebrated effects of stress has been post-traumatic stress disorder (PTSD), discussed briefly in the previous chapter. An immediate or delayed aftermath of trauma, the disorder involves recurring dreams, memories and even flashbacks of the traumatic events sometimes triggered by a sight, sound, smell or situation with some relationship to the original crisis. PTSD typically involves emotional detachment from loved ones, extreme suspicion of others and difficulty concentrating.

At least 500,000 of the 3.5 million American soldiers who served in Vietnam have been diagnosed as suffering from PTSD.[25] An estimated 30 percent suffer from such a severe version of the disorder that they will never lead a normal life without medication and/or therapy.[26] A quarter of those who saw heavy combat were involved in criminal offenses after returning to the United States. Only 4 percent of those had had prior psychological problems.[27] A 1997 study of Vietnam veterans with PTSD found that they were also more likely to be suffering from serious physical ailments, such as heart disease, infections, and digestive and respiratory disorders.[28]

It is testimony to the basic soundness of the human mind that the horrors of war often create psychological wounds in combat troops. In the early 1980s,

Israeli soldiers suffered nearly a quarter as many psychiatric casualties as physical casualties during the invasion of Lebanon. Israeli Army psychologists claim that the rate of psychiatric to physical casualties was even higher during the 1973 Middle East war, some 40 to 50 percent.[29]

The diagnostic manual of the American Psychiatric Association indicates that PTSD is induced by events that lie "outside the range of usual human experience." But there is evidence that PTSD may be triggered by experiences that are not nearly as unusual as the manual implies. A study of young adults in metropolitan Detroit revealed that some 40 percent of the more than one thousand randomly selected subjects had experienced one or more of the traumatic events that were defined as "PTSD stressors." These included sudden serious injury or accident, physical assault or rape, seeing someone seriously hurt or killed, and receiving news of the unexpected death of a friend or close relative. About a quarter of those who reported these experiences (93 of 394) developed PTSD.[30] If these data are generalizable, PTSD would rank fourth among the most common psychiatric disorders troubling young urban adults.[31]

Trauma can have a very long reach. A 1991 report of psychologists at the VA Medical Center in New Orleans that studied 22 U.S. soldiers who had been taken prisoner during the Korean War found that as many as 19 of them (86 percent) were still suffering from PTSD and other mental problems more than 35 years after their release. More than half of the former POWs who developed PTSD also suffered from other forms of anxiety disorder, such as panic attacks, and about a third experienced severe depression. Prisoners of war appear to be far more likely to develop PTSD than combat veterans.[32] The traumas that they experienced were both prolonged and extreme, involving random killings, forced marches, months of solitary confinement, torture and the like. Yet trauma need not be this extreme to have very long-lasting effects on emotional health and mental stability.

From the perspective of human reliability in dangerous-technology systems, two characteristics of trauma-induced disorders are particularly relevant. First, since the onset of the problems that result from stress disorders may be delayed days, months or even years, someone who appears to be completely recovered from trauma and untouched by such disorders may still harbor them. Psychiatrist Andrew Slaby used the general term "aftershock" to describe "any significant delayed response to a crisis, whether this reaction is anxiety, depression, substance abuse or PTSD."[33] According to Slaby, "everyone, even the calmest, most levelheaded person, has a breaking point that a trauma, or a series of traumas, can set off and bring on aftershock."[34] Second, even if someone who has been severely traumatized in the past does not appear to be dysfunctional or unreliable, his/her ability to cope with stress can be severely compromised. Either chronic stress or the acute stress of a future crisis might overwhelm him/her well before it became severe enough to render a person without such problems unreliable. Because these disorders are common in the human population and can be difficult or impossible to detect, it is impossible to completely avoid them when recruiting large dangerous-technology work forces.[35]

CIRCADIAN RHYTHMS:
DISRUPTING THE BIOLOGICAL CLOCK

The behavior and metabolism of most biological organisms seems to be partly regulated by an internal biological clock. For many centuries, it was believed that this rhythmic time pattern was simply the result of plants or animals responding passively to natural cycles in their environment (such as the day-night cycle). But even when an organism is deprived of all environmental time cues (by keeping light, sound, temperature and food availability constant), the majority of its time patterns continue—with a period close to but not exactly 24 hours.[36] Such patterns are called "circadian" rhythms (from the Latin for "approximately one day"). Apparently, most organisms on this planet have internalized the naturally occurring 24-hour period of an earth day.

The part of the human circadian pattern most important to reliability is the sleep-wake cycle, or, more generally, the variation in alertness and psychomotor coordination over the course of a day.[37] When time patterns are abruptly shifted, the internal biological clock is thrown out of phase with the external time of day. For example, flying at jet speed across a number of time zones causes "jet lag," which disrupts sleep, dulls awareness, reduces attention span and produces a general feeling of disorientation and malaise. It takes several days for most people to completely adjust to the simple twice a year, one-hour time shift between standard time and daylight saving time.[38] It is no surprise then that rapidly crossing five, six or more time zones can throw our biological clocks out of balance.

The fundamental problem with being out of phase with external time is that the external world may be demanding highest alertness and capability just when the internal cycle is at lowest ebb. A businessperson flying from New York to London crosses five time zones. When it is 9:00 A.M. in London, he/she will want to be bright and alert to deal with typically high demands at the beginning of a new business day. But his/her internal clock will be set at 4:00 A.M., a time when the level of alertness and psychomotor performance tends to be at or near its daily minimum.[39]

Many dangerous-technology workers must staff all critical duty stations throughout the 24-hour day, every day. That kind of round-the-clock shift work inevitably plays havoc with the biological clock. There appears to be an underlying circadian rhythm that reaches its lowest levels at night, regardless of sleep-wake schedules. Thus, night-shift workers inherently tend to perform less well than day-shift workers. Swedish studies showed that the normal performance of night-shift workers was similar to that of day-shift workers who had lost an entire night's sleep.[40]

Rotating the work schedules of shift workers both aggravates the problem and spreads it to the day shift. Yet a survey by the National Center for Health Statistics in the U.S. showed that by the late 1970s, more than 27 percent of male workers and 16 percent of female workers rotated between day and night shifts. Over 80 percent of these shift workers suffered from insomnia at home and/or sleepiness at work, and there is evidence that their risk of cardiovascular problems and gastrointestinal disorders also increased.[41]

Even if it were possible to keep the same workers on the night shift permanently, that would not solve the problem. Night-shift workers usually try to function on something approaching a day schedule on their days off, so they won't be completely out of step with the world around them. Consequently, their circadian rhythms are in a continual state of disruption. If they don't try to follow a more normal schedule on their days off, they will be much more socially isolated, and that will subject them to increased stress, which will also degrade their performance.

In the early 1980s, a group of applied psychologists in England studied sleep and performance under real-life conditions.[42] They monitored and recorded the sleep of a dozen male factory shift workers in their own homes, then measured their performance at the factory where they all worked. The men were followed over one complete three-week cycle during which they rotated between morning (6:00 A.M. - 2:00 P.M.), afternoon (2:00 P.M. - 10:00 P.M.) and night (10:00 P.M. - 6:00 A.M.) shifts. The researchers found that, compared to night sleep, sleep during the day was lighter, more "fragile." The normal pattern of sleep stages was disrupted. As a result, not only were reaction times significantly slower on night-shift work, but performance tended to get worse as the week progressed. Performance of workers on the morning and afternoon shifts remained nearly stable. Even if night-shift workers took more or longer naps, they could not compensate successfully for the lower quantity and poorer quality of day sleep.[43] "The night-shift worker must sleep and work at times when his or her body is least able to perform either activity efficiently. The body is programmed to be awake and active by day and asleep and inactive by night, and it is extremely difficult to adjust this program in order to accommodate artificial phase shifts in the sleep-wake cycle. . . ."[44]

Circadian rhythms also play a role in disease. The timing of death of both surgical and nonsurgical patients follows a circadian rhythm. Studies of ongoing disease processes in animals show real circadian variation in the average dose of toxins and the severity of injuries that prove fatal. Thus, shift workers may also be more vulnerable to health-related reliability problems. Further, the effectiveness or toxicity of a variety of drugs has been shown to follow a circadian rhythm as well, apparently because of underlying circadian rhythms in drug absorption, metabolism and excretion.[45] Drug and alcohol abuse might therefore reduce the reliability of dangerous-technology shift workers even more than it reduces the reliability of workers on a stable day schedule.

The timing of the accident at the Three Mile Island nuclear power plant in 1979 strongly suggests that circadian factors played a significant role. The operators at Three Mile Island had just passed the middle of the night shift when the accident occurred, at 4:00 A.M. They had been on a six-week slow shift rotation cycle.[46] In general, the incidence of work errors seems to be much higher in the early morning hours for rotating shift workers whose circadian rhythms have not been fully synchronized to their work schedules. According to a study by the National Transportation Safety Board, truck drivers falling asleep at the wheel are a factor in 750 to 1,500 road deaths each year. Fatigue is a bigger safety problem for truckers than drugs or alcohol.[47] According to another study, they are three times as likely to have a single-vehicle accident at 5:00 A.M. than during usual daytime hours.[48]

Poor circadian adjustment is also an important problem in aviation. Late one night, a Boeing 707 jetliner whose crew had filed a flight plan to land at Los Angeles International Airport passed over the airport at 32,000 feet headed out over the Pacific. The aircraft was on automatic pilot and the whole crew had fallen asleep! When local air traffic controllers could not get a response from the aircraft, they managed to trigger a series of alarms in the cockpit. One of the crew woke up. The plane, which had flown a hundred miles over the Pacific, still had enough fuel to turn around and land safely in Los Angeles. The pilot of another jetliner coming in early one morning after a night flight from Honolulu was in the process of landing the plane when he fell asleep only 200 feet above the ground. The copilot realized what had happened and was able to land the plane safely. Not all such incidents have a happy ending. Pilot error was cited as the main cause of the 1974 crash of a Boeing 707 in Bali that killed 107 people. The crew had flown five legs on this flight since it had begun in San Francisco, combining night and day flying across 12 time zones. It is likely that disrupted circadian rhythms were a major contributor to the "pilot error" officially listed as the cause of this accident.[49]

Because most circadian rhythms follow a more or less 24-hour day, forcing workers into an average day length radically different from 24 hours can also interfere with the normal functioning of the biological clock. Yet American nuclear submarine crews normally operate on an 18-hour day. Each sailor is at work for 6 hours, off duty for 12 hours, then back at work for another 6-hour shift. Short-term studies have shown that the 18-hour day can cause insomnia, impaired coordination and emotional disturbance. This schedule is probably one reason for the extremely high turnover rate of American submarine crews. After each voyage, as much as 30 to 50 percent of the enlisted crew does not sign up for another tour. Only a small number of sailors undertake more than two or three of the 90-day submarine missions.[50] The high turnover rate means that a large fraction of the crew on any given voyage is not fully experienced in the operation and maintenance of the ship on which they are sailing, and not used to being confined to a tube sailing under water for three months. This too is a potential source of unreliability.

Finally, disturbances of circadian rhythms have long been associated with certain forms of mental illness.[51] Waking up early in the morning, unable to fall back asleep, for example, is one of the classic symptoms of depression.[52] There is evidence that sleep-wake disorders in manic-depressive individuals may result from misfiring of their circadian pacemakers. It is even possible that those malfunctions may help cause manic-depressive syndrome. In general, studies suggest that disturbances of the biological clock caused by abnormalities in circadian pacemakers may contribute to some forms of psychiatric illness. If so, the circadian disruptions so common among dangerous-technology workers may create reliability problems through this route as well.[53]

LEADERS AND ADVISORS

There is nothing in the exalted positions of political leaders (or the advisors on whose counsel they depend) that makes them immune to any of the reliability problems we have already discussed. There is also nothing in the process of achieving those exalted positions that insures that failures of reliability will not occur. Quite the opposite; in nearly all modern governments, most political leaders follow a long and arduous path to the pinnacles of power. By the time they get there, they have been exposed to a great deal of physical and mental stress, and have often reached an advanced age. By the time they relinquish power, they are older and usually have been subjected to even greater stress. On the positive side, that means that those who reach high political positions have lived enough years to accumulate valuable experience and have shown that they can cope with stress. On the negative side, stress and advanced age certainly do take their toll.

Some people show significant physical or psychological effects associated with aging before they leave their fifties, while others suffer little or no deterioration in intellectual and creative ability well into their seventies or eighties. Bernard Baruch authored the American plan for the control of atomic energy after World War II at age 76; Goethe completed *Faust* at age 80; Michaelangelo was still creating extraordinary sculpture in his 80s.[54] With the right lifestyle, nutrition and exercise, it is possible for individuals to maintain their critical physical capacities much longer as well. Still, the mental and physical capability to cope successfully with the pressures of an acute crisis does tend to diminish with age.[55]

A variety of physical and psychological problems relevant to reliability are more common among people in the later stages of life. And when these problems occur, the process of aging tends to aggravate them. Psychiatrist Jerrold Post cites a series of psychological difficulties that tend to grow worse with age "once the march of symptomatic cerebral arteriosclerosis or other presenile cerebral degeneration has begun": (1) thinking tends to become more rigid and inflexible, with things seen more in terms of black and white, right or wrong; (2) concentration and judgement are impaired, and behavior becomes more aggressive and less tolerant of provocations; (3) there is less control of emotions, with anger, tears and euphoria more easily triggered and a greater tendency to depressive reactions; (4) rather than mellowing, earlier personality traits can become exaggerated (for example, someone who has generally been distrustful can become truly paranoid); (5) the ability to perform mental tasks is degraded, but wide day-to-day fluctuations in mental function can lead others to underestimate the seriousness of the deterioration that has taken place; (6) there is a marked tendency to deny the seriousness and extent of disabilities—a failing leader may therefore "grasp the reins of power more tightly at the very time when he [or she] should be relinquishing them."[56]

Political leaders who have manifested some of these difficulties are as diverse as Joseph Stalin and Woodrow Wilson. Stalin was never a particularly trusting soul. That trait became exaggerated with time, and according to Post, "Joseph Stalin in his last years was almost surely in a clinically paranoid state."[57] Woodrow Wilson became quite ill while president, yet refused to acknowledge the extent of his illness. He "suffered a major cerebrovascular accident [a stroke] in September 1919 which left him paralyzed on the left side of his body and was manifested by severe behavioral changes. The manner in which Wilson stubbornly persisted in fruitless political causes was in part to sustain his denial of disability."[58] The European political leaders of the 1930s have often been condemned for failing to stop the rise to power of aggressive dictators in Germany and Italy short of World War II. But "All the evidence suggests that . . . [the leaders of Europe] were sick men rather than sinners."[59]

The remarkable career of Franklin Delano Roosevelt was filled with physical challenges. He had scarlet fever at school, typhoid fever at age 30, and was stricken by crippling poliomyelitis nine years later. Throughout his life he suffered from serious nose, throat and sinus infections. In 1937 a member of Roosevelt's cabinet said, "the President . . . looks all of fifteen years older since he was inaugurated in 1933."[60] After late 1943, Roosevelt's physical and mental condition continually deteriorated. It was dangerous, and probably irresponsible, for a seriously ill Roosevelt to run for a fourth term in 1944. Aside from the personal consequences, it was not a good idea to have a man as ill as FDR negotiating with Stalin and Churchill over the political future of Asia and Europe and the lives of hundreds of millions of people.

The World War II leader of Britain, Winston Churchill, became prime minister at age 65. Churchill has been described as "a medical textbook in himself." Lord Allanbrooke was chief of the Imperial General Staff, and in daily contact with Churchill. Referring to the prime minister's exhaustion and deterioration in 1944, Allanbrooke noted: "He seems quite incapable of concentrating for a few minutes on end, and keeps wandering continuously"; and "Winston had been a very sick man with repeated attacks of pneumonia and frequent bouts of temperature."[61] In May 1944, the Polish ambassador observed, "I began to wonder whether Churchill . . . really grasped all that was going on. . . . Perhaps, however, he has his own reasons for repeating certain things to us over and over again."[62] After Churchill's second serious stroke in 1952, his personal physician noted, "he was not doing his work. He did not want to be bothered by anything." His physician also revealed that Churchill apparently suffered from bouts of severe depression. By the time he resigned in April 1955, his various disabilities caused him to spend most of the day in bed.[63]

At age 65, in his third year as president, Dwight Eisenhower suffered a myocardial infarction (heart attack), apparently as a result of both chronic and acute stress. Fortunately, it came at a quiet time, domestically and internationally. But Eisenhower later commented that if a dangerous situation, such as the Lebanon crisis of 1958, had arisen early in his illness, "the concentration, the weighing of pros and cons, and the final determination would have represented a burden . . . which the doctors would likely have found unacceptable for a new cardiac patient to bear. . . . [However] had there been an emergency, such as the detection of incoming enemy bombers, on which I would have had to make a rapid decision regarding the use of United States retaliatory might, there could have been no question, after the first forty-eight hours of my heart attack, of my capacity to act."[64] Clearly, even in the President's own view, his ability to make critical decisions *within* the first 48 hours after his heart attack was at least questionable.

Sitting at his desk on November 25, 1957, Eisenhower felt a brief giddiness. He could not pick up a piece of paper or grip his pen. It fell to the floor. He could not read or find the right words to express his thoughts or needs. The president had suffered a mild stroke. Understanding that another stroke or heart attack could happen again at some unpredictable time, Eisenhower wrote a detailed letter to Vice President Nixon early in 1958, specifying the procedures under which Nixon would temporarily or permanently take control, in the event he became medically disabled.[65]

The youngest man to be elected to the U.S. presidency (in 1961, at age 43), John F. Kennedy was also troubled by serious medical problems. He suffered from Addison's disease, a condition of the adrenal cortex that results in decreased production of steroid hormones over time. Symptoms include tiredness, weakness, anemia, bouts of diarrhea and indigestion with nausea and vomiting. In a patient with Addison's, even a mild infection can create sufficient stress to cause acute adrenal failure resulting in dehydration and loss of consciousness (among other things) in the absence of careful medical attention.[66] Because of this condition, Kennedy was treated almost continually with steroids from 1947 on. He had injured his back while playing football in 1937, an injury later aggravated during his stint in the Navy in World War II. In 1944, and again in 1954, he underwent back surgery. He used braces and crutches periodically beginning as early as 1952. Kennedy's back problems caused him to suffer from recurrent pain and disability throughout his life.[67]

During the final days of his presidency, Richard Nixon was under enormous pressure as a result of the Watergate scandal. Investigative journalists Bob Woodward and Carl Bernstein reported that General Alexander Haig, White House chief of staff, said Nixon was a battered man, strained to his limit, and he was afraid the president might try to kill himself. Woodward and Bernstein described a meeting between Nixon and Secretary of State Kissinger on August 7, 1974—two days before Nixon resigned in disgrace: "The President was drinking. He said he was resigning. . . . The President broke down and sobbed. . . . Nixon got down on his knees . . . prayed out loud. . . . He was weeping. And then, still sobbing, Nixon leaned over, striking his fist on the carpet, crying, 'What have I done? What has happened?' Kissinger touched the President, and then held him, tried to console him, to bring rest and peace to the man who was curled on the carpet like a child. The President of the United States."[68] Later that night, Nixon telephoned Kissinger. According to Woodward and Bernstein, "The President was slurring his words. He was drunk. He was out of control. . . . He was almost incoherent."[69]

Whatever one's opinion of Richard Nixon and his political career, it is possible to empathize with the specter of a man whose life was coming apart at the seams. Nevertheless, that same man was at this very time commander-in-chief of the world's most powerful armed forces, the only person authorized to order the use of American nuclear weapons on his own judgement. He was at the head of the nuclear chain of command as his life slowly descended into chaos. As long as there are nuclear weapons, it is impossible to insure that no leader of a nuclear-armed nation will ever get into an emotionally tortured state like this again.

On March 30, 1981, three months into his administration, Ronald Reagan was wounded when one of six bullets fired at his entourage ricocheted off his limousine and buried itself in his chest. The Twenty-Fifth Amendment to the Constitution (passed in 1967) provided for the orderly temporary or permanent transfer of power in the event of presidential illness or incapacity. Yet there was no attempt to transfer power when Reagan was shot.[70] Even as he lay in the operating room, undergoing surgery under anesthesia, Reagan still held all presidential powers.

The public was led to believe that Ronald Reagan was not badly wounded. However, according to Stanford University medical professor Herbert Abrams, Reagan's condition was much more serious than was admitted at the time.[71] By the time he was wheeled into the operating room to have the bullet removed from just behind his heart, he had lost 35 percent of his blood, though the bleeding had been slowed and his blood pressure kept up by transfusions. By the time he entered the recovery room he had lost about 50 percent of his total blood volume. He continued to cough up blood the next day. Speaker of the House Tip O'Neill visited Reagan that day, and, shocked by his appearance, later commented, "in the first day or two after the shooting he was probably closer to death than most of us realized." Yet that day a White House staffer was quoted in the *Wall Street Journal* as saying, "If a really grave crisis occurs, Mr. Reagan would be on top of it."[72]

Reagan was still seriously debilitated nearly two weeks after the attack: "A visitor describes him as 'pale and disoriented, walking with the hesitant steps of an old man.' Entering a room, Reagan starts to sit down and 'falls the rest of the way, collapsing into his chair.' He can concentrate for only a few minutes at a time and is able to work and remain attentive for only an hour or so a day."[73] In the opinion of his personal physician, Reagan was not fully recovered until October, more than half a year after the attack.[74] Then in 1994, six years after leaving the presidency, Ronald Reagan was diagnosed as having Alzheimer's disease, a disease characterized by progressive deterioration of mental faculties. Alzheimer's can proceed so slowly that it can significantly affect the kind of mental functions critical to decision making for years before it is finally diagnosed.[75]

Fourteen of the eighteen U.S. presidents who held office during the twentieth century had significant illnesses during their terms. Four had strokes; five suffered from various kinds of chronic respiratory illness; six underwent major surgery at least once; seven suffered from serious gastrointestinal disorders; and nine had heart disease. Wilson, Franklin Roosevelt, Eisenhower, Johnson and Reagan were medically incapacitated while president; Harding, FDR, McKinley and Kennedy died in office (the latter two by assassination). Four of the seven leaders of the USSR suffered from serious heart conditions and five died while in office.[76]

After the collapse of the Soviet Union, Boris Yeltsin became the first president of Russia. He had been first hospitalized for heart trouble in the late 1980s. In July 1995, he was rushed to the hospital with "acute heart problems." It appears that he had yet another heart attack that was kept secret in the spring of 1996, and subsequently had major coronary-bypass surgery. In the first six months after his re-election, he was able to work in his office for only two weeks.[77] As of early 1999, it was clear that Yeltsin remained a very sick man, still holding the reins of power in an economically and politically deteriorating nation, a nation armed with the world's second-largest nuclear arsenal.

The kinds of illness and trauma that have so frequently plagued political leaders do not simply affect physical function. They can impair psychological function as well. Heart attacks, for example, are often followed by anxiety, depression and difficulties in sleeping and concentrating. In more than half the patients, some of these psychological disturbances persist for months after the attack. According to one study, more than 30 percent suffer from irritability, fatigue, impaired memory, inability to concentrate and emotional instability for six months to two years after their heart attack.[78]

Strokes, another common problem in aging leaders, cause many patients to suffer from depression, anxiety and emotional volatility; 40 to 60 percent are cognitively and emotionally impaired. Inability to sleep and feelings of hopelessness are also common. Severe depression may persist for 6 to 24 months. Major surgery also produces important psychological side effects, including confusion serious enough to make it hard to think clearly (especially in elderly patients). It can produce disorientation and an inability to grasp concepts and use logic.[79]

Depression and anxiety are both common effects of serious physical illness and trauma in general. People who are depressed have a hard time focusing their attention, concentrating and remembering. They tend to overemphasize negative information; their analytic capabilities can be seriously impaired. Anxiety also degrades learning and memory, as well as interfering with the ability to reason. These are extremely serious problems for any leader having to make crucial decisions, or any advisor upon whose counsel a leader must depend in a crisis.

FAMILIARITY

There is one more issue of individual fallibility that is often overlooked. When put into a novel work situation, especially one which involves expensive, dangerous or otherwise critical systems, people tend to be very careful of what they do—for a while. But no matter how expensive or dangerous the systems might be, if things go well and all is calm for a long time, most people begin to assume that nothing will go wrong. The cutting edge of their vigilance begins to dull. Even if familiarity does not breed contempt, it does breed sloppiness.

There is no reason to expect that it is any different in dangerous technological systems. Military personnel assigned to duty that brings them directly in contact with nuclear weapons undoubtedly feel a sense of awe and danger at first. But after months of guarding them, loading them on ships or planes, etc., nuclear weapons are just another bomb, if not just another object. If this seems exaggerated, consider how careful most people are when they first learn to drive. Being aware that they could get hurt or killed in a car crash tends to make beginning drivers more careful—for a while. But once they get comfortable with the act of driving, most people pay much less attention its inherent dangers. The car is just as deadly, but the act of driving has become much more routine.

The tendency to relax once we become familiar with a task is not only a common human trait, it is useful in most situations. But when it causes vigilance to fail in dealing with critical dangerous technological systems, it can lead to catastrophe. There is no way to completely avoid this or any of the other fundamental problems of individual fallibility we have discussed, no way to be sure that we can completely avoid recruiting workers whose reliability has been compromised by the vulnerabilities, traumas and afflictions that are part of every human life.

Because we humans are, after all, social animals, programmed to interact with each other, it is not enough to consider our behavior as individuals. When we function as part of a group, as we so often do, the behavior of the group can be very different from the sum of our individual behaviors. It is now time to consider just how dramatically that difference can affect the reliability of those who interact with dangerous technologies.

CHAPTER EIGHT

Bureaucracy, Groupthink and Cults

Most of the time we are at work or play, we are part of a group. Interacting with other people is such a normal part of life that we take it for granted and sometimes fail to appreciate just how much the presence of others affects our own behavior, for better or for worse.

When we think about it at all, we tend to think of groups in a work setting as constraining individual behavior and increasing reliability by protecting us against the mistakes and foolish, crazy or malicious acts of any one person. The nuclear military, for example, uses groups to counter the possibility that an individual will become unreliable. The nightmarish scenarios of fiction writers—from the crazy general who manages to trigger a nuclear war to the distraught missile officer who launches nuclear weapons against an unsuspecting city—are thought to be rendered impossible by the "two-key" rule, which requires that two or more people always act together to carry out critical activities. But groups are not always reliable. Despite the high skill of surgical teams, an estimated 75 to 80 percent of anesthesia mishaps are linked to human error, and many operating room mistakes result from interpersonal communication problems rather than technical deficiencies.[1] Even normally functioning groups carrying out their day-to-day activities sometimes behave in ways that create serious problems. And under certain conditions, group behavior can degenerate into pathological forms that actually foster risky or pernicious behavior rather than preventing it.

BUREAUCRACY, ORGANIZATIONS AND RELIABILITY

Bureaucratic organizations link the activities of pigeonholed subgroups, each with its own specialized set of functions, into a grand organization with lines of communication, authority and responsibility clearly specified. They often take the shape of hierarchical pyramids so that "the buck" is guaranteed to "stop here" somewhere within the organization for every conceivable type of decision. In theory, hierarchical bureaucracy combines the expertise of many individuals in an organizational form with a certain clarity of function and decisiveness of action. In fact, real-world bureaucracies do not work nearly as well as theory would lead us to believe.

Information is the lifeblood of any management-administrative organization. It flows up to top decision makers, then down, in the form of directives, to those responsible for carrying them out. The impact of those directives must then be monitored, evaluated and sent back up the organization to decision makers in a feedback loop that is critical to judging how well the directives are working and whether or not they should be changed. Without an efficient flow of accurate and relevant information, neither managers nor administrators can perform their jobs effectively. Yet bureaucracies systematically distort the flow of information in both directions.

Subordinates *must* edit the information they send up the hierarchy and present it in a manageable form. They cannot faithfully transmit all the information gathered by those on the lower levels of the pyramid or those at the top would be overwhelmed. They could not conceivably digest it all, let alone act on it. Too much information can paralyze decision makers. But as soon as there is editing, there is room for distortion. Some criteria must be used to decide what to pass on and what to throw away. It is not always easy for subordinates to judge which information is most important to those higher in the organization, but, unfortunately, importance is often not even the prime criterion they use. Holsti and George argue that in foreign-policy-making organizations, incoming information is filtered to fit into "existing images, preconceptions, preferences, and plans." When the information gathered does not fit, it is distorted by those with a stake in a particular viewpoint to support their theory. They "tend to screen out at least some of the data adverse to their own interests and to magnify data that are favorable."[2]

The Good-News Syndrome

Subordinates sometimes try to avoid superiors who might report facts they want suppressed, even when it means not reporting potentially dangerous situations.[3]

Personal beliefs, rigid world views and concepts of loyalty have also been shown to inhibit the communication of accurate, unbiased information to those higher up in the organization.[4] There is a "good-news syndrome," a tendency for subordinates to edit out information that highlights their own errors or failures of judgement, or the errors or misjudgements of their superiors. No one likes to be the bearer of bad news. Sometimes the "creative editing" of the good-news syndrome crosses the line into outright, blatant deception. In June 1996, Miami-based U.S. immigration officials reportedly ordered the sudden, unscreened release of some detainees, including criminals, and the temporary movement of others so that a visiting congressional delegation would not see how terribly overcrowded their immigration detention center was.[5]

Combined with the necessary filtering at each level of the organization, the good-news syndrome creates a major distortion in the upward flow of information, a distortion that unfortunately tends to get worse when there is more at stake. As organizational-communications expert Chris Argyris puts it, "the literature suggests that the factors that inhibit valid feedback tend to become increasingly more operative as the decisions become more important. . . . This is a basic organizational problem . . . found not only in governmental organizations, but also in business organizations, schools, religious groups, trade unions, hospitals, and so on."[6]

The internal politics of bureaucracies can also bias information flows. The various subgroups in a bureaucracy often have different and conflicting interests, values and viewpoints. Conflicts arise over budget, status and influence with and access to top decision makers. Parochialism can cause subgroup leaders to consciously or unconsciously edit information so that their own particular division is presented in the best possible light, believing (or at least rationalizing) that this is good for the organization as a whole.[7]

The result of these distortions in the upward flow of information is that top decision makers may find themselves living in an unreal world. It is hard to make good decisions when you don't really know what is going on. Worse yet, being this out of touch can lead to catastrophic loss of control. That is why Arthur Schlesinger, Jr., a close confidant of the president in the Kennedy White House, argued for the importance of what he called "passports to reality" in upper levels of government, especially in the presidency.[8]

The special presidential commission set up to investigate the explosion of the space shuttle *Challenger* on January 28, 1986, concluded that one of the prime causes of the accident was a breakdown in the upward flow of information at NASA. The immediate technical problem was the failure of an O-ring seal between two segments at the rear of the right-side solid-fuel rocket booster. *Challenger* was launched in weather colder than any previous shuttle flight, even though there was clear evidence that the O-ring was only about a fifth as flexible in weather that cold as at normal room temperatures. It was not flexible enough to seal the joint properly. Escaping fumes ignited, causing the explosion. But the deeper problem was that the serious misgivings expressed before the launch by engineers from Morton Thiokol (builder of the booster) failed to reach the top management at NASA.[9]

In January 1990, the Department of Energy (DoE) permanently halted nuclear weapons manufacturing at the Rocky Flats plant near Denver. A year and a half earlier, Secretary of Energy James D. Watkins had publicly expressed grave concern about his subordinates' performance, as well as their ability to run the nation's nuclear weapons complex. In his view, these managers and supervisors did not have the technical skills to run the weapons production system. They were giving him unreliable information about problems at the nuclear weapons plants, and some were too undisciplined to safely operate nuclear reactors.[10] "When I get the briefing, I only get one side, so I have to dig in myself. . . . I don't have the database coming to me that I need. I have omissions in the database. So I am making decisions today on a crisis basis. . . . It has been a nightmare for me to try to unravel the background sufficient to make some decision. . . . It's been very confusing."[11] Eventually, Watkins said he had uncovered "serious flaws" in the procedures intended to insure that reactors in the nuclear weapons complex were safe to operate.

In March 1992, memorandums filed by the U.S Department of Justice and Rockwell International Corp. in the federal district court in Denver detailed the conflicting policies and sloppy environmental procedures that had prevailed for decades at Rocky Flats. These documents "described how the Department of Energy established a 'prevailing culture' that put production of plutonium triggers for nuclear weapons above any other concern, including care for the environment and public safety."[12] Rampant bureaucratic problems and poor management by DoE had led to a loss of control over this key nuclear weapons facility.[13]

The DoE's Savannah River Plant, near Aiken, South Carolina (the site of plutonium- and tritium-producing nuclear reactors), had its share of these problems as well. There have been quite a few unpublicized incidents at the plant since the 1950s, among them two in which fuel elements nearly melted. "The problems were recorded in log books, monthly reports and semiannual briefings. But they were not well studied. . . . Nor were they clearly reported to top officials in Washington. . . . Many reveal faults in operation, not hardware."[14]

In the late 1980s, DoE hired inspectors from the Nuclear Regulatory Commission (NRC) to evaluate the operation of the nuclear weapons-related reactors. One of them insisted on a full review of the Savannah River Plant's ability to withstand earthquakes, arguing that previous seismic analyses done by

DoE were inadequate and failed to use state-of-the-art methods. The review reportedly turned up "hundreds of potential structural weaknesses" at Savannah River's so-called P reactor. The inspector reported that "design documentation was 'grossly inadequate'. . . . *[P]lant managers could not put their hands on a set of complete blueprints showing the reactor as it exists now* and that some vulnerable components thought to have been removed long ago are still in place"[15] (emphasis added). In August 1988, the chair of DoE's safety group learned belatedly and indirectly that Du Pont (who operated the plant for DoE) and local DoE officials were not keeping him informed on what action was being taken on the recommendation of his inspectors that the reactor should be shut down. DoE and Du Pont subsequently decided to take the inspectors' advice. By late 1988, all 14 of the reactors in the DoE's nuclear weapons complex had been closed, many of them permanently.[16]

Charged with assuring the safe operation of America's *commercial* nuclear power plants, the NRC is hardly a paragon of effective operation itself. In 1996, it was reported that "Higher-level NRC managers sometimes downgrade the severity of safety problems identified by on-site inspectors without giving reasons for the change," and that "NRC inspectors who persist in pressing safety issues have been subject to harassment and intimidation by their supervisors."[17] In 1997, a General Accounting Office review found that "NRC has not taken aggressive enforcement action to force licensees to fix their longstanding safety problems on a timely basis," and argued that this laxity was so embedded in the organization that "changing NRC's culture of tolerating problems will not be easy."[18]

Without timely and accurate information, bad news problems that would not have been difficult to handle early on can spin out of control by the time they become obvious. It is simply not wise for upper-level managers to assume that the information that flows to them on a daily basis is correct and unbiased. According to Irving Shapiro, former chief executive of Du Pont, "Getting the right information is a substantial part of the job. . . . You get lots of information and most of it is totally unnecessary. The organization tends to want to give you the good news and not cough up the bad news. . . . But to manage well, you have to get the message across, that whatever the story is, let's get it on the table."[19] Many organizations have a long way to go in their efforts to overcome this problem. Robert Bies, professor of organization at Northwestern University, argues, "Most corporations still punish those who deliver the bad news and it's more the exception that managers will tolerate and even want bad news."[20]

What can be done? There appear to be two main schools of thought among experts in organization theory. The first is that the solution is cultural: an environment must be created that makes managers seem more accessible to subordinates. Probably the best way to achieve this is to manage by "wandering around"—get out of the office and be seen in the corridors, the workplaces of subordinates, the cafeteria, etc., so that informal contacts are encouraged. Most subordinates are inhibited about making an official visit to their boss's office, or, even worse, to the office of their boss's boss to report a problem. That is especially true when they want to point out an error or misjudgement that either they or higher-level managers have made. Repeated informal contacts help employees to feel more at ease about talking to higher-ups, and make it less threatening to bring up bad news. Problems can then be dealt with early, before they have become compelling enough to justify a formal visit. It is a tremendous advantage for managers to be made aware of problems while they are still small.

The other main school of thought is that the answer lies in reshaping the organization, replacing the sharply rising pyramid with a much flatter structure. Hewlett-Packard, for example, divided the company into small units whose managers are therefore more visible and informally accessible to their subordinates. But the best structure and the most peripatetic managers in the world will do no good if the people who fill key management positions do not behave in a way that encourages open communication. Bad news will still not reach the top.

More than once, military commanders have blundered into major disasters because they have been sold an overly optimistic picture of what is actually happening on the ground. Their advice, based on that rosy view, can cause political leaders to compound earlier mistakes and multiply the disaster. During the Vietnam War, top-ranking American commanders, looking at inflated "body count" data, kept reassuring the president and his advisors that Viet Cong and North Vietnamese military forces were depleted and nearing collapse. They could see "the light at the end of the tunnel." Thinking their efforts were just about to succeed, American military and civilian leaders plunged the country deeper and deeper into the morass that was Vietnam.

In February 1997, the Pentagon reported it could find only 36 pages of an estimated 200 pages of logs detailing detections of chemical/biological weapons that it had kept during the 1991 war in the Persian Gulf. Since paper and computer-disk copies of the logs had been placed in locked safes at two different locations in the United States after the war, the report fueled speculation by veterans that the military was either grossly incompetent or engaged in a cover-up. General Norman Schwartzkopf, the commander of American forces during the war, testified before Congress that he was shocked when the Pentagon announced that thousands of troops might have been exposed to nerve gas. "We never had a single report of any symptoms at all on the part of the 541,000 Americans over there."[21]

Distortions are much more troubling in organizations in which crisis decisions with potentially catastrophic consequences must be made in a matter of minutes. Yet it is clear that distortions in the upward flow of information persist in these organizations as well. A former Air Force missileman, describing life in the hardened silos of the U.S. land-based missile forces, put it this way: "Crew members dare not tell higher command that the regulations are flouted. *Noncommunication with higher command is endemic in the missile field,* with the result a gap between regulations and what is really done in the capsule [that is, the missile silo]"[22] (emphasis added).

There isn't much potential for "management by wandering around" in the armed forces. By its nature, the military does not encourage casual contact between lower and higher ranks.[23] Failing to respect the chain of command is not taken lightly. A soldier who tries to report information critical of his/her superiors is not likely to get a warm reception, as many women in the military found in the late 1990s when their complaints of sexual harassment were reportedly "met with ridicule, retaliation or indifference."[24] The system fosters strict obedience to orders and puts severe constraints on questioning authority.[25] Not only are soldiers supposed to automatically obey those of higher rank, but superiors run the risk of being seen as weak and indecisive if they do not "stick to their guns."

Flattening the hierarchy holds even less promise for the military. The armed forces must ordinarily be ready to operate in a coordinated fashion from a central command to carry the day. It is not inconsistent with military principles to break forces into small units capable of operating autonomously (in guerilla warfare, that is key), but there still must be a way of coordinating forces and pulling them together at the critical moment. The system of rigidly enforced hierarchy with many ranks is deeply embedded in traditional, large-scale military organizations. It not clear they could operate effectively without it. There does not appear to be any effective solution to the good-news syndrome that is compatible with the basic structure and operation of military forces.

Directives

The ability to routinize many specialized functions and rely on standard operating procedures (SOPs) is the greatest alleged advantage of bureaucratic organizations. Yet widespread adherence to SOPs creates an inertial atmosphere. Change is not particularly welcome and not readily incorporated. SOPs help to establish an organizational culture, style and even ideology biased toward incremental rather than fundamental change. When directives that require substantial departure from SOPs flow down from above, they are not received with enthusiasm. Instead, they are often reinterpreted to better fit more familiar patterns. Sometimes subordinates make higher-level managers aware of these alterations. More often, they do not. Coalitions form among interest groups within the bureaucracy, and negotiation and compromise, rather than analysis, come to dominate policy making. Behavior is driven by what is familiar and comfortable rather than what is best.[26]

Both bureaucratic inertia and internal politics can cause directives to be diverted, distorted or ignored, compromising the position and flexibility of top decision makers at critical moments. Probably the most dramatic case on record involves the Cuban Missile Crisis, the confrontation that brought the world to the edge of global nuclear war. Fearing the massive U.S. advantage in strategic nuclear forces (and wanting to protect Cuba against another possible U.S.-supported invasion), the Soviets tried to balance the scales quickly by sneaking nuclear-armed missiles into Cuba, 90 miles from the American mainland. The United States caught them at it, threw up a naval blockade around the island and demanded that the USSR remove the missiles immediately. The Soviets refused, and their warships, obviously carrying additional missiles toward Cuba, continued to steam toward the blockade line. According to Robert MacNamara, then secretary of defense, the United States did not realize at the time that Soviet forces in Cuba already had 36 nuclear warheads for the two dozen intermediate-range missiles targeted on U.S. cities. That is easily enough firepower to kill tens of millions of Americans and lay waste to a good part of the United States. The CIA had told the administration that they did not believe there were any nuclear warheads in Cuba. This was only one of a number of ways, in MacNamara's words, "that each nation's decisions immediately before and during the crisis had been distorted by misinformation, miscalculation and misjudgment."[27]

After more than a week of crisis and confrontation, President Kennedy received a formal letter from Soviet Premier Khruschev proposing that the USSR withdraw its missiles from America's doorstep in exchange for U.S. withdrawal of its nuclear-armed missiles from Turkey, on the Soviet's doorstep. According to Robert Kennedy,

> On several occasions over the period of the past eighteen months, the President had asked the State Department to reach an agreement with Turkey for the withdrawal of Jupiter missiles in that country. They were clearly obsolete. . . . At the President's insistence, Secretary [of State] Rusk had raised the question . . . in the spring of 1962. The Turks objected and the matter was permitted to drop. . . . In the summer of 1962 . . . President Kennedy raised the question again. He was told by the State Department that they felt it unwise to press the matter with Turkey. But the President disagreed. He wanted the missiles

removed. . . . The State Department representatives discussed it again with the Turks, and finding they still objected, did not pursue the matter.

The President believed he was President and that, his wishes having been made clear they would be followed and the missiles removed. Now he learned that the failure to follow up on this matter had permitted the same obsolete Turkish missiles to become hostages of the Soviet Union. . . . He was angry. . . . He pointed out to the State Department . . . that our position had become extremely vulnerable, and that it was our own fault.[28]

And so the international situation was gravely aggravated at a crucially dangerous point in human history, not by a conspiratorial plot, not by a mentally deranged or drug-addicted military officer, not even by a stress or monotony-induced failure of vigilance, but simply by the politics and inertia of bureaucracy.

Thirty years later, a frightening postscript to the Cuban Missile Crisis appeared in the form of a *New York Times* Op-Ed piece written by Fedor Burlatsky, a former advisor to Nikita Khrushchev. He writes that once Khrushchev got the idea to put nuclear-tipped missiles in Cuba, the group of Soviet advisors and decision makers somehow managed to convince themselves that the United States would fail to notice the movement of a hundred ships, dozens of planes, 42 nuclear-armed missiles, more than a hundred antiaircraft weapons and 40,000 Soviet soldiers before their installation in Cuba was complete. But it is unlikely that even they were ready for the telegram subsequently sent by Castro to Khrushchev, which Burlatsky claims he saw at the time. He says the telegram read, "I propose the immediate launching of a nuclear strike on the United States. The Cuban people are prepared to sacrifice themselves."[29]

The failure of subordinates to carry out management directives also played a key role in the world's worst nuclear power accident, the Chernobyl disaster of April 26, 1986.[30] The design of the RBMK reactors at Chernobyl made them particularly vulnerable to instability. The design errors were so serious, and apparently unnecessary, that most Western designers who had reviewed Soviet reports on the reactors prior to the accident found them difficult to believe. The Russians were aware of these problems and knew that the reactors were dangerously unstable, especially at low power. Rather than going about the difficult and costly business of fixing them, the managers issued directives with strict rules for operation. But they apparently made little or no attempt to educate the plant operators or to strictly oversee their activities. The day before the accident, those rules were being violated. The reactor was intentionally being run below 20 percent power, a dangerously low level. And *six* major safety devices had been *deliberately disconnected*. After the accident, a new rule was instituted requiring that a senior person—whose primary function is to see that the rules are obeyed—be present whenever RBMK reactors were started up or shut down. According to the Soviet minister of atomic energy, "this by itself would not have prevented the accident at Chernobyl, because it was the deputy chief engineer who was most responsible for breaking the rules."[31]

Hierarchical bureaucratic organizations coordinate and control the behavior of those who work within them to avoid disruption and reliably achieve established goals. Those constraints can interfere with reliable performance by creating troubling conflicts between responsibility to the organization and personal and/or social responsibility. "Whistle-blowers" are those who ultimately try to resolve this conflict by violating the organization's norms, stepping outside their bureaucratic role to expose its behavior to wider public scrutiny. Unfortunately, social norms of conformity are so strong that whistle-blowers are often scorned and harassed for their trouble, rather than being praised for their courage.

Professionals, such as medical doctors, are placed in precisely such a conflictual situation when they work for a bureaucratic organization. One of the defining characteristics of a "profession" is that its practitioners are subject to a code of ethics to which their behavior is expected to conform, regardless of how or where they practice. Yet this independent set of values may come into conflict with the bureaucracy's goals, creating a stressful dilemma for those who take their professional obligations seriously. The common assumption is that the professional, whether school teacher, public defender or medical doctor will be guided primarily by the ideals of public service and responsibility to the client embedded in ethical codes and professional training. Yet,

> studies suggest that it cannot be assumed that organizational goals will give way to professional goals. Professionals who wish to succeed may well respond to the specific immediate pressures of their employing organization before the more abstract and distant expectations of their profession. . . . The changes or modifications in professional practice which result from adaptation to organizational requirements . . . often affect the core functions of the profession.[32]

This issue became increasingly important in the late 1990s, as more medical doctors in the United States went to work for health maintenance organizations (HMOs). They found themselves caught between their obligations to their patients and the pressures and incentives that came from the business interests of the HMO that employed them.

One species of "captive" professional with particular relevance here is the military psychiatrist.[33] Psychiatrists in the military are there to assist in the control or management of behavior that the military defines as being deviant. From the point of view of the military bureaucracy, the psychiatrist's primary responsibility is to the organization, not the patient. Regulations specify the diagnostic standards that are to be followed. The psychiatrist's job is to evaluate and make recommendations about a patient in accordance with those standards and not to freely apply unrestricted professional judgement. Recommendations are made to the patient's superior officers, not to the patient. The psychiatrist is primarily a decision-making tool for the organization, not a therapist for the patient. The interest of the patient is subordinated to the interest of the organization.

It might be argued that this is appropriate to the critical business of assuring reliable behavior in the nuclear and conventional military. There are surely circumstances in which it might be dangerous to allow judgement about what is best for the patient to supersede what is necessary for safe and reliable performance of military duty. But that isn't always true. There are also times when loyalty to the organizational role might be more dangerous. For example, a particular commander might believe that a discharge or even a change of duty for a troubled soldier is too softhearted and would set a bad example. The judgement of a psychiatrist trying to be a good soldier might be unduly influenced by what the psychiatrist believes the local commander wants to hear. That might not be in the best interest of either the soldier or the wider society. Whatever policies may be made at the highest levels of authority within the military, "commanders of hospitals, posts or departments sometimes have powers similar to those of feudal barons in their fiefs. Current 'official' policy about regulations may be overturned or 'reinterpreted' by local commanders or their key administrators."[34]

When the values, goals and expectations of the organization diverge from those of the profession, professionals are all too likely to give in to organizational pressures. It is frightening to think that military psychiatrists are just as vulnerable to these pressures as any other professional—even when giving in might compromise their vital role in assuring the reliability of people dealing with exceedingly dangerous technologies.

GROUPTHINK

There are times when groups are not only unreliable, but are actually less reliable than individuals. The tendency of members of a decision-making group to take greater risks than the individuals in the group would take acting alone is called "risky shift." Risky shift has been observed in dozens of controlled experiments carried out over decades.[35] One plausible explanation for this odd behavior is that the presence of others sometimes stimulates those who are part of the group to try to prove to each other how brave, tough and daring they are.

Laboratory experiments are one thing. But there is strong evidence that this and other forms of behavior that render groups less reliable than individuals also occur in critical real-world decision-making situations. In his landmark 1972 book, *Victims of Groupthink: A Psychological Study of Foreign-Policy Decisions and Fiascos,* psychologist Irving Janis presents a fascinating series of case studies of the deeply flawed group decision processes that produced some of the worst foreign-policy disasters of the twentieth century. Janis coined the term "groupthink" to refer to "a deterioration of mental efficiency, reality testing and moral judgment that results from in-group pressures."[36] It is what happens when the members of a decision-making group become more focused on maintaining good relations with each other and achieving unanimity than on realistically analyzing the situation and critically evaluating all available alternatives. As the reputation of the FBI's famous crime lab came under fire in the late 1990s, one FBI technical expert argued that part of the problem was that "You get an inadvertent bonding of like-minded individuals supporting each other's false conclusions."[37] Unfortunately, this state of affairs can occur even in the most critical decisions, when thousands or millions of lives are at stake.

According to Janis, there are eight major symptoms of groupthink:

> (1) an illusion of invulnerability . . . which creates excessive optimism and encourages taking extreme risks; (2) collective efforts to rationalize in order to discount warnings which might lead the members to reconsider their assumptions . . . ; (3) unquestioned belief in the group's inherent morality, inclining the members to ignore the ethical or moral consequences of their decisions; (4) stereotyped views of enemy leaders as too evil to warrant genuine attempts to negotiate, or as too weak and stupid to counter whatever risky attempts are made to defeat their purposes; (5) direct pressure on any member who expresses strong arguments against any of the group's stereotypes, illusions or commitments . . . ; (6) self-censorship [by each group member] of deviations from the apparent group consensus . . . ; (7) a shared illusion of unanimity concerning judgments conforming to the majority view . . . ; (8) the emergence of self-appointed mindguards—members who protect the group from adverse information that might shatter their shared complacency about the effectiveness and morality of their decisions.[38]

Korea

On June 24, 1950, North Korean troops invaded South Korea, touching off the Korean War. The United States led UN military forces into the field to drive the Northerners back above the thirty-eighth parallel, which divided communist and noncommunist Korea. Within a matter of months, the UN forces had achieved that objective.

During the difficult early days of the war, the Truman administration's key decision-making group had developed a real sense of mutual admiration and had grown insulated, self-congratulatory and perhaps a bit intoxicated with success. In early November, the administration decided to escalate the war, authorizing a large-scale American military action to pursue the defeated North Korean forces above the thirty-eighth parallel, conquer North Korea and unite it with the noncommunist South. They chose to ignore repeated warnings by the Chinese that they would enter the war if UN troops invaded North Korea.

In late November, the Chinese attacked the American invasion force en masse. Throwing hundreds of thousands of troops into the battle, they trapped entire American units and forced the rest out of North Korea. The rout continued as the Chinese pushed into South Korea, nearly driving the UN forces out of the entire Korean peninsula. The disastrous foreign-policy decision to escalate the conflict came close to snatching defeat out of the jaws of victory. What had led to that blunder?

It certainly wasn't lack of information. The Chinese began issuing warnings in September. In the first few days of October, they directly and explicitly warned that they would not stand idly by if the U.S.-led UN troops poised on the North Korean border crossed the thirty-eighth parallel. The American policy-making group dismissed these repeated strident and belligerent statements as a bluff. Then the Chinese strengthened the warning by sending forces into North Korea, and strengthened it further when those forces actually engaged South Korean and U.S. troops. Yet despite the fact that U.S. policy makers strongly desired to avoid war with China, they still did not believe that the Chinese would enter the war in force. They were somehow able to convince each other that although the United States had not hesitated to resist aggression against an ally eight thousand miles away, the Chinese would not do the same for an ally on their border.

The State Department's own policy staff opposed crossing the thirty-eighth parallel. Assuming the role of "mindguard," Secretary of State Acheson made sure that the dissenters were kept far away from the policy-making in-group. In classic groupthink style, careful consideration of the alternative views of outside experts was sacrificed on the altar of group cohesion. They knew that risking war with China on the Asian mainland was risking disaster. Yet they decided to take actions that made it virtually impossible for the Chinese to stay out of the conflict.

A few days after the Chinese made good on their oft-repeated threats, President Truman received a call direct from General MacArthur's headquarters informing him that the massive Chinese assault was forcing the U.S.-led forces into a full-scale retreat. Truman consulted key members of his policy in-group, who still reported unanimity and refused to admit their blunder.[39] Truman's decision to try to unify Korea by military force under South Korean rule almost succeeded in Korea being unified by force under North Korean rule. It was a horrible mistake that cost millions of lives, nearly led to ignominious defeat and, given the tenor of the times, might even have precipitated World War III.

Pearl Harbor

The Japanese assault on Pearl Harbor in December 1941 is often cited as the ultimate in treacherous "sneak attacks." Yet the lack of vigilance at this American military fortress, despite repeated warnings of imminent attack, is just as good an example of how groupthink fosters complacency.

In 1940, American cryptographers had managed to break nearly all of the Japanese communication codes. That gave America the enormous advantage of being privy to almost all of the secret messages that the Japanese transmitted. With this advantage, the U.S. military knew that Japan was preparing for major military operations, although it was not clear exactly where. Two weeks before the attack, the Navy commander at Pearl Harbor received a strong warning from the chief of naval operations in Washington that war with Japan could break out at any moment. It read: "Chances of favorable outcome of negotiations with Japan very doubtful. This situation coupled with statements of Japanese government and movements their naval and military forces indicate in our opinion that a surprise aggressive move in any direction . . . is a possibility."[40] A few days later, on November 27, 1941, naval operations in Washington strengthened the message: "This dispatch is to be considered a war warning. Negotiations with Japan . . . have ceased and an aggressive move by Japan is expected within the next few days. . . . Execute appropriate defensive deployment preparatory to carrying out [the Naval War Plan]."[41]

The Navy commander in Hawaii and his advisory group took seriously the warning that war was imminent, but persisted in believing that Pearl Harbor could not be the target. They not only didn't increase airborne reconnaissance, they continued to completely neglect reconnaissance over the sector to the north of Hawaii. Had American planes been patrolling that sector, they almost certainly would have spotted the Japanese aircraft carriers approaching Oahu. If they had, there would probably have been no battle. The Japanese contingency plan was to turn around and head back to Japan without attempting an attack if they were spotted.

No alert was sounded when the Japanese planes arrived over Pearl Harbor until *after* the bombs started exploding. How could the commander and his advisory group have been so convinced that Pearl Harbor would not be Japan's target, despite those strong and urgent warnings from higher command? They supported each other in maintaining an illusion of invulnerability—not that uncommon in military history—based on a series of assumptions that were never subjected to serious scrutiny.

One assumption was that a power as small as Japan would never initiate a war against a nation as powerful as the United States, a war they could not win. But the United States had imposed a blockade against Japan in July 1941 that was cutting them off from supplies of vital raw materials. "No one discussed . . . how the Japanese would view the risks of *not* attacking the United States, of allowing themselves to be relegated to the status of a third- or fourth-rate power, deprived of all their hard-won territories gained from years of fighting and sacrifice, divested of all national honor."[42]

The Navy group also believed that the concentration of military power at Pearl Harbor was so great that it alone would deter any possibility of attack, an important historical example (among many) of the failure of deterrence. An admiral stationed in Hawaii later testified, "If we had ten minutes warning everybody would have been there [shooting the planes down], and we didn't anticipate that they could get in without ten minutes warning."[43] But why not? Even hours before the attack, when patrols *did* encounter hostile submarines and radars *did* detect unidentified aircraft, these obvious warnings that something was afoot were completely missed or mishandled.

U.S. forces had been tracking the Japanese aircraft carriers across the Pacific, and the commander at Pearl Harbor was aware of a mysterious loss of radio contact with the carriers during the week before the attack. (Of course, the carriers were observing radio silence—they were approaching their target.) On December 2, 1941, the commander was discussing the "lost" Japanese carriers with a subordinate who informed him that they were still out of contact and the U.S. Navy didn't know where they were. The admiral joked, "What, you don't know where the carriers are? Do you mean to say that they could be rounding Diamond Head [10 miles from Pearl Harbor] and you wouldn't know it?"[44] Self-assured laughter at a clear danger signal is a classic manifestation of groupthink. Here, as is often the case, the joke did not turn out to be very funny.

The British decision to appease Hitler (in which the Sudetenland was ceded to Germany in exchange for "peace in our time"), the Kennedy administration's Bay of Pigs fiasco (in which Cuban exiles were to invade Cuba and overthrow Fidel Castro), the long series of bad decisions in the 1960s and 1970s that marked America's disastrous experience in the Vietnam War, the foolish decisions that led the Soviet government into the morass of the War in Afghanistan during the

1980s—all of these are cases in which groupthink played a major role. Those members of the decision-making in-group that knew better simply held their tongues or were silenced by group pressures. Outside experts with dissenting opinions were kept away. Treating silence as agreement and believing "if we all agree, it must be right" reinforced questionable policies favored by key decision makers and kept them locked into policies about which even they may have had serious doubts. In the grip of groupthink, groups can easily be less effective and less reliable than individuals.

Groupthink can be prevented, or at least minimized, by direct and explicit action. Specific members of the group can be assigned to criticize *every* proposal that anyone makes. Group members can be encouraged by their leader to actively debate the pros and cons of each proposal, and to search for and explain any misgivings they might have, without fear of ridicule. Outside experts of very different background and beliefs than the in-group can be sought out and consulted, rather than excluded.

Having learned from the Bay of Pigs fiasco, the Kennedy administration took such actions when the Cuban Missile Crisis erupted. As a result, there was much higher quality decision making during the Missile Crisis than during the Bay of Pigs, even though many of the same people were involved in both decision-making in-groups.[45] Still, it is important to remember that critical errors and miscalculations were made during the Missile Crisis that nearly led the world into the terminal disaster of general nuclear war.

GROUP PSYCHOSIS

Members of cohesive decision-making teams caught up in groupthink may be in denial about some elements of what is going on in the outside world, but they are still basically in touch with reality. They may be deluding themselves about their own power, wisdom or morality, but they are not living in a world of delusion. There are, however, conditions under which the behavior of groups can become truly psychotic.

The difference, it is said, between a psychotic and a neurotic is that "the neurotic builds castles in the air, and the psychotic lives in them." Although they are disconnected from reality in important ways, psychotics may still be functional enough to avoid isolation or even detection by those around them. Yet because they are living in a world of delusion, they are not only unreliable, but their behavior can be both bizarre and extremely dangerous.

It is easy to understand why a group of mentally deranged individuals might exhibit psychotic behavior. It is very difficult to understand—or even believe—

that a group of otherwise normal people can together behave if they were psychotic. How can such a thing happen?

Under the right conditions, a seriously deluded but very charismatic leader can draw a group of basically normal though perhaps vulnerable people into his/her own delusional state. The most critical elements appear to be: (1) the charisma of the leader; (2) the degree of control he/she can exert over the conditions under which the group's members live; and (3) the extent to which the leader is able to isolate group members from outside influences, especially those of nongroup friends and family. Four of the most striking cases of group psychotic behavior in twentieth century America are briefly described below. They provide a frightening insight into the character and potential consequences of this most extreme form of group pathology.

The Manson Family[46]

On the morning of August 9, 1969, the police were called to the scene of a grisly series of murders at 10050 Cielo Drive in Los Angeles. Five bodies were discovered, including that of the pregnant movie star Sharon Tate, wife of Hollywood director Roman Polanski. The three men and two women had not so much been murdered as butchered. There were large pools of blood in front of the house, and more blood inside. The killers used Tate's blood to print the word "PIG" on the front door. The very next day, the Los Angeles police again confronted a similarly bizarre and bloody murder scene in another house only a few miles away.

After a long, complicated and well-publicized trial, Charles Manson, Patricia Krenwinkel, Susan Atkins and Leslie Van Houten were each convicted of committing these brutal, vicious and apparently senseless random murders. Somehow the charismatic Manson had managed to draw these three young women (and other members of the so-called Manson Family) into his own psychotic fantasy world. Just who is Charles Manson, and how did he accomplish this?

Born in 1934, Manson had a very difficult childhood. His mother would leave him with obliging neighbors who had agreed to watch him for an hour, then disappear for days or weeks. At twelve, the court sent him to a caretaking institution. Shortly after, he began a life of crime, committing his first armed robbery at age 13. By the time he was 32, Manson had spent 17 years, more than half his life, in institutions. In all those years, he had been examined by a psychiatrist only three times, and then superficially. There was nothing in his record to foreshadow the butchery that lay ahead.

Manson had learned how to manipulate people at a very early age. He honed his skills over the years, becoming especially adept at controlling females. In the spring of 1967, Manson moved to Haight-Ashbury in San Francisco, at the time in its heyday as the center of hippie life. The Haight was filled with young and naive people eager to believe in something and needing to belong, to feel wanted. In the midst of that free-floating, "do your own thing", drug-oriented yet nonviolent culture, he used his charm and hypnotic style to create the Family.

In the almost completely isolated world of the Spahn Ranch, where the Family lived in the months before the brutal killings, Manson created an atmosphere of acceptance and affection, punctuated with fear and intimidation. He used drugs (especially LSD), sex, sleep deprivation and constant harangues to deprive Family members of most if not all of their individuality, sense of personal responsibility and capacity for critical thought or independent moral judgement. Manson had taken a group of young, vulnerable people and brainwashed them.

Because he is such an extraordinary con artist, it is impossible to be sure if he actually believed the bizarre, paranoid fantasy into which he drew his followers. Manson claimed that the English rock group the Beatles were speaking personally to him in the songs they recorded, telling him that there was about to be a race war. In the words of a Family member, "He used to explain how it would be so simple to start out. A couple of black people . . . would come up into the Bel Air and Beverly Hills District . . . up in the rich piggy district . . . and just really wipe some people out, just cutting bodies up and smearing blood and writing things on the wall in blood . . . all kinds of super-atrocious crimes that would really make the white man mad."[47] These heinous acts would cause mass paranoia among whites, who out of fear would enter black ghettos and begin shooting everyone. The slaughter would create a split in the white community that would set them against each other. After they had mostly killed each other off, the Black Muslims would come out of hiding, destroy all the remaining whites and take over. The blacks would be unable to run things, so they would ultimately seek out Manson and his Family, who had been hiding in a bottomless pit in the desert. The Family, by that time grown to 144,000 (as predicted in the Bible) would emerge, a pure white master race for whom all the remaining blacks would become servants. The Family would rule the world—a neatly wrapped, racist, paranoid psychotic fantasy. The brutal murders the Family committed were intended to get things going, because Manson believed that blacks were too slow to get the message.

It certainly cannot be said that the members of the Manson Family were a representative cross section of American society. But, prior to becoming wrapped up in Manson's psychosis, it is extremely unlikely that any of them would have even imagined becoming involved in acts of savagery. Yet, as part of that psychotic group, they not only committed those brutal and senseless crimes, but they did so virtually without emotion at the time or remorse afterwards. Years later, only one of those convicted, Leslie Van Houten, seemed to have been able to break free and regain control of her own mind.

Jonestown[48]

It was November 18, 1978. Nine hundred twelve human corpses lay in the clearing in the jungle that was Jonestown, Guyana. Row after row of men, women and children, most of them face down, their faces twisted into violent contortions by the terminal agony of cyanide poisoning, blood streaming from their noses and mouths. On the primitive stage they faced, their leader, the Reverend Jim Jones, was toppled over a podium, a bullet in his head. This was the last act in a tragic play that began in great hope, two decades earlier in Indianapolis, Indiana, when the very first People's Temple was opened.

The original Temple was a model of progressive thought and compassionate action. Racially integrated when racial integration was not common, the church ran a soup kitchen that fed anyone who was hungry, a free employment agency that helped people in need find jobs, and a nursing home that provided health care. Jones and his wife had one child of their own and adopted eight more children of various racial backgrounds. The mayor of Indianapolis was so impressed by the People's Temple that he appointed Jones director of the city's Human Rights Commission in 1961.

From an early age, Jones was fascinated by preachers and used to practice giving sermons. His small-town life and his stable upbringing could not have been more different from that of Manson. Yet he too became adept at manipulating people and fascinated by the power that implied. A fellow minister commented, "I've never seen anyone relate to people the way he could. He would build them up, convince them that anyone as intelligent and sensitive as they were ought to do whatever it was that he wanted them to do."[49]

Like Manson, Jones also had a vision of catastrophic war—not of race war, but of nuclear holocaust. Every sane person fears nuclear war, but for him the fear was grossly exaggerated. One neighbor reported that "There were times when just the sound of an airplane flying overhead would start him crying." He began to look outside the United States for a place that would be safe from both bombs and bigotry. In 1963, he visited Guyana and dreamed of establishing an isolated utopian settlement there.

In 1965, Jones moved the People's Temple to California. He convinced many of its members that the Ku Klux Klan, the CIA or some other external evil force would kill them if it weren't for him. He encouraged them to inform on spouses or children who violated any of his rules or expressed doubts. Severe public paddlings, suicide rituals with fake poison and sexual abuse became part and parcel of life in the Temple. Every woman close to him was required to have sex with him regularly. He used sexual activity with males in the Temple as a tool to humiliate or blackmail them, having them observed or photographed in the act.

Ignorant of this, the outside world acclaimed Jones for establishing effective drug rehabilitation programs, clinics and nursing homes. The mayor of San Francisco appointed him chairman of the city's Housing Authority. More and more, Jones associated with high-level local and national politicians. But his growing celebrity status contained the seeds of his downfall. The press began to get curious about the secret inner world of the People's Temple. The imminent appearance of an investigative exposé in the August 1, 1977, issue of *New West* magazine prompted Jones to move the People's Temple to Guyana, taking more than 800 members with him to build Jonestown on 27,000 acres of land he had leased.

Some of Jones's followers were young college graduates from socially progressive upper-middle-class backgrounds whose parents were educated professionals or executives. Some were middle-class blacks and whites from fundamentalist religious backgrounds. More were young blacks with limited education from poor ghetto neighborhoods. The greatest percentage were elderly, mostly blacks, also from the San Francisco ghetto. They were much more of a cross section of American society than the members of the Manson Family.

Once ensconced in their clearing in the Guyanese jungle, Jones became even more controlling, paranoid and psychotic. A master of mind control, he employed a variety of brainwashing techniques, some of which paralleled those used by Manson.[50] Life in what was originally supposed to be a peaceful, caring, nonracist utopian community became a living hell. By November 18, Jones's followers had largely lost their ability to exercise independent judgement. They were no longer fully in control of their own minds. When he commanded them to begin the oft-rehearsed ritual of mass suicide, they were primed for mindless obedience. One by one, they came up to the vat of flavored water laced with cyanide and drank a cup of the poisonous brew. In a few minutes they began to gasp, convulse and vomit. And soon, it was all over.

The tragedy at Jonestown seems like nothing so much as a bizarre, ghoulish nightmare. It is hard to believe that it really happened. But it did.

David Koresh and the Branch Davidian

On February 28, 1993, more than a hundred armed federal agents of the Bureau of Alcohol, Tobacco and Firearms (ATF) swarmed onto the compound of a fringe religious sect known as the Branch Davidian and led by David Koresh. The agents wanted to arrest Koresh for illegal possession of firearms and search the compound for what they had been told was an impressive arsenal of illegal weapons. Things did not go well. They encountered a barrage of high-powered gunfire that left 4 of them dead and 15 more wounded. The failed raid marked the beginning of an armed standoff that made headlines for months.[51]

David Koresh was 33 at the time of the raid. Born in Houston, and raised in the suburbs of Dallas-Fort Worth, Koresh referred to the schools he attended as "special schools" for slow learners and said his parents and teachers never expected him to accomplish much. A ninth-grade dropout, he worked at a number of jobs. Eventually, Koresh found his "calling" as a preacher. He was a very charismatic speaker who combined religious fervor with a talent for weaving a spell with words.[52] According to one near-convert, his style was "almost schizophrenic and dissociative. . . . The words would all be there, the syntax would be correct, but when you put it together, it didn't make logical sense."[53]

Like Jim Jones and Charles Manson, Koresh lived in a world that was an odd combination of fantasy and reality, wrapping his followers in his delusion that he was Jesus Christ incarnate, come to earth to interpret the truths contained in the biblical Book of Revelation, which could only be understood by opening the Seven Seals described in that book.[54] First, Koresh believed, he would become a rock star. Then, he would go to Israel and show the rabbis the biblical truths they were unable to see because of their limited wisdom and understanding, truths to which only he was privy. His presence in the Middle East would create havoc. The United States would be forced to send in troops, triggering a war that would destroy the world. He would then be crucified.[55]

While waiting for the drama to begin, Koresh lived a decidedly un-Christlike existence. At age 24, he had married the 14-year-old daughter of a high-ranking Branch Davidian. Former Davidians allege that Koresh routinely abused even the cult's youngest followers, physically and psychologically. After a long investigation, the *Waco Tribune-Herald* reported complaints that Koresh boasted of having sex with underage girls in the cult. He had at least 15 "wives" of his own and claimed the divine right to have sex with every male cult member's wife. Koresh claimed that the children he fathered with these women would rule the earth with him after he and the Davidian men had slain the nonbelievers.[56]

He kept his followers isolated at the group's compound and spellbound with his oratory. They believed that he was their key to entry into heaven after the soon-to-come end of the world. He controlled all the details of their daily lives, putting them through lengthy and rigorous Bible study every day, sessions sometimes lasting as long as 15 hours.[57]

Provoked by an assault on the compound by federal forces after months of armed standoff, the main building at Mt. Carmel was set on fire and nearly 80 Branch Davidians (including 21 children) died together in mid April 1993. Unlike the grotesque group suicide-murder discovered only after the fact in the remote jungles of Jonestown, Guyana, the events in Waco developed before an awestruck public who watched as flames engulfed the compound on national television. Once again a deluded, charismatic leader had led the group of people who trusted him to self-induced disaster.

Heaven's Gate[58]

In late March 1997, as the comet Hale-Bopp drew ever closer to earth, 39 men and women dressed in loose, androgynous black clothing, swallowed a potent mixture of phenobarbitol, and alcohol and quietly lay down on their beds to die. They believed they were shedding their earthly bodies to join aliens trailing the comet in a spacecraft that had come to take them to another plane of existence, the Level Above Human.

Unlike those who died in agony in the jungle at Jonestown or were burned to death in the spare compound at Waco, the members of the cult called Heaven's Gate drifted peacefully off to their final sleep in the rooms of a large, comfortable house near San Diego. These quiet, hardworking, clean-cut, celibate and religious people had become enveloped in the ultimately deadly science-fiction fantasy of their charismatic leader, Marshall Herff Applewhite.

The son of a minister, Applewhite grew up in South Texas, a likable, energetic man. The first 40 years of his life were unremarkable. Then he went through a painful divorce in 1968 and lost his father in 1971. By 1972, his life was coming apart. He was in debt, his career was in trouble, and in March of that year, he was hospitalized with heart trouble. According to his sister, he had a "near death" experience in the hospital, and one of the nurses there, Bonnie Nettles, told him that God had kept him alive for a purpose. "She sort of talked him into the fact that this was the purpose—to lead these people—and he took it from there."[59]

Within a year, Applewhite and Nettles began travelling together; by the summer of 1973, both were convinced that they were the "two lampstands" referred to in the biblical Book of Revelation and that God had directly revealed their "overwhelming mission" to them. Unable to handle the reality of his life, Applewhite had constructed his own reality, restoring his self-esteem with the idea that he was God's special agent. Such fantasies are easier to sustain if others can be drawn into them. So Nettles and Applewhite set out to recruit followers, with considerable success.

Once convinced to join the cult, members had to give away their material possessions, change their names and break all contact with friends and family. They were not to watch television or read anything other than the "red letter" edition of the Bible. Each member was assigned a partner and they were encouraged to travel always as a pair. Control of sexual behavior was again part of the routine, in this case in the form of complete celibacy. Camped in Wyoming in the late 1970s, cult members had to wear gloves at all times, limit their speech to "Yes," "No" and "I don't know" and otherwise communicate almost solely through written messages. They woke up and ate meals on a rigid schedule, sometimes wore hoods over their heads and changed their work chores every 12 minutes, when there was a beep from the command tent.

When Nettles died in 1985, Applewhite kept the cult in deep seclusion, living in various southwestern cities until moving to California in the 1990s. There the group did free-lance computer work designing Internet Web pages. On their own Web site, they published documents outlining their beliefs: Two thousand years ago beings from the Level Above Human sent an "Older Member" (Jesus Christ) to earth to teach people how to enter God's Kingdom, but demonically inspired humans killed him. In the 1970s, the Level Above Human gave humanity another chance, this time sending two "Older Members" (Applewhite and Nettles) to resume the teachings.[60]

Another document published on their Web site, ironically called "Our Position Against Suicide," explained their belief that they would soon board an alien spacecraft, referred to the destruction of David Koresh and the Branch Davidian, and alluded to the conditions under which they were prepared to take their own lives.[61] But Heaven's Gate was not the subject of a pending media expose or congressional investigation, as was the Jonestown cult, nor were they surrounded by heavily-armed law-enforcement agents, as were Koresh and his followers. In the end, they took their own lives not because of threat, but simply so that they could ascend to the spacecraft they believed would take them to their long-sought Level Above Human.

How did a small group of young wayward souls become trapped in the paranoid fantasies of Charles Manson and molded into heartless, vicious, emotionless killers? How was a much larger group of essentially normal people drawn so completely into the crazy world of a disintegrating psychotic like Jim Jones that they took their own lives and murdered their own children? How did a twisted fanatic like David Koresh take a group of deeply devoted believers in the teachings of a loving God to the point where they betrayed the most central principles of those teachings? How was Marshall Applewhite able to convince a group of clean-cut, well-educated adults to believe so strongly that an alien spaceship had come to take them to a better world that they all calmly and willingly committed suicide? Under what conditions can groups of basically sane people be led to behave in such insane ways?

Brainwashing

In the early 1960s, Yale psychiatrist Robert Jay Lifton published a landmark study of the brainwashing methods used by the Chinese communists during their takeover of mainland China more than a decade earlier.[62] The essence of the process was depersonalization, achieved through repeated attacks on the individual's sense of self. They created a totally controlled, extremely stressful living environment, then applied enough physical and psychological pressure to break the victim's will, and finally offered leniency in return for complete cooperation. It was all quite effective. A relatively short but intense period of brainwashing overrode years of experience and belief.

The brain is an extraordinarily complex information-processing system. Just as muscles must be exercised to function properly, it appears that the brain must have a continual flow of internally or externally generated information to process. Sensory-deprivation experiments have shown that drastic distortions of mental function can be produced by halting the flow of external information. Visual and aural hallucinations may result after as little as 20 minutes. Prolonged total deprivation may even damage brain function in ways that are difficult if not impossible to reverse. Yet the alterations of consciousness that are produced by relatively short periods of sensory deprivation can include drug-like ecstatic highs and feelings of spiritual bliss. In any case, it is clear that information is vital to normal brain function.[63]

Because people are capable of learning both facts and ways of interpreting them, experience must affect the way in which the brain processes information. It is therefore logical that new and intense experiences can alter not only what we know, but also how we think. The brain must both take in facts and organize them around frameworks of interpretation. The key to "brainwashing" or "mind control" lies in creating intense new experiences that disrupt existing ways of thinking enough to throw the brain off balance in its search for order, then to present new ways of thinking that offer relief from the mental chaos that has been created. Because chaos is so painful to the order-seeking mind, the brain will tend to lock onto the new way of thinking, even if it is quite foreign and bizarre. Since it plays off the basic way our brains work, we are all vulnerable to brainwashing. Some of us are much more vulnerable than others, and we are all more vulnerable during particularly stressful times than when life is flowing smoothly.

Conway and Siegelman use the term "snapping" for the point at which individuals stops thinking and feeling independently, break their ordinary connection to the outside world in terms of awareness and social relationships, and give up their own mind to automatic or external control.[64] Brainwashing has succeeded when snapping occurs. The more control can be exerted over a person's conditions of life, the easier it is to create new experiences of sufficient intensity to induce snapping. That is why prisoners are more vulnerable to brainwashing than those who are free to come and go as they please. Prisoners of war may be at special risk, because it is so easy to rationalize mistreatment of "the enemy."

Sleep disruption, sleep deprivation, starvation and violations of personal physical integrity (from rape to beatings to control over bodily functions such as bathroom habits) are effective mind-control tools because they create enough mental disorientation and emotional chaos to throw the brain into disorder. Constant ideological harangues thrown into the mix help to further disrupt the mind's accustomed way of thinking and to provide new ways of thinking that, if accepted uncritically, promise to end the awful noise. An individual already primed for unquestioned obedience to authority by early exposure to dogmatic religion or abusive and authoritarian parenting is easier to brainwash. Perhaps that is why so many of the cults that program their followers' minds arise in the context of distorted, extremist religious faith.

Cults are not the only groups that manipulate participants to gain control over their minds. So-called mass therapies like est (Erhard Seminars Training) and Lifespring also use these or similar techniques. Extremely popular during the 1970s, est begins with a 60-hour-long course, training 250 people at a time in marathon sessions typically held on two consecutive weekends in a large meeting room. In order to participate, est trainees must sign an agreement at the beginning of the training not to get out of their seats without permission or speak unless they are asked to speak. No one is allowed to eat, smoke, drink, or use drugs during the sessions. Bathroom breaks are also severely limited.

A series of est "processes" are woven around many hours of lectures on the nature of reality, perception and belief systems. These are mental exercises intended to change the trainees' patterns of thought and feeling. There are many direct verbal assaults in which trainees are called "assholes" and "turkeys" and directly confronted by the est trainer. "During the course of the weekend, many trainees cry, faint, vomit, or lose sphincter control. At the end of the training, . . . each trainee is supposed to 'get it' in a moment of sudden realization that he alone is responsible for creating everything that happens to him[sic]."[65]

Founded in 1973, Lifespring presents itself as a self-help group that sells intensive seminars which are supposed to help their participants come to terms with their problems and lead a more fulfilling life.[66] Like est, it uses powerful brainwashing techniques, such as high-pressure confrontation and exhausting marathon sessions. The basic course consists of 50 hours over a period of only five days. Doors to the training rooms are locked, the temperature of the room is manipulated, food is limited, trips to the bathroom are restricted, etc.

Lifespring searches out and highlights each participant's greatest fear. Calling that intense personal fear a "Holy Grail", the group calls on the person to confront it directly—not under the care of trained mental-health professionals or even among caring friends and family, but out in the world at large. In the words of one recruit, "you have to get this Grail. If you don't get this Grail, you're not a good person, you gotta get your Grail."[67] Programmed with this sort of determination, getting the Grail can be extremely dangerous. Urged to confront his paralyzing fear of heights, one man put on a business suit and jumped off a bridge high over rocks in a dangerous river. He nearly died. Another, who was terribly afraid of water and had never learned to swim, jumped into a river and drowned trying to swim the quarter mile to the other side.

During one Lifespring session, trainers suggested to a young man who was experiencing an intense pain in his head that his pain was purely psychological. Actually, he had suffered a broken blood vessel in his brain, which put him in the hospital for months, paralyzed. Another Lifespring trainer refused to give any assistance to a young woman who suffered a severe asthma attack during training, telling her that her gasping and wheezing was also purely psychological. She died. Why didn't she just leave? According to her Lifespring training partner, "She told me that she . . . didn't want to leave; she felt very intimidated by the trainers."[68] John Hanley, Lifespring's founder, admits, "there are people who freak out in the training. And sometimes they can control it and handle it . . . and sometimes they just can't. . . . We had a veteran who freaked out reliving his experiences in Vietnam. . . . [H]e was screaming and yelling in the parking lot and he was gonna tear everything up."[69]

Brainwashing, Group Psychosis and the Nature of Military Life

In order to mold an effective fighting force, military training must condition essentially normal people to do things that the vast majority of them would otherwise not think of doing. They must be trained to put themselves in grave danger and be ready to inflict great harm on others who for the most part have never really done anything to them personally. Militaries must condition ordinary people to put themselves in the unnatural position of killing or being killed, and to do so for relatively abstract principles, such as patriotism or ideology.

The many wars with which the human species has been plagued to date make it clear that militaries have become quite good at this conditioning.

Military training (or perhaps we should call it military conditioning) begins with a process of depersonalization. First, each recruit is made to look as much like every other recruit as possible. Heads are shaved, clothing and accessories that might carry signs of individual personality are removed and replaced by uniforms to make everyone's appearance "uniform." All conditions of the recruit's life are controlled as well: when, where and what to eat; when and where to sleep (in barracks that are also uniformly equipped and arranged); and virtually all activities during the day. These activities are prearranged and ritualized. There are endless drills, from marching on parade grounds to inspections of clothing, living quarters and equipment. There is constant repetition, and continuing verbal and physical harassment. Punishment for even minor infractions is swift, and often includes language or tasks that are intended to humiliate and degrade.

Why does it matter if recruits can march precisely on a parade ground? It has been at least two hundred years since marching in formation has had anything to do with the conduct of war. What difference does it make if their shoes are shiny? Why all of the harassment and humiliation? The central purpose of basic training is much less to give recruits the military skills they will require than to deprive them of their individuality. The idea is to wipe away pre-existing patterns of thought and behavior and replace them with the military way of thinking and behaving. The recruits are to be turned into components of the military machine that will do what they are told, when they are told to do it, without question. In other words, military basic training itself is primarily a form of brainwashing.

From the very beginning, members of the armed forces are conditioned to operate within a rigid, authoritarian system in which orders are to be obeyed, not debated. They are thus primed for further, perhaps pernicious manipulation by those who so completely control the conditions of their life and work. It is important to understand that this cannot be avoided, given the mission military organizations are asked to perform. The only way to get large numbers of people to reliably do what militaries must get them to do is to establish a rigid, authoritarian system and condition the troops to follow orders.

Some American military personnel, including some in the nuclear military, have also undergone training by cults or pop-psychology mass therapies. At one point, the military itself reportedly invited Lifespring to train military personnel on several American bases in the United States and abroad. According to ABC-TV News, Lifespring held training sessions for hundreds of military personnel and their families at Vandenburg Air Force Base in California in late 1979 and again in March 1980. Among other things, Vandenburg was a Strategic Aerospace Command training base for the missile silo crews that tended the land-based missile component of the U.S. strategic nuclear forces. An Air Force captain involved in security brought these Lifespring trainings to the attention of ABC. He was concerned that "we don't know if this person that we certified as being combat-ready prior to Lifespring training is the same person in terms of reliability after that Lifespring training."[70] According to a Lifespring consultant, they had also been invited to give training at Moumstrum Air Force Base in Montana, one of the bases at which American land-based strategic nuclear missiles are located.

According to an investigative reporter who secretly attended one of the Lifespring sessions at Vandenburg, "At several points during the training they were so vulnerable that they were . . . talking about their most intimate problems, and they were ranting and raving and screaming and cursing . . . and spitting; it was like bedlam. It was like being at a mental institution, and I think at that point they would've answered any question put to them."[71] When Harvard psychiatrist John Clark was asked if he thought Lifespring graduates in the military and other sensitive positions might be unreliable enough to be security risks, he replied, "I believe they are. . . . I don't think they can be trusted because something has happened to their mind. . . . They are loyal to Lifespring and dependent on it, as though they were dependent on a drug."[72]

Group Psychosis in the Military

Primed by military training and the obedience-oriented, authoritarian character of military life, members of the armed forces may be particularly vulnerable to being drawn into the fantasy world of a charismatic but deeply disturbed commanding officer. The bonding of military men and women to each other and their often fierce pride in their unit, so necessary to cohesiveness and effectiveness under life-threatening conditions, also predisposes them to becoming part of a group loyal to a leader who may have departed from firm contact with reality.

The ability of a charismatic leader to involve followers in acting out a script that the leader has written is greatly enhanced if the followers can be isolated from outside influences. An appreciation of this is obvious in the tactics of both the cults and the "mass therapies." Isolation is a normal part of life in the armed forces. Ordinarily, groups rather than individuals are isolated from the outside world for periods ranging from hours or days (in the land-based missile forces) to months (in nuclear submarines). The longer the isolation, the greater the control the commanding officer has over the environment in which subordinates live and work.

It would be relatively simple for a charismatic military commander, capable of inspiring extraordinary loyalty in troops already primed for obedience, to segregate them from the outside world. Once the group is isolated, the commander, with control over so many elements of their lives, could easily create the conditions that would draw them ever more tightly into a fantasy world. The basic elements of such a scenario are illustrated in a 1960s-vintage film, *The Bedford Incident.*[73] The charismatic captain of an antisub-marine-warfare ship, the *Bedford,* does not appear to be psychotic. But he is so totally wrapped up in his fierce sense of mission and determination to humiliate the enemy that his judgement is terribly distorted. Once at sea, he drives the men beyond reason and good sense. Disrupting their sleep and keeping his crew almost constantly at high levels of alert during their tour at sea, he draws them into his grand obsession. Ultimately, one of his most loyal men breaks under the strain, misunderstands a command and launches a nuclear attack that could begin a nuclear war. Though the script is purely fiction, the story is frighteningly realistic.

Probably the most dangerous plausible place for a *Bedford*-like scenario in the real world is a nuclear missile submarine. The crew is completely cut off from outside contact for months at a time, their world confined to the cramped quarters of a large metal tube sailing below the surface of the ocean. Encased day and night in a totally artificial environment from which there is no respite, the captain and ranking officers aboard have nearly complete control over the living and working conditions of the crew. The chain of command on the ship begins with the captain and is clearly specified. Obedience is expected from the crew, which has, of course, been thoroughly trained to obey orders. There is no external physical control over the arming and launching of the nuclear-tipped missiles the submarine carries. Though they are not authorized to do so without orders from the highest command authority, the captain and crew of a missile submarine are capable of launching all the nuclear weapons on board by themselves, at any time. And each single submarine carries enough offensive nuclear firepower to destroy any nation on earth.

Under these conditions, a charismatic captain who commands the loyalty and trust of both officers and crew is in an almost ideal position to lead them wherever he or she wants them to go. If the captain has become deluded, paranoid or otherwise enveloped in an internal fantasy world (like Manson, Jones, Koresh or Applewhite), while still maintaining enough contact with reality to appear sane, the stage is set for disaster on a scale that could dwarf even the horrors unleashed by a genocidal maniac like Adolf Hitler.

Transferring responsibility from individuals to groups does not make dangerous-technology systems proof against the limits of safety imposed by human error or malevolence. From the banalities of bureaucracy to the arrogance of groupthink to the nightmare of group psychosis, groups not only fail to solve the problem of human fallibility, they add their own special dimension to it.

There is no way to avoid the fallibility that is an essential part of our human nature. Individually or in groups, we have accidents, we make mistakes, we miscalculate, we do things that we would be better off not doing. If we continue to create, deploy and operate dangerous and powerful technological systems, we *will* eventually do ourselves in. That is the bottom line. After all, we are humans, not gods. The only way we can permanently prevent human-induced technological disaster is to stop relying on technologies that do not allow a very, very wide margin for the errors we cannot help but commit.

PART IV

Why Improving Technology Won't Save Us

CHAPTER NINE

The Failure
of Technical Systems

More than ever before, we are dependent on a web of interconnected technical systems for our most basic needs and our most fleeting whims. Technical systems are integral to providing us with water, food and energy, to getting us from place to place, to allowing us to communicate with each other and coordinate all the activities on which our physical and social lives depend. The more technical systems have become central to our way of life and critical to the normal functioning of society, the greater the disruption caused when they fail.

Most of us don't really understand how any of these complex and sophisticated technical systems work, and no one understands them all. Most of the time, that isn't much of a problem. We don't have to know how they work to use them, and often don't need detailed technical knowledge to maintain or repair them. Knowing which pedal to push to speed up or slow down, when and how much to turn the steering wheel, how to back up and so on is all you need to know to drive a car. You don't have to understand what actually happens inside. Even those who repair cars don't need engineering or scientific knowledge of the electrical, chemical, and mechanical processes that make the car work. All they need to know is what each part does and how to adjust, repair or replace it.

Technical expertise is also not necessary to understand why technical systems fail and why the possibility of failure cannot be completely eliminated. It is enough to generically comprehend the inherent problems involved in developing, producing and operating them, as well as in assessing their costs, benefits and risks. There is great power in technology, but there are also inherent limits to that power.

COMPLEXITY AND RELIABILITY

The reliability of a technical system depends on both the reliability of its parts and the complexity of the system. Complexity, in turn, depends upon the number of parts and how they interact with each other. All other things being equal, the more parts there are that must perform properly for the system to work, the less reliable the system will tend to be.[1] Multiplying the number of parts can make the whole system less reliable even if each part is made more reliable.[2] Greater interdependence among components also tends to make the system less reliable.[3]

In a system with more parts, there are more ways for something to go wrong. Given the same quality of materials, engineering and construction, a system with more parts will thus fail more often. When the parts of a system are tied together more tightly, the failure of any one part is more likely to overload or otherwise interfere with the other parts. One failure tends to lead to others, dragging the system down. At first this may only degrade performance, but if it produces a cascading series of failures, it can cause the whole system to break down completely.

The complex of satellites orbiting the earth has, in effect, become just such an increasingly interdependent technical system. Some near-earth orbits are so full of active satellites, dead satellites, discarded rocket booster stages, and an enormous amount of small, miscellaneous debris that even a small scrap of very high speed space junk smashing into a large orbiting object could shatter it into hundreds of pieces, which would then shatter other objects in a continuing chain reaction. Although it would take decades for this slow-speed chain reaction to run its course, it could destroy many billions of dollars worth of vital satellites. Some experts believe we are already near the critical point at which a random collision could start this expensive disaster in motion.[4]

Sometimes complexity is unavoidable. Achieving the required performance might require a complex design. The more complex version might also be faster, safer, cheaper to operate, or higher precision than a simpler design. Still, increasing the number of and interdependence of components rapidly overwhelms attempts to make the system more reliable by improving the reliability of each of its parts. Despite our best efforts, complicated technical systems tend to fail more often than simpler ones.

The Therac 25 linear accelerator, an electron-beam and X-ray radiation therapy machine, was designed to destroy tumors deep inside a cancer patient without damaging skin tissue. On three separate occasions in 1985 and 1986, the machine failed, delivering a dose 100 times larger than the typical treatment dose. Two patients died, the third was severely burned. Therac 25 had a metal target designed to swing into place and convert its high-energy electron beam into lower energy X-rays. A minor error in the machine's software made it unable to keep up when instructions were typed at unusually high speed. The critical target apparently failed to swing into place. It is not even clear that complex and sophisticated computer controls were needed for this function. A simple on-off switch and timer might have done the job just as well, with a much lower probability of failure.[5]

In an internal briefing for the Air Force in 1980, military analyst Franklyn C. Spinney documented the reliability-reducing effects of complexity.[6] Using several different measures, he compared the reliability of aircraft of varying complexity in both the Air Force and Navy arsenals. Data drawn from his analysis are given in Table 9-1.

All modern combat aircraft are very complex, interdependent, high-performance technical systems. Even the most reliable of them breaks down frequently. None of the aircraft in the sample averaged more than 72 minutes of flying time between failures (problems that require maintenance), averaged fewer than 1.6 maintenance events per flight or had all of the equipment essential for its mission operating properly more than 70 percent of the time. Anyone whose car broke down anywhere near this often would consider that car a world-class lemon.

Even so, substantial differences in complexity do give rise to major differences in reliability. While the complexity designations in the table are necessarily general, there is a clear pattern of reduced reliability with increasing complexity, by any of the measures used. The contrast between the most and least reliable aircraft is striking. The highly complex F-111D averaged only 12 minutes of flying between failures, while the much simpler A-10 flew six times as long between problems. At least one piece of equipment essential to the F-111D's mission was broken nearly two-thirds of the time, while the A-10 was fully mission capable twice as often. And the F-111D averaged more than six times as many maintenance problems as the A-10 every time it flew.

Complexity-induced failures of reliability are an inherent feature of all technical systems, be they single machines or large interconnected networks of equipment. The primary air traffic control system is based on computerized radar tracking. Computers in the regional air route traffic control centers receive a continuous flow of data from radars transmitted via telephone lines. The computers are designed to diagnose and correct a variety of malfunctions. Nevertheless, in 1980 the system was experiencing an average of one interruption

TABLE 9-1

COMPLEXITY AND RELIABILITY IN MILITARY AIRCRAFT

(FY 1979)

AIRCRAFT (AF=AIR FORCE; N=NAVY)	RELATIVE COMPLEXITY	AVERAGE PERCENTAGE OF AIRCRAFT NOT MISSION CAPABLE AT ANY GIVEN TIME	MEAN FLYING HOURS BETWEEN FAILURES	AVERAGE NUMBER OF MAINTENANCE EVENTS PER SORTIE
A-10 (AF)	Low	32.6%	1.2	1.6
A-4M (N)	Low	31.2%	0.7	2.4
A-7D (AF)	Medium	38.6%	0.9	1.9
F-4E (AF)	Medium	34.1%	0.4	3.6
A-6E (N)	High	39.5%	0.3	4.8
F-14A (N)	High	47.5%	0.3	6.0
F-111D (AF)	High	65.6%	0.2	10.2

Source: Spinney, F.C., Defense Facts of Life (December 5, 1980: Department of Defense unreviewed preliminary staff paper distributed in typescript by author).

of service of a minute or longer per center per week. That doesn't sound like much . . . until you realize that a modern jetliner flies about nine miles in one minute (at normal cruising speed). On average, these disruptions lasted seven minutes.[7] For most of us, a one-minute telephone outage or switching error is little more than a nuisance. We just hang up and dial again. But even short-lived interruptions in transmission or switching failures can create serious, even deadly problems in critical systems such as those used for air traffic control.

The complex, interconnected telephone system fails fairly often, though most of these failures are so fleeting and trivial we scarcely notice them. Every once in a while, though, we get a spectacular illustration of just how wrong something can go when something does go wrong. At 2:25 P.M. on January 15, 1990, a flaw in a single AT&T computer program disrupted long-distance service for nine hours. Roughly half the national and international calls made failed to connect.[8] Robert Allen, AT&T's chairman, called it "the most far-reaching service problem we've ever experienced."[9]

The program involved was part of a switching-software update designed to determine routings for long-distance calls. Because of the flaw, a flood of overload alarms was sent to other computers, stopping them from properly routing calls and essentially freezing many of the switches in the network.[10] Ironically, the system had been designed to prevent any single failure from incapacitating the network.[11] Furthermore, there was no sign that anything was wrong until the problem began, but once it did, it rapidly spun out of control. In the words of William Leach, manager of AT&T's network operations center, "It just seemed to happen. Poof, there it was."[12]

Ten years earlier, a forerunner of the Internet called "Arpanet," then an experimental military computer network, failed suddenly and unexpectedly. Its designers found that the failure of a small electronic circuit in a single computer had combined with a small software design error to instantly freeze the network.[13] In 1987, a complex network of hundreds of computers TRW had created for U.S. intelligence in Europe began to behave in peculiar and unpredictable ways. On careful investigation, engineers could not find anything wrong with the way it had been designed. Yet it was clearly not performing as intended.[14]

Opacity

As technical systems become more complex, they become more opaque. Those who operate ever more complex systems usually cannot directly see what is going on. They must depend on readings taken from gauges and instruments, and this can be very misleading. During the buildup to the massive power failure New York City experienced in 1977, one of the operators checked a current-flow reading on a particular line and saw it was zero. Since that line normally carried little if any current, that part of the system seemed to be operating normally. But what the operator was really seeing was the combined result of two switching failures, one of which would have sent current surging through that line if the second failure hadn't blocked the flow of any current to the line. The indirect information on which the operator was relying created a false sense of confidence. When the lights finally went out in the operator's control room, it became clear enough that something was very wrong.[15]

When a system becomes so complex that no one—including its designers—can really visualize how the system as a whole works, patchwork attempts to fix problems or enhance system performance are likely to create other hidden flaws. Just such an attempt seems to have caused the great AT&T crash of 1990. In 1976, AT&T pioneered a system called "out of band" signalling, which sent information for coordinating the flow of calls on the telephone network as each call was made. Engineers who were updating the out-of-band system in 1988 inadvertently introduced the software flaw that caused the system to crash two years later.

Those who modify very complex systems often do not understand enough about how they work to completely analyze all the ways in which changing one part will affect the rest of the system under all conceivable conditions. If they are careful and do the job properly, they may avoid creating problems in the part of the system they are changing. But it is virtually impossible for them to see all the subtle ways in which what they are doing will alter the overall system's characteristics and performance.

If patchwork change can open a Pandora's box, why not redesign the whole system when it fails or needs updating? Fundamental redesign may be a good idea from time to time, but it is much too time consuming and expensive to do whenever a problem arises or a way of improving the system occurs to someone. And frequent fundamental redesign has a much higher chance of introducing more serious problems than does patchwork change.

Although it sounds unbearably primitive, trial and error is still very important in getting complicated systems to work properly. "Bugs" are inevitable, even in the most carefully designed complex systems, *because* they are complex systems. Only by operating them under realistic conditions can we discover and correct unexpected (and sometimes unpredictable) problems and gain enough experience to have some confidence in their reliability. That is why engineers build and test prototypes. How confident would you feel flying in an airliner of radically new design that had never actually been flown before?

On April 24, 1990, the National Aeronautics and Space Administration (NASA) successfully launched the Hubble Space Telescope. Nearly a month later, Hubble produced its first blurry light image. Euphoria soon turned to concern, as the telescope just would not come into perfect focus, despite repeated commands from ground controllers at the Goddard Space Center outside Washington. Two months after launch, it became clear that the telescope suffered from spherical aberration, a classic problem covered in basic optics textbooks. A very slight flaw in the curvature of the Hubble's 2.4-meter primary mirror resulted in the optical system's failure to focus all incoming light at precisely the same spot. More than a decade of painstaking development had failed to prevent this crippling defect, and an extensive program of testing had failed to detect it. Scientists estimated that up to 40 percent of the scientific experiments the $1.6 billion space telescope was designed to carry out would have to be completely given up, and most of the remainder would be negatively affected. It would be years before the Hubble could be fixed.[16]

In December 1993, the telescope was finally repaired by a team of astronauts riding the space shuttle *Endeavor* into orbit. After more than a year of painstaking earthbound rehearsals, the astronauts were able to replace some defective equipment and install carefully designed corrective devices during an unprecedented series of space walks.[17] The next month, NASA jubilantly announced that the long-awaited repairs had been successful. Although Hubble still did not meet its original design specifications, it had finally been brought into clear focus and was producing remarkable images of the heavens.[18] By late March 1999, the space telescope was once more in trouble. Only three of Hubble's six gyroscopes were working properly. If one more failed, the telescope would become too unsteady to do observations and would shut down. NASA prepared to launch an emergency repair mission by October, but there was an estimated 20 percent chance that another gyroscope would fail before then.[19]

Sometimes even careful design, testing *and* early experience leaves critical problems hidden until a substantial track record has been built up. The De Havilland Comet was the first commercial jet aircraft. In May 1953, a year after the inauguration of jetliner service, a Comet was destroyed on takeoff from Calcutta during a severe thunderstorm. The weather was blamed. Eight months later, a second Comet exploded soon after takeoff from Rome, Italy, in clear and calm weather conditions. The Comets were withdrawn from commercial service for ten weeks, then reinstated even though investigators were still unable to find any flaw in the planes. In April 1954, a third Comet exploded in mid air. Only when one of the remaining Comets was then tested to destruction was it determined that the plane had a critical flaw. Repeating the normal cycle of pressurizing and depressurizing the cabin again and again caused a fatigue crack to develop in the corner of one of the windows that soon tore the plane's metal skin apart. The Comet's designers had been confident that such a problem would not develop until many more flights than the plane was capable of making during its estimated service life. They were wrong. Only in the trial and error of continuing operation was their tragic mistake exposed.[20]

Backup Systems and Redundant Design

Reliability can usually be improved by creating alternative routes that allow a system to keep working by going around failed components. When these alternate routes remain unused until there is a failure, they are called backup systems. For example, hospitals typically have their own backup generators to keep critical equipment and facilities operating when normal power supplies are disrupted. Emergency life vests on airliners are designed to inflate automatically when a tab is pulled, but they have a backup system—tubes through which a passenger can blow to inflate them if the automatic system fails.

Alternate routes used during normal operation can also operate as backups when failures occur. For example, calls can be routed over the telephone network from point A to point B over many possible pathways. In normal times, this allows the system load to be distributed efficiently. But the same design also allows calls to go through even when part of the system fails, by routing them around failed components.

"Voting" systems are another means of using redundancy to achieve reliability. "Triple modular redundancy" (TMR) is sometimes used in the design of fault-tolerant computers. Three components (or "modules") of identical design are polled and their outputs compared. If one disagrees with the other two, the system assumes the disagreeing module is faulty, and acts on the output of the two that agree. In this way, the system can continue to operate properly even if one of the modules fails.[21]

Backup systems and redundant design can make the system as a whole more reliable than any of its parts.[22] Still, backup systems themselves can also fail. In 1976, the failure of both the main audio amplifier and its backup left Gerald Ford and Jimmy Carter speechless for 27 minutes before an estimated audience of 90 million viewers during the first presidential campaign debate in 16 years.[23] Three separate safety devices designed to prevent chemical leaks all failed during the Bhopal disaster in 1984.[24] Many backup systems failed during the nuclear power plant accident at Three Mile Island on March 28, 1979, and the much more serious accident at Chernobyl on April 26, 1986. In October 1980, when technicians entered the containment building at New York's Indian Point nuclear power plant, they discovered areas flooded with nine feet of cold, brackish water that had leaked into the building from the Hudson River. A safety device had failed to detect the flooding because it was designed to detect only hot water. Two sump pumps designed as a backup should have triggered automatically and removed the water, but both failed—one because of blown fuses, the other because of a stuck mechanism.[25]

Boeing 747 jetliners have three sophisticated navigational systems. Yet on December 20, 1989, a navigational system failure led a Thai Airways 747 carrying 391 people to mistakenly reverse its course over the northern Pacific and begin flying east instead of west. The plane flew 600 miles off its flight path before it was notified of the problem by air traffic controllers. When they finally convinced the pilot that he was going the wrong way, he reported that all of the navigational devices had failed.[26]

Adding backup systems or redundancies tends to make a system design more complex, creating potential reliability problems that offset at least some of the advantages of having those backups or redundancies. One of the nine Ranger spacecraft flights intended to survey the moon before the Apollo mission failed *because* of extra systems designed into it to prevent failure. To be sure that the mission's TV cameras would come on when the time came to take pictures of the moon's surface, redundant power supplies and triggering circuits were provided. A testing device was included to assure that those systems would work properly. But the testing device short-circuited and drained all the power supplies before the spacecraft reached the moon.[27]

Common-Mode Failure and Sensitivity

When several components of a system depend on the same part, the failure of that part can disable all of them at the same time. This is known as a "common mode" failure. For example, if a hospital's backup generator feeds power into the same line in the hospital that usually carries electricity from the power company, a fire that destroyed that line would simultaneously cut the hospital off from the outside utility and make the backup generator useless.

At 4:25 A.M. on March 20, 1978, an operator replacing a burned-out light bulb on the main control panel at the Rancho Seco nuclear power plant near Sacramento dropped the bulb. This trivial event led to a common-mode failure that almost triggered disaster. The dropped bulb caused a short circuit that interrupted power to key instruments in the control room, including those controlling the main feedwater system. The instrument failures not only caused the main feedwater system to malfunction, but also cut off information the operators needed to know what to do. Equipment designed to control the reactor automatically didn't take proper corrective action because it also depended on the malfunctioning instruments. Worse yet, the instruments sent false signals to the plant's master control system, which then caused a rapid surge in pressure in the reactor's core, combined with falling temperatures.[28] In an older reactor, this is very dangerous.

The resistance to fracture of the steels from which pressure vessels are made depends on the metal's temperature. If it falls below the "reference temperature," the metal becomes very brittle and prone to break. Given enough time, neutron irradiation from the reactor's core can raise the reference temperature high enough so that rapid cooling of the reactor vessel may bring the metal close to this critical threshold. A sharp increase in pressure could then cause any pre-existing cracks in the vessel to grow quickly, producing a potentially catastrophic fracture. If the Rancho Seco power plant had been operating at full power for 10 to 15 years instead of 2 or 3, the vessel might have cracked.[29] With no emergency system left to cool the core, that could have led to a core meltdown and a nightmarish nuclear power plant accident.

Just after 4:00 PM on July 19, 1989, a United Airlines DC-10 jumbo jet carrying nearly 300 passengers and crew crashed short of the runway at Sioux City, Iowa, and burst into flames, killing 112 people. The problem was a loss of control due to failure of all three of the plane's hydraulic systems. The failure of any one would make it harder to fly the plane. With two gone, "the plane is like a drunk elephant," but it can still be flown.[30] In the United accident, an explosion in the rear engine showered the tail with shrapnel. All three hydraulic systems run through the tail section of the plane, and so all three of the hydraulic lines were severed at once. With no shut-off valve available, hydraulic fluid began to leak from the damaged lines, causing the pilot to lose control.[31]

On January 9, 1995, construction crews at Newark Airport using an 80-foot pile driver to pound 60-foot steel beams into the ground drove one beam through a foot-thick concrete wall of a conduit 6 feet under ground and severed three high-voltage cables serving the airport's terminal buildings. Hundreds of flights were cancelled and tens of thousands of people had to scramble to rearrange travel plans, as the airport was forced to shut down for nearly 24 hours. Local power company officials said that if one cable or even two had been knocked out, there still would have been enough power to keep the airport operating. If the main power cable and auxiliary power cables ran through separate conduits rather than lying side by side in the same conduit, this expensive and disruptive common-mode failure would not have occurred.[32]

Unless every part of a system has an alternative to every other part on which it depends, common-mode failures can undo some or all of the advantages of backup systems and redundancy. As long as there are any unique common connections, there can be common-mode failures that render the system vulnerable. But duplicating every part of every system is simply not workable. In a hospital, that would involve duplicating every power line, every piece of equipment, even to the point of having a spare hospital available.

High-performance technical systems are often more sensitive, as well as more complex. Greater sensitivity also predisposes systems to reliability problems. The high-performance aircraft of today are much more sensitive to variations in fuel, collisions with birds, and the quality of materials used to manufacture them than were the planes of 50 years ago. And for all their disadvantages, electronic devices that relied on vacuum tubes were much less sensitive to voltage surges and other disruptions of current than modern solid-state electronics.

More-sensitive devices are more easily overloaded, and thus more prone to failure. They are also more likely to react to irrelevant and transient stimuli. A highly sensitive smoke alarm will go off because of a bit of dust or a burned steak. False alarms make a system less reliable. They are failures in and of themselves, and if they are commonplace, operators will be too likely to assume that the next real alarm is also false.

Highly sensitive, complex interactive systems can also behave unpredictably, undergoing sudden explosive change after periods of apparent stability. When the Dow Jones average plummeted by 508 points at the New York Stock

Exchange on October 19, 1987, losing 22 percent of its value in one day, many were quick to point an accusing finger at computerized trading practices. Programmed trading by computer, it was argued, had made the market less stable by triggering buying and selling of large blocks of stock in a matter of seconds to take advantage of small movements in prices. A stock market of program traders approximates the key assumptions underlying the theory of nonlinear game dynamics (including rivalry). When systems like this are modeled on a computer, strange behaviors result. The system may be calm for a while and seem to be stable, then suddenly and unpredictably go into sharp nonlinear oscillations, with both undershooting and overshooting.[33]

DESIGN ERROR

Engineering design results in a product or process that never existed before in that precise form. Mathematical verification of concepts, computer simulation, and laboratory testing of prototypes are all important and useful tools for uncovering design errors. But until the product or process works properly under real-world conditions, it has not really been put to the test.

In 1995, after 14 years of work and 6 years of test flights, the B-2 bomber still had not passed most of its basic tests, despite an astronomical price tag of more than $2 billion a plane. The design may have looked fine on the drawing board, but it wasn't doing so well in the real world. Among other things, the B-2 was having a lot of trouble with rain. GAO auditors reported, "Air Force officials told us the B-2 radar cannot distinguish rain from other obstacles."[34] Two years later, they reported that the plane "must be sheltered or exposed only to the most benign environments—low humidity, no precipitation, moderate temperatures."[35] Not, on the whole, the kind of conditions military aircraft typically encounter. The B-2 was first used in combat in March 1999, during the NATO air campaign against Yugoslavia. Since they were too delicate to be based anywhere that didn't have special facilities to shelter and support them, two B-2s were flown out of their home base in Missouri, refueled several times in the air each so that they could drop a few bombs and then hurry back to Missouri, where they could be properly sheltered and cared for.[36]

Designers of the space shuttle booster rockets did not equip them with sensors that could warn of trouble because they believed the boosters were, in the words of NASA's top administrator, "not susceptible to failure. . . . We designed them that way."[37] After many successful launches, the explosion of the right-side booster of the space shuttle *Challenger* in 1986 proved that they were very wrong. That same year, a design flaw that had made the RBMK-1000 reactor unstable at low power from the beginning finally led to disaster, as two explosions at Chernobyl released "hundreds of times more radiation than was produced by the atomic bombings of Hiroshima and Nagasaki."[38]

Sometimes a design fails to perform as desired because the designers are caught in a web of conflicting or ambiguous design goals. The designer of a bridge knows that using higher-grade steel or thicker concrete supports will increase the load the bridge is able to bear. Yet given projected traffic, a tight construction schedule and a tight budget, the designer may intentionally choose a less sturdy, less expensive design that can still bear the projected load. If someday a key support gives way and the bridge collapses under much heavier traffic than had been expected, it is likely to be labelled a design error. But whose error is it: the engineer who could have chosen a stronger design, the person who underestimated future traffic, or the government officials who insisted the bridge be built quickly and at relatively low first cost?

Designers always try to anticipate what might go wrong when their design is put to the test in the real world, so they can prevent problems from developing. But there are so many ways things can go wrong that even the best designers can never think of them all. In June 1995, launch of the space shuttle *Discovery* had to be indefinitely postponed because a flock of "lovesick" male woodpeckers intent on courting pecked at least six dozen holes (some as big as four inches wide) in the insulation surrounding the external fuel tanks.[39] That would have been a hard problem to foresee. The odds are good that if one of the shuttle's designers had raised the possibility that passionate woodpeckers might someday attack the fuel tank insulation en masse, the rest of the design team would have laughed out loud.

Complexity, Interactions and Design Error

Failure to understand or pay attention to the way components of a complex system interact can be disastrous. The Northern California earthquake of October 17, 1989, brought down a mile and a quarter stretch of the elevated Nimitz Freeway in Oakland, killing and injuring dozens of motorists. Reinforcing cables installed on columns supporting the two-tiered, elevated roadway as part of the state's earthquake-proofing program may actually have made the damage worse. The quake sent rolling shock waves down the highway, and when some of the columns and cables supporting the roadbed collapsed, they pulled adjoining sections down one after another, like dominoes.[40]

On January 4, 1990, the number three engine fell off a Northwest Airlines Boeing 727. Investigators didn't find anything wrong with the design or construction of the engine, the way it was attached to the fuselage, or the fuselage itself. Instead, they reported that the loss of the engine was the result of a peculiar interaction. Water leaking from a rear lavatory was turned into ice by the cold outside air. The ice built up and then broke loose, striking the engine and causing it to shear off. In 1974, a National Airlines 727, and in 1985, an American Airlines 727 had lost the same engine in flight, for the same reason.[41] Designers just had not anticipated that a water leak in a lavatory could threaten the integrity of the aircraft's engines.

Even the designers may not fully comprehend the workings of complicated systems. It is impossible to enumerate, let alone to pay attention to, all of the ways things can go wrong. After-the-fact investigations often determine that failure was due to a simple oversight or elementary error that causes us to shake our heads and wonder how the designers could have been so incompetent. But mistakes are much easier to find when a failure has focused our attention on one particular part of a complex design.

One of the worst structural disasters in the history of the United States occurred on the evening of July 17, 1981, when two crowded suspended walkways at the Hyatt Regency Hotel in Kansas City collapsed and fell onto the even more crowded floor of the lobby below. More than a hundred people were killed and nearly two hundred others were injured. When investigators finally pinpointed the problem, technical drawings focusing on a critical flaw were published on the front page of the *Kansas City Star,* as well as in technical journals. An ambiguity in a single design detail had led the builder to modify the way that the walkways were connected to the rods on which they were suspended from the ceiling. That change made them barely able to support their own weight, let alone crowds of people dancing on them. Yes, the problem should have been caught during design or construction, when it could have been easily solved. But the fatal flaw was much easier to see once the collapse forced investigators to carefully examine how the walkways were supported. At that point, the critical detail was no longer lost in the myriad of other details of the building's innovative design.[42] Unless a great deal of attention is paid to interactions, knitting together the best-designed components can still produce a seriously flawed system design. Before computers, thorough analysis of complex designs was so difficult and time consuming that good designers placed a high value on simplicity. When computers became available, designers gained confidence that they could now thoroughly evaluate the workings of even very complicated designs. This degree of confidence may have been unwarranted. After all, a computer cannot analyze a design, it can only analyze a numerical model of a design. Any significant errors made in translating an engineer's design into a computer model render the computer's analysis inaccurate. So do any flaws in the software used to analyze the numerical models.

The piping systems of nuclear power plants are so complex, it is hard to imagine designing them without the use of computers. Yet one of the computer programs used to analyze stresses in the piping system was reportedly using an incorrect value for pi (the ratio of the circumference to the diameter of a circle); according to civil engineer Henry Petroski, in another "an incorrect sign was discovered in one of the instructions to the computer. Stresses that should have been added were subtracted . . . leading it to report values that were lower than they would have been during an earthquake. Since the computer results had been employed to declare several nuclear power plants earthquake proof, all those plants had to be rechecked. . . . This took months to do."[43] Had a serious earthquake occurred in the interim, the real world might have more quickly uncovered the error—with disastrous results.

Writing the programs that model engineering designs, and writing the software the computer uses to analyze these models, are themselves design processes that involve complex, interactive systems. The same cautions that apply to designing any other complex, interactive system apply here as well. A computer cannot hope to accurately appraise a design's performance in the real world unless it is given realistic specifications of system components and the way they interact. That is not an easy thing to do. Abstract numerical models are much easier to build if idealized conditions are assumed. For example, it is much easier to model the performance of an aircraft's wing in flight if it can be assumed that the leading edge is machined precisely, the materials from which the wing is made are flawless, the welds are perfect and uniform, and so on. Taking into account all of the complications that arise when these idealized assumptions are violated—and violated in irregular ways at that—makes building the model enormously more difficult, if not impossible.

The problem of spherical aberration that crippled the Hubble Space Telescope was due to an error in the curvature of its primary and/or secondary mirror of between 1/50 to 1/100 of the width of a human hair.[44] Yet the error still might have been detected if the mirrors had ever been tested together. Each was extensively tested separately, but their combined performance as an optical system was only evaluated by computer simulation.[45] The simulation couldn't have detected an error in curvature unless that error was built into the computer model. That could have been done only if the engineers knew the flaw was there. If they had known that, they wouldn't have needed the computer to tell them about it.

No computer can provide the right answers if it is not asked the right questions. Computers are extremely fast, but in many ways very stupid. They have no "common sense." They have no "feel" for the design, no way of knowing whether anything important is being overlooked. They do what they have been told to do; they respond only to what they have been asked. It is up to the designers and users of a program that analyzes a model to ask the correct questions.

In January 1978, the roof of the Civic Center in Hartford, Connecticut, collapsed under tons of snow and ice only a few hours after thousands had attended a basketball game there. The roof was designed as a space frame, supported by a complicated arrangement of metal rods. After the collapse, it was discovered that the main cause of failure was insufficient bracing in the rods at the top of the truss structure. The bars were bending under the unexpectedly heavy weight of snow and ice, and when the rods that were bent the most finally folded, the part of the roof's weight that they were bearing was shifted to adjoining rods. The unusually heavy load those rods now had to bear caused them to fold, setting up a kind of progressive collapse of the support structure which brought down the roof. A computer simulation finally solved the problem of why and how the accident had happened, but only after investigators had directly asked the right question of a program capable of answering that question. The original designers had used an unduly simplified computer model and had apparently not asked it all the right questions. But their analysis had given them such confidence in their design that when workers pointed out that the new roof was sagging, the designers assured them that nothing was wrong.[46]

Computers are very seductive tools for designers. They take much of the tedium out of the calculations required for routine design. They allow designers to reach more easily into unexplored territory. But as with most things that are seductive, there are unseen dangers involved. Exploring new domains of complexity and sophistication in design means leaving behind the possibility of understanding a design well enough to "feel" when something is wrong or problematic. It is not clear whether or not this will increase the frequency of design error, but it is almost certain to increase the severity of errors that do occur. Undue confidence and opaque designs will make it difficult for designers to catch some catastrophic errors before the real world makes them obvious.

The Pressure for New Design

If we were content to use the same designs year after year in the same operating environment, design error would be much less common. "Tried and true" designs are true because they have been repeatedly tried. Using the same design over many years allows evolutionary correction of flaws that come to light, and more complete comprehension of how the design is likely to be affected by minor variations. As long as new products, structures and systems are replicas of old and operating conditions remain the same, there is less and less scope for catastrophic error as time goes by. But that is not the world in which we live.

In our world, there is constant pressure to look for designs that work better, are more cost effective, more aesthetically pleasing and so on. Design change is driven by our creativity, our need for challenge, our confidence that better is possible, and our fascination with novelty. That means that the technical context within which all of the systems we design must operate is changing constantly. The social context may be changing as well, on its own or directly because of technological change. Automobile technology, for example, greatly increased suburbanization, changing the pattern of land use, our use of time (increased commuting), even the degree of our social interactions with neighbors. It also increased our flexibility of travel, became a major new source of death and injury, and sharply increased environmental pollution and the rate of depletion of nonrenewable resources. The environmental impacts alone tightened the constraints imposed on other technologies. Automotive pollution and resource depletion affected the design of other energy-using systems by raising the priority attached to reducing their own polluting emissions and increasing their fuel efficiency. Thus, the development and diffusion of new designs changes the context within which the design process takes place, which reinforces the need for new designs.

Reaching beyond existing designs brings with it a higher likelihood of error. In military systems, the pressure for new designs is intense, because even a small performance advantage is believed critical in combat. It is therefore not surprising that the designers of weapons and related systems make more than their share of significant design errors. Even when they try to "play it safe" by avoiding radical changes in a proven design, the constant pressure to improve performance opens the door to serious error. The Trident II missile was to be a submarine-launched ballistic missile (SLBM) with greater range and accuracy than Trident I. The first Trident II test-launched at sea exploded four seconds into its flight. The second test went all right, but the third test missile also exploded. The nozzles on the missile's first stage failed. They were damaged by water turbulence as the missile rode the bubble of compressed gas that propels it through 30 to 40 feet of water to the point where it breaks the surface of the sea and its engines can fire. The Navy has been launching SLBMs this way for decades. Trident I had passed this test. But Trident II was much longer and nearly twice as heavy. Its designers expected that it would create more turbulence in the water, but miscalculated in extrapolating previous experience. According to Rear Admiral Kenneth C. Malley, chief of the Navy's ballistic missile program, engineers using computer simulations seriously underestimated both the effect of water jets and how much pressure there would be on the missile as it surged upward through the sea.[47]

Decades earlier the designers of the De Havilland Comet jet airliner made a similarly fundamental miscalculation. Metal fatigue created in the skin of the plane when the cabin was repeatedly pressurized and depressurized was much greater than expected. Aircraft designers of the day were quite familiar with the problem of metal fatigue. But De Havilland was pushing into uncharted waters and, despite extra precautions, made a fatal error. Using "well-established methods," the designers thought that "a cabin that would survive undamaged a test to double its working pressure . . . would not fail in service under the action of fatigue."[48] As it turned out, they were very wrong.

Proponents of relatively new complex technical systems are frequently over-optimistic in projecting their experience with early successes to second- and third-generation versions. Over-optimism comes easily as the enthusiasm that surrounds an exciting new technology combines with a rapid rate of progress in its early stages of development. But later-generation systems are often more sophisticated and very different in scale. Being too ready to extrapolate well beyond previous experience is asking for trouble. This problem has been endemic in the trouble-plagued nuclear power industry. The first commercial plant was ordered in 1963, and only five years later orders were being taken for plants six times as large as the largest then in operation. There had only been 35 years of experience with reactors the size of Unit 2 at Three Mile Island when the partial meltdown occurred. That is very little experience for a technological system so big and complex.[49]

Thirty-five years is forever compared to the operating experience we have with many complex military systems. Design changes are so frequent, introduction of new technologies so common that extrapolation from previous experience is particularly tricky. New military technical systems are frequently not thoroughly tested under realistic operating conditions. It is not surprising that they often don't behave the way we expect them to at critical moments. Even when they are tried out under special test conditions, performance aberrations are sometimes over-looked in the pressure to get the new system "on line." During flight tests beginning in 1996, one of the wings on the Navy's $70 million F/A-18 Super Hornet fighter would sometimes suddenly and unpredictably dip when the plane was doing normal combat maneuvers. Engineers and pilots struggled without success to figure out what was causing this unpredictable "wing drop," which could prove fatal in combat. Though the flaw had still not been fixed, the Pentagon authorized purchase of the first 12 production-model F/A-18s in March 1997.[50]

Acts of God and Assumptions of People

The fruits of engineering design are real products that must be able to withstand the stresses, loads, temperatures, pressures, etc. imposed by their operating environments. Engineers must therefore build assumptions about that environment into the design process. Unfortunately, there is often no way of knowing all of its key characteristics precisely in advance. Consider the design of a highway bridge. It is possible to calculate gravitational forces on the bridge with great accuracy, and to know the load-bearing capabilities, tensile strength and other relevant characteristics of the materials used to construct the bridge (provided they are standard materials). Calculating wind stresses is less straightforward, though still not that difficult under "normal conditions." But it is much harder to calculate the strength and duration of the maximum wind stress the bridge will have to bear during the worst storms it will experience in its lifetime. Or the greatest stress it will have to bear as a result of flooding or earthquake. We simply do not know enough about meteorological or geological phenomena to be able to accurately predict these occasional, idiosyncratic but critical operating conditions. For that matter, we don't always do that good a job projecting future traffic loads either.

The subtle effects of slow-acting phenomena like corrosion can also interact with the design in ways that are both difficult to foresee and potentially catastrophic. In 1967, after 38 years of service, the Point Pleasant Bridge, spanning the Ohio River between West Virginia and Ohio, suddenly collapsed. Seventy-five vehicles were on the bridge, and 46 people were killed. The bridge was suspended from two giant chains made of links 50 feet long, rather than the more standard round wire cables. The design made thorough inspection of the links in the chain difficult, at the same time it encouraged greater than normal corrosion. Over the years, undetected corrosion created cracks that eventually weakened one link to the point where it broke. That shifted the load on the rest of the supports and triggered a rapid progressive collapse of the bridge.[51] Like corrosion, the embrittlement of nuclear reactor vessels (discussed earlier) is also a slow-acting, subtle process that threatens structural strength. Since we have far less experience with embrittlement than with corrosion, it is easier to see why its effects have been dangerously underestimated by some reactor designers.

Because making assumptions about operating conditions is risky but unavoidable, it is common practice to design for "worst case" conditions. That way, any errors made are less likely to cause the design to fail. But what is the worst case? Is it, for example, the worst earthquake that has ever been recorded in the area in which a bridge is being built? Or is it the worst earthquake that has ever been recorded in any geologically similar area? Or perhaps the worst earthquake that has ever been recorded anywhere?

As the assumptions escalate, so does the cost and difficulty of building the bridge. Does it make sense to bear the cost for making all bridges able to withstand the same maximum earthshaking, knowing that few if any of them will ever experience such a severe test? The definition of "worst case" is therefore not purely technical. It almost inevitably involves a tradeoff between risk, performance and cost. Even if the most extreme assumptions are made, there is still no guarantee that more-severe conditions than had been thought credible (such as a worse earthquake than anyone had predicted) won't someday occur. During the great Kobe earthquake of 1995, at least 30,000 buildings were damaged or destroyed, 275,000 people were left homeless and the death toll passed 4,000.[52] Japanese earthquake engineers, among the best in the world, were shocked by the extent of the damage. But ground motions in the quake were twice as large as had been expected. In the words of an American structural engineering expert, the Japanese structures "will perform well during an earthquake that behaves according to their design criteria. But . . . [this] quake did not cooperate with the Japanese building codes."[53]

Engineers also try to insure proper performance by building safety factors into their designs. The safety factor is the demand on a component or system just great enough to make it fail, divided by the demand that the system is actually expected to face. If a beam able to bear a maximum load of 10,000 pounds is used to bear an actual load of 2,000 pounds the safety factor is five; if an air traffic control system able to handle no more than 60 flights an hour is used at an airport where it is expected to handle 40 flights an hour, the safety factor is 1.5. Safety factors are the allowed margin of error.

If we know how a component or system performs, and we have an established probability distribution for the demands it faces, we can calculate the factor of safety that will provide any given degree of confidence that the design will not fail. But where the component or system design is innovative and the demands it may face are unknown or subject to unpredictable variation, there is no science by which it is possible to calculate exactly what the safety factors should be. Just as in worst-case analysis, we are back to projecting, estimating, guessing how much is enough. Unfortunately, that is the situation in constantly evolving, complex high-tech systems. When the design must face unpredictable environments *and* rivals actively trying to make it fail—as is the case in military systems—the problem of preventing failure is that much more difficult.

Peculiar confluences of circumstances can also defeat a design. Engineers examining the remains of the sections of the Nimitz Freeway destroyed in the Northern California earthquake of 1989 found evidence of just such a possibility. The frequency with which the ground shook might have matched the resonant frequency of the highway. In other words, after the first jolt, the highway began to sway back and forth. By coincidence, the subsequent shocks from the earthquake may have been timed to give the highway additional shoves just as it reached the peak of its swing. Like pushing a child's swing down just after it reaches its highest point, the reinforcing motion caused the highway to sway more and more until it collapsed.[54] It would have been hard for engineers to foresee this odd coincidence when the highway was being designed.

On July 13, 1977, New York City suffered one of the worst power failures in its history. It began when lightning struck a transmission tower north of the city and short-circuited two transmission lines. Transmission towers are designed to ground most strikes. This time, the grounding was ineffective. Circuit breakers tripped, as they were supposed to, to isolate the power surge. But they failed to close again after it passed. A remarkable series of failures of switches and circuit breakers then occurred, resulting in the loss of three major transmission lines and the utility's most heavily loaded generator—all from one lightning strike. Then another lightning bolt hit another transmission tower, short-circuiting two more lines. Further failures ensued, compounded by errors made by operators in the utility's control room. The problems snowballed. An automatic load-shedding mechanism began disconnecting customers as it was designed to do, but it also caused an unexpected surge in voltage that knocked another major generator out of service. And that was that. Almost exactly one hour after the first lightning strike, New York City went dark.

"How could the . . . automatic load-shedding system . . . produce such unexpected, and disastrous, results? Largely because . . . [the utility's] engineers never dreamed their system would be reduced to such a small island. So they never bothered to analyze what would happen to system voltages after automatic load shedding on an isolated system."[55] In other words, as in the Kobe quake, the real situation exceeded the worst-case scenario engineers imagined likely enough to be worth considering. After all, two towers both with faulty grounding struck by lightning within 20 minutes, an astonishing series of equipment failures, multiple operator errors and other serious control-room problems—at some point the whole thing does begin to sound wildly implausible. But of course, this bizarre, implausible scenario is exactly what happened.

In the United States, big earthquakes are much less common in the East than in the West, so those who design most structures built in the East are not required to include severe earthquakes in their worst-case scenarios and typically do not do so. Yet the most powerful earthquake ever known to hit the United States occurred in the eastern half of the country, at New Madrid, Missouri, near the Tennessee-Kentucky border in 1811. It was later estimated to have had a magnitude of about 8.7 on the Richter scale, and was one of three quakes in that area in 1811-12, all stronger than magnitude 8.0.[56] The Richter scale is logarithmic (base 10). Thus, those nineteenth century eastern quakes were more than ten times as powerful as the magnitude 6.9 earthquake that brought down part of the Nimitz Freeway and did so much damage to San Francisco in the fall of 1989. (The New Madrid quake was nearly 100 times as strong.) They temporarily forced the Mississippi River to run backward, permanently altered its course, and were felt as far away as Washington, Boston and Quebec.[57]

All along the eastern seaboard there are geological structures similar to those responsible for a 7.0 earthquake that devastated Charleston, South Carolina, in the late nineteenth century. Milder quakes of magnitude 5.0 are not all that rare in the East. Such a quake hit New York City in 1884. Seismologists have estimated that there is about a 50-50 chance of a much stronger 6.3 quake on the New Madrid fault by the year 2000.[58] A study of New York City's vulnerability to earthquake found that a magnitude 6.0 temblor centered within five miles of City Hall would do about $25 billion worth of damage; even a quake more than three times as far away could cause an estimated $11 billion in direct damage to buildings. Estimates of damage in Northern California from the 6.9 quake in 1989 ran from $4 billion to $10 billion.[59] So a New York City–area earthquake about one-tenth as powerful as the California quake could cause up to six times as much property damage.

The reason is partly geological: the earth's crust is older, colder and more brittle in the eastern United States. But it is also because designers of structures in New York do not typically include severe groundshaking in their assumptions. By contrast, designers of structures in California have been compelled to take earthquakes into account. The modern skyscrapers in San Francisco swayed during the 1989 quake, but sustained little or no damage. Virtually every structure that suffered major damage there had been built before stringent earthquake-resistance requirements were incorporated in building codes over the last 15 to 20 years.[60] The assumptions about operating environment made by designers really do make a striking difference in how products perform.

Five large nuclear reactors used to produce plutonium and tritium for American nuclear weapons were designed and built without strong steel or reinforced-concrete "containments."[61] The last line of defense against the accidental release of dangerous radioactive materials, containments are built as a matter of course around civilian nuclear power reactors in the United States (and most other developed nations). The containment surrounding the Three Mile Island reactor prevented a large release of radioactivity such as the one that occurred at Chernobyl where there was no containment. Why were America's nuclear weapons production reactors designed and operated without this key safety feature? Ignorance may be a partial explanation. All of these reactors are "old" in terms of the nuclear business, dating back to the 1950s. The dangers of radiation were much less well understood then, and exposure to radioactive materials was often treated much too cavalierly. For example, it was during the 1950s that American soldiers were ordered to crouch in ditches during nuclear bomb tests in the atmosphere and then made to march or drive right through the point of detonation. But the 1950s was also the era of McCarthyism and of intense Cold War fears and hatreds. It is likely that the pressure to get bomb production moving, to keep ahead of the "godless Communists," also played an important role. Technical matters are not the only considerations that enter into the designers definition of worst case.

There are many sources of error inherent in the process of designing complex technical systems. With great care, the use of fault-tolerant design strategies and thorough testing, it is possible to keep such flaws to a minimum. But even the most talented and careful designers, backed up by the most extensive testing programs cannot completely eliminate design errors serious enough to cause catastrophic technical failures. This is especially true where complicated, innovative or rapidly changing technologies are involved. Many dangerous technologies are of just this kind.

MANUFACTURING AND TECHNICAL FAILURE

Even the best designs for technical systems remain only interesting ideas until they are made real by manufacturing. The process of fabricating major system components and assembling the systems themselves creates ample opportunity for error. Flawed manufacturing can translate the most perfect designs into faulty products. The more complex the system, the more sensitive and responsive it must be, and the more critical its function, the easier it is to introduce potential sources of failure during manufacture. Subcomponents and subassemblies must be checked and rechecked at every step of the way. Still serious errors persist.

The nuclear power industry provides innumerable examples of how easy it is for slips to occur during manufacturing. On October 5, 1966, there was a potentially devastating meltdown at the Enrico Fermi Atomic Power Plant, only 30 miles from Detroit. In August 1967, investigators discovered a piece of crushed metal at the bottom of the reactor vessel that they believed had blocked the coolant nozzles and played a key role in the accident. In 1968, it was finally determined that the piece of metal was one of five triangular pieces of zirconium installed as an afterthought by the designer for safety reasons. They did not even appear in the blueprints. This particular shield had not been properly attached.[62] Only a few weeks after the accident at Three Mile Island (March 1979), the Nuclear Regulatory Commission reported that they "had identified thirty-five nuclear power plants with 'significant differences' between the way they were designed and the way they were built."[63]

The multibillion-dollar nuclear power plant at Diablo Canyon in California sits close to an active fault. During construction in 1977, the utility hired a seismic-engineering firm to calculate the stresses different parts of the plant would have to withstand in an earthquake. A little more than a week before the plant was due to open, a young engineer working for the utility that owned the plant discovered a shocking error. The utility had sent the diagrams of the wrong reactor to its seismic consultants! Guided by the faulty stress calculations that resulted, the utility reinforced parts of the plant that did not need reinforcing, while vulnerable parts were reinforced too little, if at all. More than a hundred other flaws in the reactor's construction were subsequently discovered.[64]

In an attempt to stop the continuing release of radiation after the accident at Chernobyl, the Soviets built what amounted to an after-the-fact containment around the burned-out reactor—a ten-story concrete and steel "sarcophagus" with walls some 20 feet thick. Designed to last 30 years, it was constructed so quickly and under such dangerously radioactive conditions that the quality of work was poor. Only ten years later, the metal had rusted and the concrete was riddled with cracks. The whole structure was in serious need of repair.[65]

By mid 1990, the U.S. Air Force had over 1,700 nuclear warhead–carrying air-launched cruise missiles in its arsenal. These 3,000 pound, 21-foot-long, state-of-the-art missiles are designed to be carried under the wings of high-flying strategic bombers. When launched, they drop down close to the ground to avoid radars as they fly toward their targets. High-speed, ground-hugging flight requires a highly accurate navigation system to keep the missile on its preset path. Northrop manufactured the guidance system, which utilized a series of gyroscopes surrounded by a viscous fluid known as DC-200 (produced by Dow Chemical Corp.). Previously secret Pentagon and corporate documents released in late July 1990 raised doubts as to whether the guidance system, and thus the missile itself, was reliable. Apparently, prolonged exposure to extreme cold caused the fluid to interfere with the spinning of the gyros. A military-wide standard requires that the missiles must be able to function at minus 65 degrees Fahrenheit, a common temperature outside aircraft flying at or above the normal cruising altitudes of long-range commercial airliners (34,000 to 37,000 feet). An internal 1987 Northrop report stated flatly, "DC-200 does not meet the -65 ° F requirement and never did." It was corroborated by tests performed by Boeing (the missiles' prime contractor) in early 1989, in which six of nine gyroscopes failed after being kept at -65 ° F for up to two hours; only when the temperature rose to minus 40 degrees would DC-200 thaw enough to allow every gyro to spin freely.[66] The *Wall Street Journal* maintained, "The Air Force and Northrop Corporation . . . have gone to great lengths to mask the problem. . . . [I]nstead of fixing the part, the Air Force simply decided to make the test less stringent."[67]

It was not until three and a half months after the Hubble Space Telescope was launched in April 1990 that NASA finally determined the cause of the perplexing spherical aberration (curvature error) problem that had made it incapable of performing the full range of tasks for which it had been designed. When the Hubble mirror had arrived for final polishing at Perkin-Elmer's Danbury, Connecticut, plant back in 1979, it was tested with a newly developed, superaccurate tester to assure that the mirror's optical properties met NASA's exacting standards. The tester showed that the mirror had a small degree (one-half wavelength) of spherical aberration, well within acceptable limits for that stage of manufacture. The Perkin-Elmer team then began the final polishing process (which continued until 1981), polishing out the deviation their new tester had found. The only problem was that there was an undiscovered one millimeter error in the structure of the tester. By using it to monitor the polishing process, Perkin-Elmer had distorted rather than perfected the mirror's surface during final polishing, creating the spherical aberration that was later to produce such headaches in the orbiting telescope.

Interestingly, the mirror had been checked with a testing device of more standard design before it was shipped to Danbury. That device had *not* shown the degree of spherical aberration that the newly developed tester had (incorrectly) detected. The company's scientists had the results of the earlier test, but were sure that the more sophisticated tester was correct. They did not bother to conduct further tests or investigate the discrepancy.[68] *Science* reported that "astronomers experienced in making ground-based telescopes say they are appalled that NASA and Perkin-Elmer would rely on one single test. . . . [T]here are any number of simple and inexpensive experiments that could have seen the spherical aberration that now exists in Hubble."[69] In the excitement of meeting the kinds of technological challenges involved in designing and building complex, state-of-the-art systems, mundane matters, such as "simple and inexpensive" checks during the manufacturing process, are easy to overlook. When we have our eyes on the stars, it is all too easy to trip over our own feet.

COMPONENT FLAWS AND MATERIALS FAILURE

It is impossible to make high-quality, reliable products out of poor-quality, unreliable parts or seriously flawed materials. Semiconductor chips have become almost a raw material to the electronics industry. Their performance and reliability is the bedrock on which the performance and reliability of modern electronic equipment is built. In the latter part of 1984, officials at Texas Instruments (the largest chip manufacturer in the United States at the time) disclosed that millions of integrated circuits it had manufactured might not have been tested according to specification. The chips had been built into more than

270 major weapons systems by 80 different contractors. The attendant publicity—and a Department of Defense investigation—led officials at Signetics Corporation (the nation's sixth-largest microchip manufacturer) to audit their own chip-testing procedures. They concluded that as many as 800 different types of microcircuits they had supplied to military contractors might not have been tested properly. At least 60 different types of microchips sold to the Pentagon by that bulwark of the computer industry, IBM, were also determined to have "confirmed problems."[70] In the spring of 1985, the Pentagon's inquiry found that irregularities in testing military-bound microchips were pervasive in the electronics industry. The director of the industrial productivity office at the Department of Defense put it this way: "What we found was that it was common practice for the microcircuit makers to say 'Yeah, we'll do the tests,' and then for them never to conduct them."[71]

The fact that microchips are not properly tested does not mean that they are faulty. What it does mean is that the reliability of those components, and therefore of every product that contains them, cannot be assured. Because dissassembly and retesting costs would have been high, reports at the time indicated that there was very little, if any, retesting of the suspect microchips that had already been built into weapons systems. But the fact that such a basic part of so many military systems might be unreliable also has its costs. If these systems fail at critical moments, the consequences can be disastrous. On June 3, 1980, the failure of a single, 46-cent microchip generated a major false warning that the United States was under land- and sea-based nuclear attack by Soviet missiles. Three days later, the same faulty chip did it again.[72]

Early in the summer of 1994, Intel Corporation, by then the standard-setting computer chip manufacturer, discovered a flaw in its much touted and widely used Pentium chip.[73] The chip could cause computers in which it was embedded to give wrong results in certain division problems that used the chip's floating-point processor.[74] Intel waited until November to publicly disclose the problem, provoking an avalanche of angry messages on the Internet from engineers and scientists disturbed by what they considered to be the company's cavalier attitude. A computer that gives the wrong results "silently" (that is, without any indication that anything is wrong) is no small thing. It is not just annoying or misleading, it is potentially dangerous. It does not take a great deal of imagination to see how a computer that gives the wrong results because of a flawed processor could cause a lot of damage if it were used to design, analyze, control or operate a dangerous-technology system. The June 1980 false warning of nuclear attack is only one of many frightening possibilities.

In September 1989, the Pentagon was forced to delay deployment of new rockets for Air Force fighters and replace components in rockets carried by Navy fighters because of defects in circuit boards costing $12 each. The boards, manufactured by Asher Engineering, were built into "stator switches" that help arm the warhead before the missile is fired. According to Congressman John Dingell of Michigan, "In our first raid on Libya, we used both HARM and Harpoon missiles which contain Asher stator switches. . . . During that raid, at least 25 percent of the HARM's and Harpoons did not detonate." Since those missiles were never recovered, there is no way to know whether the substandard switches were the reason for the warheads' failure to explode.[75]

Metallurgical defects are also a major potential source of systems failure. They seem to have played a key role in the July 19, 1989, crash of United Airlines Flight 232 near Sioux City, Iowa, discussed earlier. Apparently, a flaw in a 270-pound cast-titanium-alloy disk in the rear engine grew into a crack that broke the disk apart, shattering the plane's tail section. Metallurgists working for the National Transportation Safety Board found a tiny cavity in the metal that grew into the fatal crack. The cavity was large enough to be "readily visible with the unaided eye," raising questions as to why it was not detected either at the factory or during routine maintenance.[76] It was also discovered that the manufacturer had mistakenly given two disks made at the same time the same serial number. One had failed to pass inspection, and one was destroyed. Investigators thought that the good disk might have been destroyed, and the faulty one installed in the DC-10. The company managed to convince them that that was not true.[77]

Flaws in even the simplest of components can cause very complex technical systems to fail. On April 28, 1989, ABC News reported that "every year in this country, companies buy some $200 billion of nuts and bolts . . . and put them in everything from jet planes to children's amusement rides. . . . We now know that billions of bad bolts have come into this country." Peterbilt Trucks issued a recall because of a number of its customers reported that the steering mechanism on their vehicles would suddenly stop working, causing a virtually total loss of control. Defective bolts were found. They were brittle because they had not been properly heat treated during manufacture. According to ABC, some bolt distributors had become aware of the problem, but few said anything about it to anyone, including their customers. ABC reported, "Twice in the past year and a half, bolts holding jet engines on commercial airliners broke in flight, and the engines then fell off the airplanes. . . . The State of Louisiana noticed that the bolts that hold the Calcasieu River Bridge together were breaking off. It temporarily closed the bridge and replaced the bolts."[78]

Of course, bolt manufacturers also supply the U.S. military. Pentagon officials reassured ABC in writing that they were not aware of any "death or injury attributable to substandard fasteners installed in military equipment." But Army documents ABC obtained showed that defective or broken nuts and bolts were involved in 11 aircraft (mostly helicopter) accidents over the previous decade in which 16 people had died. Over a thousand M-60 tanks were temporarily taken out of service because of defective bolts.[79] ABC News also reported that NASA had spent millions of dollars removing suspect bolts from key systems, like the space shuttle engines. There were reports that bad bolts had been supplied for the Air Force's MX missile.[80]

THE CRITICALITY OF MAINTENANCE

There is no more mundane issue in high-technology systems than maintenance. Yet without proper maintenance, the best-designed and most carefully built system can slowly turn into a useless piece of high-tech junk. The "if it ain't broke, don't fix it" attitude is a prescription for endless, expensive trouble. Complex, sensitive systems often require extensive and painstaking preventive maintenance, not just after-the-fact repair.

More than half a decade after planning for its $30 billion space station began, NASA uncovered a serious problem with the project's design. In January 1990, Richard Kohrs, the head of the space station program told a gathering of contractors, "This program has too much EVA [extra vehicular activity, that is, space walks] . . . they're talking about 1,700 man-hours of EVA a year just for maintenance. . . . If that's true, we don't have the right design." A few months later, a special NASA team (working with estimates of failure rates derived mainly from NASA agencies and contractors) projected that more like 2,200 hours of space walks each year would actually be needed to maintain the station's 6,000 parts.[81] NASA subsequently boosted the maintenance estimate further to 3276 hours per year—about 9 astronaut-hours per day (including preparation time).[82] Space walks are dangerous. There are radiation risks and possibilities of injury from fast-moving space debris or micrometeorites. They are also expensive and time consuming. From the beginning of the American space program to 1990, astronauts had only accumulated about 400 hours of EVA. The NASA team argued that the enormous amount of maintenance EVA required each year would divert so much time from key construction tasks that it might never be possible to complete the station.[83]

It is clear enough that the problems of space station maintenance are not merely theoretical. The USSR orbited the core module for its much simpler *Mir* space station in the mid 1980s. By 1990, construction was two years behind schedule, at least in part because Soviet astronauts had to spend an inordinate amount of time maintaining and repairing the 80-ton, 100-foot-long orbiting laboratory.[84] In June 1997, a collision between *Mir* and an unoccupied cargo vessel punctured part of the space station and knocked out much of its power supply. *Mir* limped along throughout that summer, with repeated power outages, computer failures and other problems focusing the attention of ground controllers and its occupants more on maintenance and repair than on its mission.[85]

In April 1988, the mechanics responsible for maintaining Aloha Airlines's fleet were routinely inspecting one of the airline's Boeing 737s. They failed to note that a section of the upper fuselage was starting to come loose, and that the overlapping metal skins in that section were beginning to develop fatigue cracks around the rivets. On April 28, in flight, the cracks suddenly began to grow, connecting to each other and literally ripping the top off the body of the plane. A flight attendant was killed, but the pilot was able to bring a plane full of very frightened passengers down for an otherwise safe landing.[86]

Boeing's 737 jet airliner has a long record of effective service. But it wasn't designed to last forever. It was supposed to fly for 20 years, 51,000 flight hours or 75,000 takeoffs and landings. The Aloha Airlines plane had taken off and landed nearly 90,000 times—20 percent beyond its design life. By mid 1990, almost 20 percent of the 737s in use across all airlines were more than 20 years old. Airline officials argued that careful maintenance procedures, along with upgrades to the planes, could keep them flying well beyond their original design life.[87] In May 1998, the FAA issued an emergency order grounding dozens of older Boeing 737s because of a possible maintenance problem involving their fuel pump wiring.[88] The wiring on some 35 of the planes inspected in the first few days after the FAA found the problem showed some wear, and in 9 or more aircraft the insulation was worn at least halfway through.[89] Worn insulation can lead to sparks, and sparks and jet fuel are a deadly explosive combination.

The 1979 crash of a DC-10 on takeoff from Chicago's O'Hare Airport was traced to a tear in the pylon connecting one of the aircraft's engines to its wing. Rough handling that resulted from shortcut procedures used during routine maintenance was apparently the culprit.[90] The 1989 explosion of the United Airlines DC-10 near Sioux City, Iowa, discussed earlier was also attributed to a maintenance problem: flawed inspection that failed to detect a visible cavity in the rear engine's titanium-alloy disk. And faulty maintenance—poorly done repair work—was cited as the primary culprit in the worst single aircraft disaster in history, the death of 520 people in the crash of a Japan Airlines 747 northwest of Tokyo in August 1985.[91]

Nuclear power plants are designed to "fail safe." Any major problem, including a loss of power to the controls, is supposed to trigger an automatic shutdown, or "scram," of the reactor. Industry analysts have argued that the odds of a failure of this system are no more than one in a million reactor years. Yet on February 22, 1983, the Salem-1 reactor failed in just this way, refusing to halt the fission reaction in its core when ordered to scram by a safety control system. Three days later, the "one in a million" event happened again—at the same reactor.[92]

A key problem lay in a huge pair of circuit breakers, known as DB-50s (manufactured by Westinghouse) in the circuit supplying power to the mechanism that raises and lowers the core control rods. When the power is flowing and the breakers are closed, the rods can be held up out of the core to speed the fission reaction. But when the automatic system orders the reactor to scram, the DB-50s break the circuit, and gravity pulls the rods down into the core, shutting down the reaction. Investigators of the incidents at Salem-1 found that a UV coil inside the DB-50s had failed. As early as 1971, Westinghouse had issued technical bulletins warning of problems with the UV coil, and in 1974 had sent out letters emphasizing the importance of properly cleaning and lubrication of the coils twice a year. The utility did not heed the warnings. "Maintenance of the breakers at Salem was poor. . . . They never got the critical attention they deserved. . . . [T]here was *no* maintenance of the UV coils between their installation in the 1970s and August 1982, when they began to fail repeatedly."[93]

At the Maine Yankee nuclear power plant, high-pressure radioactive water is pumped through metal tubes 3/4 inch in diameter and 1/20 inch thick after being heated by the reactor core. Heat conducted through the walls of the tubes turns "clean" water into "clean" steam that drives the turbine, generating electricity. Proper inspection and maintenance of the tubes is critical, since any cracking or rupturing could allow the radioactive water from the reactor to leak into the otherwise "clean" steam and possibly escape to the environment. In early 1995, it was disclosed that about 60 percent of the plant's 17,000 tubes had severe cracks. Furthermore, the reactor had been operating in that dangerous condition for years.[94]

NRC Commissioner James Asseltine expressed grave concern that weaknesses in preventive maintenance, equipment reliability problems and other related difficulties are pervasive in nuclear industry. In his view, "The bottom line is that, given the present level of safety . . . we can expect to see a core meltdown accident within the next twenty years, and it is possible that such an accident could result in off-site releases of radiation which are as large as, or larger than, the releases estimated to have occurred at Chernobyl."[95]

Maintenance cannot be an afterthought in the kind of sophisticated, complex technological systems that a modern military expects to work well even in difficult operating conditions. Given the speed of modern warfare, equipment that fails often because it is poorly maintained or because its design is so inherently complex that it cannot be properly maintained is worse than useless. It can lead to military tactics and strategies that amount to fantasies, because they are built around equipment that won't perform as advertised—if it works at all—in the real world of combat.

Air Force analyst F. C. Spinney's appalling data on the reliability of fighter planes (given in Table 9-1) make it clear that this is not just a theoretical concern. These are not even data for planes in combat. Focusing on the F-15, Spinney makes the point that there is a difference between the way maintenance problems look "on the drawing board" and the way things work out in the real world. The fighter has built-in test equipment on board that lets the pilot or flight-line crew chief know that there has been a failure in a specific "line replaceable unit" (LRU). The LRU is then removed and taken to an "Avionics Intermediate Shop" (AIS) to be repaired. Meanwhile, another LRU is plugged into the plane and it is ready to fly. This approach is used to simplify flight-line maintenance. But according to Spinney, "the maintenance task is aggravated by long test times. . . . To hook up an LRU to the computer . . . can be a time-consuming task . . . sometimes taking up to 30 minutes. The computer then checks out the LRU— again a time-consuming task, averaging about three hours, but sometimes taking as long as eight hours. Since the computer is limited to hooking up and checking one LRU at a time, no other LRUs can be checked out during this period."[96] And after all that, the computer can't find any problem with the LRU 25 to 40 percent of the time. The whole maintenance process was thus useless.[97]

Maintenance, lackluster and pedestrian as it may seem, requires the closest attention. From aircraft to spacecraft to nuclear power plants, there is persuasive evidence that improper maintenance can lead to dangerous failures of technical systems. Far from being a mere footnote in the age of high technology, it is critically important to the performance of the most sophisticated technical systems.

The nature of technical systems themselves and the nature of their interactions with the fallible humans who design, build and maintain them guarantee that it is not possible to eliminate all causes of failure—even potentially catastrophic failure—of complex and critical technical systems. There is nothing about dangerous-technology systems that makes them an exception to this rule.

There are those who believe that it is possible to prevent failure of the most complex dangerous technologies by automating humans out of the system and putting computers in control. Computers are surrounded with an almost magical aura of perfection, or at least perfectibility, in the minds of many people. We sometimes think—or hope— that they can help us overcome the imperfection that is so much a part of our human nature. After all, they can bring commercial aircraft safely to the ground with remarkable efficiency and nuclear warheads to their targets with remarkable accuracy. In the next chapter, we take a closer look at these marvels of modern technology and try to understand why they are anything but a route to solving the problems of either technical failure or human error.

CHAPTER TEN

Computers
and the Technological Fix

People often think of computers as a possible solution to the unreliability of human beings. Computers don't drink or take drugs, they don't have family problems and they don't get bored repeating the same task over and over again. At certain times for certain tasks, computers *are* much more reliable than people. But they can also be much less reliable. They lack common sense, good judgement and a sense of morality. Computers can magnify rather than overcome human error. When the captain of an American Airlines flight to Cali, Colombia, typed the wrong code into the plane's navigational computer in December 1995, the computer took the plane toward Bogota instead and flew it straight into a mountainside. A computer had turned a minor typo into a death sentence for 159 people.[1]

Modern military systems depend heavily on computers. All indications are that this dependence will continue to grow. Computers control more and more of the equipment used to manufacture weapons to increasingly exacting specifications, from sophisticated metalworking machinery to high-tech inspection and quality control devices. They are an integral part of the communications, navigation and command-and-control systems critical to carrying out all aspects of military operations, from launching nuclear attack to directing forces on the battlefield to supplying them with food, fuel and ammunition.

In some cases, computers enhance the performance of weapons systems; in others, the weapons could not perform without them. Many modern missile systems could not be controlled, launched or accurately guided to their targets without computers. Some modern military aircraft designs are so aerodynamically unstable that they need constant, high-speed adjustment of their controls just to keep them flying—a task that would overwhelm an unassisted pilot, but is just what computers do best. Much touted as an ultramodern aircraft, the B-2 Stealth bomber is actually based on a decades-old design that its manufacturer, Northrop, first tried in the early 1940s. Called the "flying wing," the design was abandoned less than a decade later, when the latest jet-powered version of the plane (the YB-49) proved to be extremely unstable and difficult to fly. The design was resurrected in the 1980s when it was thought that computer control technology had advanced enough to give some confidence that the plane could actually be flown properly.

From the military point of view, computers are ideal soldiers. They follow orders without a will of their own, without raising moral questions, without exercising independent judgement. Programs are simply sets of orders, and computers follow them robotically. On the other hand, if programs are flawed or have been covertly altered to instruct computers to do undesirable things (as by a virus), they will also follow those instructions robotically, just as if they were following correct and desirable orders.

COMPUTER HARDWARE RELIABILITY

Because computers are so often used to control other systems, their reliability takes on an extra measure of significance. When control systems fail, they take the systems they are controlling with them. Failures of control can have consequences that are both serious and difficult to predict. In most machines, a worn gear or wobbling wheel might only cause performance to suffer. But when something goes wrong with a computer, it doesn't just become a little sluggish, it tends to freeze, crash or even worse—do entirely the wrong thing.[2]

Hardware failures can be caused by flawed designs or by faulty components. Design flaws include errors in both the design of the computer hardware itself and of the process used to manufacture it. When a manufacturing error results in a faulty component that causes trouble, the problem can be fixed by replacing it with a new "unbroken" copy of the same part. But when a bad design causes trouble, replacing the component with another copy won't help because the same design is used in all the copies. Only redesign will fix the problem. Because of our experience with devices other than computers, we tend to think of an equipment breakdown as the result of a faulty component rather than a design flaw. For most devices that is true. But computers are very different. While faulty components still can and do create havoc from time to time, hardware and software design flaws are much more often the culprit. Apart from the occasional recall of automobiles, the general public has little experience with significant design flaws in the products they use day to day. Until recently, most consumer products have been relatively simple. Manufacturers have been able to test them repeatedly and eliminate serious design flaws before the products ever reached their customers. The movement of computers into the consumer marketplace profoundly changed that situation. From their experience with other products, many people still seem to believe that when a computer is not working properly, some part has cracked or deteriorated. In fact, "the computer is down" rarely means that anything physical has snapped or worn out, or even that the software is somehow "broken." The computer is probably doing exactly what the programmer told it to do. Most likely, the problem is that the programmer either gave it incomplete instructions or told it to do the wrong thing.[3]

When the largest computerized airline reservation system in the United States went down on May 12, 1989, the eight IBM mainframe computers that are the heart of the system kept right on working. The problem, which disrupted operations at airline ticket counters and some 14,000 travel agencies nationwide for 12 hours, was the result of a software failure. The hardware was working just fine. The computers hadn't even lost any information on ticketing, flight schedules, and the like. A software problem simply prevented them from accessing it.[4]

While all this may be true, it should not be assumed that mundane problems like defects in materials or faulty manufacturing won't cause devices as sophisticated and "high tech" as computers to fail or become unreliable. The complexity and sensitivity of computers makes them more vulnerable to these problems than "lower tech" equipment. As with other sensitive electronics, a bit of dust in the wrong place, excess humidity, microscopic defects in materials or flaws in manufacturing microminiaturized semiconductor chips can build faults into computer components or cause them to fail unexpectedly in operation. An accumulation of lint, about as mundane a problem as they get, caused both the primary and backup computer display units in the telescope-pointing system aboard the space shuttle *Columbia* to overheat and fail in December 1990.[5]

Besides defects in the physical process of manufacturing, faulty programming may be "hard coded" into computer chips during manufacture. The unreliable chips then make the computers that contain them unreliable. It was in the summer of 1994 that leading chip manufacturer Intel Corp. discovered the flaw in its widely publicized, state-of-the-art Pentium chip that created a furor among engineers and scientists, as discussed earlier. The problem was in a part of the chip called the floating-point processor. It caused errors in long division when certain strings of digits appeared in the numbers being divided. Intel, which had shipped perhaps two million of these faulty chips, had encountered more serious problems in its earlier generation 386 and 486 chips. The storm the Pentium flaw caused was due more to the company's behavior than to the seriousness of the flaw itself--- Intel took months to make the flaw public and was slow to offer replacements. The failure to promptly notify all users could have had serious consequences if the incorrect calculations were used as the basis for designs or control systems that involved critical technologies.[6] Three years later, in November 1997, yet another serious design flaw was found in the Pentium processor. When a particular command was executed, the defect caused the processor to come to a complete halt.[7]

Hardware failures of some consequence can also be caused by apparently minor accidents—the proverbial cup of coffee spilled in the wrong place at the wrong time. On December 9, 1988, "An adventurous squirrel touched off a power failure . . . that shut down the National Association of Security Dealers automatic quotation service [NASDAQ] for 82 minutes." When power was restored, a surge put NASDAQ's mainframe computers out of action and damaged the utility's equipment so that even the backup generators could not be used.[8] The problems prevented the trading of some 20 million shares of stock and halted options trading at a number of major exchanges. This was not the first time NASDAQ's computers had serious problems. The whole system had been shut down for four hours by a hardware failure in October 1986. For that matter, it was not the first time the utility involved had to deal with rambunctious squirrels either. Squirrels had been chewing on their equipment and knocking out service two or three times a day.[9] That may have been hard on the utility, but it was probably harder on the squirrels.

Given the frequency of hardware (and software) failures and the critical nature of so many computing applications, it makes sense to try to design computer systems that can continue to operate properly despite faults. Hardware fault tolerance is simpler to achieve than software fault tolerance. Spare parts and backup systems can be combined with controlling software that brings them on-line when they are needed. Computer scientist Severo Ornstein reports, "I was in charge of one of the first computer systems that used programs to manipulate redundant hardware to keep going even in the face of hardware failures. . . . We were able to show that the machine would continue operating despite the failure of any single component. You could reach in and pull out any card or cable or shut off any power supply and the machine would keep on functioning."[10] In 1998, researchers at the University of California and Hewlett-Packard reported that an experimental "massively parallel" computer called Teramac, built by the company, was capable of operating effectively even though it contained "about 220,000 hardware defects, any one of which could prove fatal to a conventional computer."[11] That is encouraging, but it does not mean either that this success is translatable to all other computers and applications or that no hardware flaws could prove fatal to Teramac. And there is still the problem of common-mode failures.

Hardware and software *design* flaws are harder to tolerate. For one thing, the "triple modular redundancy" (TMR) technique discussed in the previous chapter would not work in the case of a design flaw. TMR involves three independent computers each of which separately gives the results of its analysis of the problem at hand. If only two of the three give the same result, it is assumed that the one that gave a different result is wrong, and the system acts on what the other two say. If there were a flaw in the design of the computers, all three identical modules would contain the same flaw. They could all agree, and all be wrong.[12] A generalized "spare parts" approach won't work either, because the part is not "broken," its design is faulty. Hardware design flaws can be very subtle and difficult to detect. They may lie hidden for a long time, until some peculiar combination of inputs, directions or applications causes them to generate a sudden, unexpected failure in what was thought to be a highly reliable, tried-and-true system. Ornstein cites a striking example:

> We were working on a program for a communications application which had to keep up with the flow of messages. We had selected a machine of which five hundred copies had been successfully used in a variety of applications . . . a mature, thoroughly tested design. Nonetheless, the machine contained a hidden design flaw. . . . [A] failure would be provoked if the random arrival of a message chanced to coincide with a very short time of a tick of the machine's clock—within about a billionth of a second, more or less. No one knew that the flaw was there because in the applications that had been programmed up to then external data arrived relatively infrequently, so that such a coincidence might occur only once every few years. But . . . we were driving the machine much harder and we observed unexplained failures once every two or three days. . . . [W]e traced that particular problem to a hardware design flaw—an improperly designed synchronizer.[13]

The same type of problems also happen with software. Different intricate programs or different parts of the same intricate program sometimes interact in surprising ways that produce completely unexpected and erratic results.

FLAWS IN THE DESIGN OF SOFTWARE

In writing software of any real complexity, it is virtually impossible to avoid programming "bugs," and nearly as difficult to find the ones you haven't avoided. So many things can happen in complex interacting systems, there are so many possible combinations of events, it is not reasonable to suppose that designers can take all of them into account.

Complexity can breed such an explosion of possibilities that the number of possible combinations could easily overwhelm even the processing speed of state-of-the-art computers. For example, suppose you tried to design a completely general chess-playing computer program by writing the software to select moves by searching through all of the possible sequences of moves in the game. Although chess is well defined and self-contained—in many ways much simpler than the mindboggling complexity of many ill-defined, loosely bounded real-world situations—there are some 10^{120} sequences of permitted moves. If a machine that was orders of magnitude faster than the fastest currently available had begun to evaluate all these possibilities when humans first began to roam this planet, it would barely have made a dent in this problem by today. Programmed in this way, the much simpler game of checkers would still require the computer to cope with 10^{40} (ten thousand trillion trillion trillion) combinations of moves.[14]

In most applications, it isn't necessary for complex programs to work perfectly. A telephone switching program that periodically but infrequently fails to handle a call properly may be annoying, but it's no great problem for most users. They can simply hang up and dial again. Even programs designed for critical applications need not be flawless, as long as whatever flaws they do have are irrelevant to the proper functioning of the system. Errors in the programming for nuclear power plants that cause trivial misspelling of words displayed in output messages fall into this category.

The problem is that there are *always* flaws, and until they are found there is absolutely no way of knowing whether they are trivial or critical. A hundred of the right kind of bugs in a complex piece of software can make little difference to the system's performance, but one critical bug buried deep within the program can cause a catastrophe. And a critical bug may not be easy to recognize even when you are looking at it. A period that should have been a comma caused the loss of the *Mariner* spacecraft.[15] During the 1991 Persian Gulf War, a software-driven clock whose programming resulted in a cumulative error of one millionth of a second per second caused a Patriot missile battery to fail to intercept the Iraqi Scud missile that destroyed an American barracks in Dhahran, Saudi Arabia.[16]

More perplexing are errors in reasoning that result in instructions that appear to do their job properly in the local part of the program which surrounds them, but interact in some unforeseen way with another part of the program to produce a serious failure. Analogous to "typos," programmers sometimes refer to simple errors in reasoning as "thinkos." They are disturbingly easy to commit.

With the approach of the year 2000, the computing community realized that a simple programming decision made decades earlier was about to cause an ungodly amount of trouble. It had become commonplace to write programs using a two-digit shorthand for the year, rather than the full four digits. So 1978 became 78, 1995 became 95, etc. The problem was that this shorthand did not allow computers to recognize any year beyond 1999. They would treat the year 2000, represented by "00," as the year 1900, with potentially catastrophic effects on payrolls, financial calculations and thousands of other programs supporting thousands of other functions. Articles from the computer industry trade press carried "images of satellites falling from the sky, a global financial crash, nuclear meltdowns . . . the collapse of the air traffic control system and a wayward ballistic missile."[17]

In fairness to the programmers of earlier decades, the decision that caused all this trouble made sense at the time. They were trying to conserve memory at a time when memory was very expensive. They were also certain, given the speed of development of the computer industry, that by the time the year 2000 rolled around, memory would be much cheaper and the programs they had written would be ancient history, long since retired. Computer technology did advance enormously, and memory did become much, much cheaper. But some of the "dinosaur" programs they had written were still being widely used and their decision to use two-digit years had become unthinking common practice. By some estimates, as the year 2000 approached, up to half of the world's computer data was still interacting with programs that could only recognize the decade of a year, but not its century.

One part of this "year 2000" (Y2K) problem that has received relatively little public attention is that an unknown number of the more than 70 billion computer chips built into everything from thermostats to pacemakers to nuclear submarines since 1972 are hard coded with programming that allows only two digits for the date. In late 1998, a report issued by the Nuclear Regulatory Commission found that the Seabrook nuclear power plant had more than 1,300 "software items" and hard-coded computer chips with Y2K problems. At least a dozen of them, including an indicator that measured the reactor's critical coolant level, had "safety implications.[18]

The problems created by the Y2K "millennium bug" are not technically challenging. But their solution requires an enormous amount of boring, pedestrian work, going through software programs line by line and methodically reprogramming them. The affected computer chips cannot be reprogrammed. They have to be ferreted out and replaced, one at a time. It is a massive undertaking. By the time the dust settles, by one estimate the worldwide cost will range from $300 to $600 billion.[19] It is an interesting question whether the early decision to save money by using only two digits for the date will actually wind up costing more and causing more trouble than it saved.

Although there are fundamental problems of fallibility that forestall the development of error-free software, it is undoubtedly true that the vast majority of software errors are the result of less-than-brilliant programming. Even the best, brightest and most careful programmers cannot be expected to write software of any real complexity perfectly. But the truth is, most programming is not written by the best, brightest and most careful programmers, nor can it be. No matter how much effort is put into training, whenever large numbers of people are needed to do a job, it is certain that only a relatively small percentage of them will do it superbly well. This is yet another unavoidable problem.

Unlike most engineering products, when software is offered for sale it normally still contains major design flaws, or "bugs." Subsequent "improved" versions may still have some of the original bugs, as well as new bugs introduced when the improvements were made. David Parnas, a computer scientist with decades of experience in software engineering explains, "Software is released for use, not when it is known to be correct, but when the rate of discovering new errors slows down to one that management considers acceptable."[20] As software is tested, the rate at which new bugs are discovered approaches, but does not reach, zero. Ornstein points out, "Eliminating the 'last' bug is . . . a standing joke among computer people."[21] Computer scientist Fred Brooks argues that the rate at which bugs are discovered *after* a new version of a program is in the hands of its users tends to follow a U-shaped curve: "Initially, old bugs found and solved in previous releases tend to reappear in a new release. New functions of the new release turn out to have defects. These things get shaken out and all goes well for several months. Then the bug rate begins to climb again . . . [when users] begin to exercise fully the new capabilities of the release."[22] Brooks adds, "My experienced friends seem divided rather evenly as to whether the curve finally goes down again."[23]

Why is it so difficult to find and fix all the significant errors in a complex program? Referring to what is euphemistically called "program maintenance" (changes/corrections in the program made after it is already in the hands of users), Brooks argues that "The fundamental problem with program maintenance is that fixing a defect has a substantial (20-50 percent) chance of introducing another. So the whole process is two steps forward and one step back. . . . All repairs tend to destroy the structure. . . . Less and less effort is spent on fixing original design flaws; more and more is spent on fixing flaws introduced by earlier fixes."[24]

The repetitive structure found in much (but not all) computer hardware makes it possible to build hardware systems made of many copies of small hardware subsystems that can be thoroughly tested. But software design is more complex because it often does not have the repetitive structure found in the circuitry itself. The relative lack of repetitive structure contributes to making software systems generally less reliable than hardware, and making programmers look less competent that they really are. Parnas argues that this "is a fundamental difference that will not disappear with improved technology."[25]

How can software design flaws that have been in a complex program from the beginning remain hidden without apparently affecting the program's performance, then suddenly appear? In designing software, a programmer creates decision trees intended to include all the cases that the program might possibly encounter. In very large and complex programs, many of the branches of these trees cover cases that are not all that likely to occur. As a result, much of the program will rarely if ever run in normal operation. Even if each part of every path were used as the program ran, some routes that travel these paths in a different order or combination might not run for a very long time. One day, after the program has been working just fine, a new situation causes the computer to follow a route through the program that has never been used before, exposing a problem that has been hidden in that particular route from the very beginning. So a frequently used program that has been running reliably for years suddenly and unexpectedly collapses.[26]

It is possible to design fault-tolerant software. Over the past 20 years programmers have developed ways of writing software that can reconfigure distributed networks of computers when part of the network crashes to work around the software (or hardware, for that matter) that is down.[27] As creative and useful as such work is, it does not and cannot cure misconceptions or oversights on the part of software designers.

Even after repeated testing, exposure and correction of errors, it is difficult to have full confidence that software does not contain some crucial hidden design flaw that will unexpectedly surface someday and take the system down. In June 1998, the billion dollar SOHO solar space probe spun out of control because a pair of undiscovered errors in its thoroughly tested software led NASA controllers to give it an errant command.[28] When software plays a key role in switching telephone calls, controlling expensive space probes or even landing commercial airliners, that is one thing. But when it is critical to controlling technologies that can cause death and destruction on a massive scale, that is another thing entirely.

ARTIFICIAL INTELLIGENCE

"Artificial intelligence" (AI) as a field of computer science first emerged in the 1950s. The very name calls forth images of a superintelligence that combines the incredible power and speed of emotionless, ultimately rational and tireless computers with the learning ability, reasoning power, intuition and creativity of the human mind. Often, when the feasibility of some grandiose technological scheme runs up against the need for computer control systems more perfect than the human brain seems capable of creating, that magical name is invoked. In the 1980s, AI was one of the deities called upon to answer the arguments of eminent scientists critical of the technical feasibility of the Strategic Defense Initiative (SDI) ballistic missile defense system proposed by the Reagan administration. SDI's originally stated goal—to render nuclear weapons "impotent and obsolete"—required the development and construction of a vast, enormously complex system of sensors and missile-attacking equipment, including perhaps ten thousand interacting computers. The system had to be coordinated quickly and accurately enough to destroy nearly every warhead launched in an all-out attack against the United States. If ordinary computer techniques were not up to the job, surely artificial intelligence could find a way to make it all work.

The field of AI is based on the assumption that the functioning of the human mind can be simulated by sufficiently clever computer programming. From its very inception, it has been a field of dreams. Perhaps the most spectacular dream was to create a computer program that would possess the full range of human problem-solving abilities. Reflecting this lofty goal, one of the seminal AI programs, written in 1957, was called the General Problem Solver. Within 20 years, the whole idea of creating a general problem-solver began to be seen as extraordinarily naive.[29] AI programming became more task specific, and some successes were achieved. Some viable "expert systems" began to emerge in the 1980s. Once again the name, which could have come out of an advertising agency, promises much more than the programs have been able to deliver.

"Expert systems" are simply programs that have built-in "rules of thumb" (called "heuristics") that have been gleaned from analyzing the procedures followed by people considered to be experts in whatever specific task (say medical diagnosis) is being mimicked. Whatever their origins, by their very nature heuristics are general guidelines, not guarantors of optimality. In essence, they are workable ways of achieving good results most of the time. They do not always result in good choices, let alone in finding and making the best choice. Expert systems are therefore *not* computer systems of superhuman intelligence and expertise capable of flawlessly performing the task at hand. In fact, they are considered to be doing well if they are anywhere near as good at it as the human experts they are designed to imitate.

Why has the dream faded? Artificial intelligence seems to be caught in an inherent tradeoff that has the flavor of Heisenberg's uncertainty principle.[30] AI programmers can take the "logic programming" approach, in which the computer breaks up all the possibilities into complex cases and subcases with rules specifying how to sort through them in deciding what to do. This is certainly thorough, but it can result in the need to consider enormous numbers of possibilities. Or, to avoid a combinatorial morass, programmers can use heuristic decision rules to dramatically narrow the range of possibilities that must be considered. Deep Blue is the remarkable IBM parallel-processing chess-playing computer system that ultimately defeated world chess champion Garry Kasparov in May 1997. It is programmed to consider possible moves for any arrangement of the pieces. But since there are many more board positions in a game of 40 moves (10^{120}) than the estimated number of atoms in the universe, even a system as powerful and fast as Deep Blue must make use of heuristics. Until there are just a few pieces still on the board, it can only partially preview possible moves.[31] Whichever approach they take, programmers cannot be sure that the software they write will find the best solution in complex, real-world decision situations.

Does this simply reflect the lack of progress to date to overcome this dilemma? Or is it a symptom of an unavoidable underlying imperfection of knowledge? We usually assume that if we just keep at it, we will eventually solve the problem. But it may also be that as we press toward the fringes of knowledge, we will increasingly encounter situations like this, in which what we are running up against is not just the boundaries of what we know, but the boundaries of what is knowable.

Artificial intelligence seems an attractive way of getting computers to do all sorts of human-like jobs. The Pentagon targeted three main options for AI development, one for each military service: an automated vehicle for the Army, a conversational cockpit computer for the Air Force to give the human pilot advice in the chaos and confusion of battle, and a similar combat-analyzing, advice-giving computer for the Navy.[32] Others have conceived of imperturbable, errorless AI devices for operating technically complex, dangerous technologies such as nuclear power plants. The advocates of such schemes may have far too grand a view of what artificial intelligence seems capable of doing, at least in the foreseeable future. The track record to date has not been all that encouraging. The Army, for example, gave up on its autonomous vehicle, after years of development. In mid 1989, the *New York Times* reported that "By 1987, it was supposed to be able to travel across six miles of open desert at speeds up to 3 miles an hour, avoiding bushes and ditches along the way. The best it has done off road so far is to travel about 600 yards at about 2 miles an hour, slower than many people walk."[33]

One of the key problems with artificial intelligence may go right to the heart of the fundamental assumption that underlies it—the ability of computers to mimic the functioning of the human mind.[34] In computing, the real world is represented in an essentially linear and rigid way. Computers move with lightning speed through the branches of complex decision trees, rigidly following a predetermined set of rules. The human brain appears to work in a much more flexible and very nonlinear way. Rather than repeatedly and "mindlessly" searching through the same network of decision trees for a solution, it is capable of stepping back and looking at a particularly baffling problem from an entirely differently perspective. Sometimes that different angle is all that is needed to clear the fog and find a solution. The brain takes leaps, doubles back and is capable of reordering and reprogramming its own thinking processes. It has proven very difficult, if not impossible, for AI researchers to come anywhere near duplicating this process.[35] Meanwhile, "AI systems are some of the least well understood and therefore most error prone systems in existence."[36]

As David Parnas once put it, "Artificial intelligence has the same relation to intelligence as artificial flowers have to flowers. From a distance they may appear much alike, but when closely examined they are quite different."[37] Artificial intelligence may indeed have great promise, despite what may be its unavoidable limitations. But it is not magic, and it offers no reason to believe that it will somehow allow us to circumvent or overcome the problems of technical (and human) fallibility that are built into the world of computers.

COMPUTER VIRUSES

Along with hardware failures and software flaws, computers can also become unreliable because they have succumbed to virus attack. A computer virus is a program that can "infect" other programs by inserting a version of itself into those programs. The virus then becomes part of the infected program. Whenever it is run, the computer acts upon the set of instructions that make up the virus in exactly the same way that it acts on all the other instructions in the program. Programs accessed by an infected program can also become infected. A virus that copies itself into the "operating system" (the master control program that supervises the execution of other programs) will have access to, and therefore can infect, every program run on that system.

Whether a virus is benign or malicious depends on what its instructions tell the computer to do. Viruses can be helpful. For example, a compression virus could be written that would look for uninfected "executables," compress them and attach a copy of itself to them. When an infected program ran, it would decompress itself and execute normally. Such a virus might be extremely useful and cost effective when space was at a premium. A compression virus could save more than half the space those files usually take up.[38] A "compiler" is a program that translates the kind of higher-level computer language used by most programmers (such as FORTRAN, Pascal and PL/1) into binary "machine language" that computer hardware can understand. A compiler that compiles a new version of itself is a virus—it is altering another program (the original compiler) to include a version of itself (the modified compiler). Most compilers incorporate this virus-like capability because it is so useful.[39]

Some viruses are neither helpful nor harmful in and of themselves. A playful programmer might, for example, create a virus that causes the programs it infects to display a mysterious message. After the message appears, the virus self-destructs, leaving the program uninfected and its users bewildered.

Malicious viruses can do a lot of damage. They can corrupt data files by manipulating the data in some unauthorized and destructive way or by erasing them entirely. They can destroy infected programs or change them in ways that make them unreliable. Having a program "killed" by a virus is bad enough, but it can be worse if the infected program appears to be operating normally when it is not. A failed program is more obvious, more easily removed and replaced. A program that is subtly corrupted may continue to be relied upon, like a steam pressure gauge that has become stuck at a normal reading. By the time the malfunction is discovered, a disaster may be unavoidable. Delayed-action viruses are particularly difficult to control. Because they do not affect the behavior of infected programs right away, those programs continue to be used, and the virus has a chance to infect many more programs before anyone realizes that anything is wrong.

Malicious viruses can also be designed to overload a single computer system or interconnected network. When the infected program is run, computers are given a rapidly expanding volume of useless work or the communications network is jammed with a proliferation of meaningless "high priority" messages. While this kind of virus does little or no long-term damage, it can temporarily take the system completely out of service. In 1996, some Internet servers were suddenly bombarded by requests for service coming from randomly generated false addresses at the rate of more than a hundred per second. For a time, service was denied to legitimate users as the attacks jammed the system.[40] Viruses can even be designed to overload a system in ways that damage computer hardware. They could speed up a disk drive, wearing it out.[41] A virus that interfered with the scan control of two video monitors in a corporate computer network in California in 1987 caused one of them to catch fire.[42]

Origins

The forerunner of the computer virus was an after-hours game called Core War, invented and played by programmers at Bell Laboratories in the late 1950s and early 1960s. They realized that since a computer's main memory (called the "core," after now obsolete magnetic core storage technology) stored both programs and the data those programs "consumed," they could create programs that would "consume" other programs in the core, rather than just data. Each player would create programs designed to multiply themselves and "kill" the opposing player's programs by destroying their instructions. The programs were triggered at a given signal, and they fought it out in the core. In the mid 1980s, what had been the private preserve of a relatively small number of expert programmers went public. Ken Thompson, one of the creators of the widely used Unix operating system, gave a public speech in 1983 describing the construction of computer viruses.[43] The next year, *Scientific American* published an article describing Core War, and offered readers "guidelines for those who would like to set up a Core War battlefield of their own" for a nominal $2 fee.[44] The article included instructions for Dwarf, a simple battle program described as "very stupid but very dangerous."[45]

In the early days, each computer was a stand-alone device, so an out-of-control self-reproducing program could at worst incapacitate one machine. But as computer networks became more and more common, the possibility of such a program spreading through the system and causing widespread damage grew. At the same time, computers became more central to all phases of life, from health care to industry to education to communications to the military. The threat posed by malicious viruses grew much more serious.

In 1983, virus expert Fred Cohen demonstrated an experimental virus at a seminar on computer security. Five carefully controlled attacks on the security system of an available computer were carried out using the virus to try to gain unauthorized access to protected system rights. "In each of the five attacks, all system rights were granted to the attacker in under an hour. The shortest time was under five minutes, and the average under thirty minutes."[46] In 1987, a researcher at AT&T's Bell Laboratories demonstrated the potential threat of virus-led infiltration to even the most secure systems. The test virus he designed secretly spread to more than 40 computers at the facility. "It succeeded in infiltrating an experimental secure operating system being designed by two other Bell Labs researchers."[47]

Real Virus Attacks

Computer viruses have ceased to be an experimental curiosity and have become instead a very real threat to computer security. As early as 1988, *Time* magazine reported that "a swarm of infectious programs . . . descended on U.S. computer users. . . . [A]n estimated 250,000 computers, from the smallest laptop machines to the most powerful workstations, have been hit."[48] There were almost a dozen major virus or virus-like attacks reported in the late 1980s.[49] One particularly interesting virus attacked Arpanet, a nationwide defense computer communications system linking more than six thousand military, corporate and university computers around the United States.

Established in 1969, Arpanet was a forerunner of the Internet of today.[50] Early in November 1988, Arpanet was invaded by a rapidly multiplying program that spread throughout the network. It made hundreds of copies of itself in every machine it reached, clogging them to the point of paralysis. Hundreds of locations reported attacks, including MIT, Harvard, the Naval Research Laboratory, University of Maryland, NASA's Ames Research Center, Lawrence Livermore (nuclear weapons) Laboratory, Stanford University, SRI International, University of California (Berkeley and San Diego campuses) and the Naval Ocean Systems Command.[51]

Designed by Robert T. Morris, a 23-year-old graduate student in computer science at Cornell University, there are conflicting reports as to whether the program was a true virus or a "computer worm" (a program able to migrate from machine to machine, using whatever resources it needs and copying itself when required).[52] The program was introduced into the network through a secret "back door" Morris discovered that had been left open years earlier by the designer of Arpanet's electronic mail program. The virus/worm entered the network through a computer at MIT in Cambridge, Massachusetts, while Morris sat at his computer in Ithaca, New York. It further concealed its point of origin by causing all copies of itself to regularly send messages to a computer at the University of California at Berkeley.[53]

Although Morris apparently meant no harm, he made a single programming error that allowed his infiltration of Arpanet to turn into what became the country's most serious computer virus/worm attack to date. Morris's mistake involved a simple reversal of logic. The program was designed to detect copies of itself in a computer so as to avoid infecting a machine more than once. This would have caused it to move slowly from machine to machine and forestalled the explosion of copies that jammed the system. But Morris thought that someone might discover the signal that prevented the program from copying itself into a computer and use it to stop the invasion. So he instructed each copy to randomly choose 1 of 15 numbers when it encountered that signal, and if the number was positive, to infect the computer anyway. One of the 15 numbers was supposed to be positive, so that the probability of overriding the "stay away" signal would have only been 1 in 15 (less than 7 percent). But the logic reversal made 14 of the 15 numbers positive, raising the probability of override to more than 93 percent.[54] Thus the program kept reinfecting already infected machines with more and more copies of itself, spinning out of control.

If the virus/worm had been intended to be malicious, the safeguard against reinfection would have been left out. More importantly, the program could easily have been designed to damage the system rather than just temporarily clog it. Morris's attack on Arpanet was not the first time the network had been penetrated. The same Arpanet program that proved vulnerable to his attack had been redesigned years earlier to correct a flaw that had allowed hackers to claim "superuser" status and do virtually anything within the computer's capabilities.[55]

In the words of Chuck Cole, deputy computer security manager at the Lawrence Livermore Laboratory, "a relatively benign software program can virtually bring our computing community to its knees and keep it there for some time."[56] In the late 1980s, scientists at Livermore were warned by government security that some of the lab's 450 computers were infected by a virus set to become active that same day. Many users stopped doing their work and frantically made backup copies of all their disks. The warning turned out to be a hoax. An attack does not even have to be real to be disruptive.[57]

In 1990, a rogue program infiltrated dozens of computer systems including those at Harvard, University of Texas, Digital Equipment Corporation and Los Alamos Laboratory. It stole passwords that allowed access to a wide variety of supposedly secure files.[58] Five years later, a federal computer security agency warned of yet another type of attack. "Internet protocol spoofing" could leave many of the 20 million computers on the global Internet vulnerable to eavesdropping and theft. According to a former FBI computer crime expert, "Essentially everyone is vulnerable."[59] In 1996, computer researchers at Princeton University uncovered flaws in the Netscape Navigator, the most popular Internet Web browser. The most serious could allow a malicious programmer to steal, corrupt or erase data on a computer connected by the Navigator to the Internet.[60] Viruses, virus-like attacks and virus hoaxes continue to proliferate. By the late 1990s, antivirus sleuths reportedly were finding about six new viruses a day.[61] Good Times (1996), AOL4FREE.COM (1997), ShareFun (1997), Join the Crew (1998), Win32/CIH (1998) and Melissa (1999) are but a few of the better-known examples of actual and "hoax" viruses.

Defending Against Computer Virus Attacks

There are basically three strategies for defending computer systems against viruses and virus-like programs: preventing exposure, "immunization," and after-the-fact cure. In the abstract, it seems that immunization would be best, since it avoids both the restrictions that might be required to prevent exposure and the problems associated with infection. Can computer systems be immunized against virus attack?

The more similar a pathogen trying to infiltrate a system is to the normal components of that system, the more difficult it is to detect, isolate and block it. Unfortunately, viruses are programs just like any other normal user program that runs on the system. They differ only in what they instruct the computer to do, not in what they are. There is no general structural way to stop them. As Cohen points out, "It has been proven that in any system that allows information to be shared, interpreted, and retransmitted, a virus can spread throughout the system."[62]

A software "vaccine" must first detect a virus program, then destroy or disable it. A vaccine program may try to detect viruses in control software or other legitimate programs by checking to see whether the amount of space those programs ordinarily occupy has changed. Or it may check whether the program has been rewritten at an unauthorized date. It might carefully watch key locations within control programs that are convenient places for viruses to hide, stop the computer and alert the operator if anything tries to change that part of the operating system.[63] It is also possible to physically "hard code" or "burn" the antiviral program and part, or even all, of a computer's master control programming onto a computer chip so that no software running on the computer could modify them.[64]

But viruses or virus-like programs can be designed to avoid signalling their presence by enlarging the programs they infect or changing "write dates." They can hide in atypical places and do not need to infect the computer's master control system. There are so many possible ways to write a virus that no one antiviral program could be effective against all of them, just as no single vaccine is effective against all the viruses that might infect the human body. The advent of computer viruses has launched a kind of "arms race" between those writing viruses and those trying to stop them. Any program written to block the current strain of viruses can be circumvented by the next type of virus, written specially to defeat that defense. Raymond Glath, a virus expert whose company produces antiviral software, has said, "No anti-virus product can offer a 100% guarantee that a virus cannot slip past it in some way."[65]

Looking at viral detection from a theoretical point of view, Fred Cohen concludes that "a program that precisely discerns a virus from any other program by examining its appearance is infeasible. . . . [I]n general it is impossible to detect viruses."[66] He argues, "I can prove mathematically that it's impossible to write a [general] program that can find out in a finite amount of time whether another program is infected."[67] Cohen goes on to say that "any particular virus can be detected by a particular detection scheme. . . . Similarly, any particular detection scheme can be circumvented by a particular virus."[68] Even if a computer were protected by a highly effective antiviral program, a serious attack might still occupy so much of the machine's resources that it could not be used for its normal purposes during the attack.[69] Thus, the virus would have taken the computer out of action anyway, at least for a time.

Hard coding part or all of the operating system onto computer chips is generally impractical. Most operating systems are complex. Complexity implies both a high probability of bugs and the need to modify the operating system over time. Hard coding greatly complicates debugging, as well as updating and testing the operating system. Rather than simply reprogramming software, new computer chips must be designed, manufactured, shipped and physically installed.[70] Immunization may seem a superior approach in the abstract, but it has many problems in practice.

Curing an infected computer can require a painstaking process of search and destroy that must work its way through all levels of the system and through all backup data and program files. This is likely to be very time consuming. Creating separately stored backup copies of all programs and data does not assure that a computer purged of a virus can be easily returned to its original state. If the infection occurred before some of the backups were made (likely in the case of delayed-action viruses), backups can be a source of reinfection rather than a solid defense.[71]

Many experts warn that the virus programs are still in a fairly primitive stage of development. More advanced viruses might, for example, change their own codes as they infect more programs. Cohen argues that "Evolving viruses are easily written and are often harder to detect than simple viruses."[72] It is not difficult to imagine a sophisticated evolving virus capable of detecting protective or curative programs and modifying itself to circumvent them. Another possibility is the so-called retrovirus, a program that communicates with the copies of itself with which it has infected other programs. If it detects that these clones have been wiped out, it assumes they have been destroyed by an attempt to disinfect the system, and shuts itself off. After lying dormant for weeks or months, it awakes and reinfects the system. The user, thinking the system has been cleared of the infection, discovers that the same virus has mysteriously reappeared.[73]

One advantage that computers infected with a virus have over infected biological systems is that it is possible to remove and replace suspect parts of the software or shut the whole system down completely to eliminate an infection. Shutting off the power long enough to insure that temporary memory has been wiped clean (often for as little as 30 seconds) will eliminate any virus that has been stored only in such a memory. If the virus has already written itself onto a more permanent storage device (such as a diskette or hard drive), this won't help. The virus won't be able to spread while the system is off, but the system won't be usable either. As soon as it is restarted, it will immediately be reinfected.

A shutdown can buy time to search for the virus and work out a method of destroying it. But by forcing a shutdown, the virus has succeeded in disabling the system, perhaps for a long while. It reportedly took programmers at the Environmental Protection Agency in Research Triangle Park, North Carolina, a week to clear an attack by the Scores virus that infected 50 Apple computers.[74]

If immunization is problematic and cure may still temporarily cripple the system, then what about preventing exposure in the first place? Protective "fire walls" can be built by using special electronic security measures such as additional layers of passwords or encryption.[75] Physical security is also possible. The system can be physically isolated. Or physical barriers can be created. Engineers at Sandia National Laboratories recently attached a tiny electromechanical combination lock to a computer chip to prevent anyone who does not have the combination from using the system.[76]

More-complex passwords and more-involved identification measures do make it more difficult for an outsider to introduce a virus. But making it harder to break in to a system can also make it harder to use. And such an approach does nothing to prevent insiders from causing trouble. According to an informal survey of corporate data security officers conducted for the FBI, more corporate computer attacks were launched by insiders in 1997 than by outsiders.[77]

Any technical system designed to permit some people access while denying it to others contains within it the seeds of its own defeat. If it is possible to figure out how it works, it is possible to figure out how to get around it. Computer security systems are no exception. After the Arpanet virus attack, a former top information-security chief at the Pentagon put it bluntly: "No one has ever built a system that can't be broken."[78] It doesn't always take an expert with long years of experience to find a way in. In 1995, a 22-year-old researcher, Paul Kocher, discovered a weakness in the "public key" data security technologies used by most electronic banking, shopping and cash systems.[79] A few years later, Kocher and his colleagues developed an ingenious scheme for breaking the coding system used in "smart cards," the credit card-like devices with special electronic encryption and data processing circuitry built into them that can be used for everything from cash cards to identification.[80] That same year, University of California researchers cracked the world's most widely used cellular phone encryption scheme, designed to prevent digital phones from being "cloned" by unauthorized users.[81]

For many computer programmers, figuring out how to break into a computer security system designed to keep them out is just the sort of intellectual challenge that got them interested in programming in the first place. The fact that this game playing can often be done by remote control with relatively little chance of getting caught makes it that much more appealing. According to one security expert, "There are enough hackers out there that someone will manage a serious compromise of security."[82] In the electronic as well as the physical world, perfect security is an illusion.

What about physical isolation? The only sure way to protect a computer system against infection is to block *all* incoming flows of information. That means no user could enter any new data or load any new programs, making the system close to useless. If the system were physically isolated to prevent any unauthorized users from even getting into the rooms where the access terminals are, someone might still be able to break in, just as burglars and spies do on a daily basis. As long as anyone is, or ever was, on the inside, that person could intentionally or accidentally infect or otherwise interfere with the proper operation of the system. Physical isolation prevents information sharing and user interaction, one of the great advantages of modern computer systems. Isolation may be workable in some unusual, highly specialized applications, but it is not a realistic solution in general.

COMPUTER FALLIBILITY
AND DANGEROUS TECHNOLOGIES

In the minimal time that dangerous technologies allow for critical decision making in a crisis, even a short-lived computer malfunction can cause disaster. The breakdown of even a single, seemingly unimportant, inexpensive piece of computer hardware—at the wrong time in the wrong place—can threaten global security. When that infamous 46-cent computer chip failed at North American Aerospace Defense Command (NORAD) on June 3, 1980, it triggered a false warning of attack by Soviet submarine-launched and land-based missiles. According to NORAD officials, the circuit board containing the chip had been installed five years earlier and was periodically tested by the Forty-seventh Communications Group at Cheyenne Mountain, NORAD's home in Colorado Springs.[83] Yet neither these periodic checks nor anything in the circuit board's behavior in *five years* of operation brought the problem to light, until the false warning that put America's nuclear military on high alert.

This dramatic incident illustrates two points made earlier: that critical flaws can remain hidden for extended periods of time and that computer problems can and do produce spectacular failures, not merely degraded performance. As it turns out, it is also an example of the deeper, more structural issue of flawed design. The communications computer used only very superficial methods for checking errors. It was the kind of mistake that might be expected of inexperienced programmers. Much more effective methods of detecting errors were commonly used in the commercial computer systems of the day, methods that would likely have prevented this dangerous false warning. Former Defense Communications Agency chief Lieutenant General Hillsman said of the false alert, "Everybody would have told you it was technically impossible."[84]

As we learned in the previous chapter, the quality control procedures used in manufacturing microchips for the U.S. military left something to be desired. Improper testing makes it difficult to be confident in this key building block of electronic hardware, and thus in all the military computers (missiles, aircraft, warning systems, etc.) of which they are a part. As the June 1980 incident makes clear, failure to assure the reliability of computer chips can have potentially dangerous consequences.

Failures of military software are legion. They have led to enormous problems in everything from the performance of particular weapons to the mobilization of the armed forces as a whole. Early in 1982, Ford Aerospace invited military brass to observe a demonstration of its Divad weapons system. Divad is an armored vehicle built something like a tank and designed for the Army with special antiaircraft capability. Here's one account of what happened at a demonstration set up to show that all of Divad's many operating problems had finally been corrected:

> The target was a drone Huey helicopter, which would hover, motionless. Divad's electronic brain was switched on. According to one of those present, the gun immediately swung at full speed away from the target and toward the reviewing stand, where the officers were sitting. Brass flashed as the officers dove for cover. Then the gun slammed to a stop, but only because an interlock had been installed the night before to prevent it from pointing directly at the stands. For a while, technicians combed over the machine. Then the Huey drone rose again, and Divad was again switched on. This time it pointed in the direction of the target but began blasting away at the ground just 300 yards out. For the rest of the day, Divad would do nothing but fire into the tumbleweeds.[85]

The test didn't even pretend to recreate realistic battle conditions in which the target would be trying to evade the attack. There was a good reason for this. The Army already knew that Divad would not be able to hit an erratically maneuvering target, not because of flaws in its technology, but because of a more basic technological limit. Using sensors, a computer can track how a target has been moving. But to shoot it down, the computer must predict where it will be when the shell launched against it catches up. Unless the target follows a smooth path, there is no way of knowing precisely where it will be. As Lieutenant General Maloney put it, "No computer can handle a jinking target."[86] A few months later, on May 27, 1982, the Defense Department awarded the contract for Divad to Ford, agreeing to buy 50 units at a total cost of nearly a third of a billion dollars, with an option to buy 226 more for about $1.4 billion extra.[87]

During the 1982 war between Britain and Argentina over the Falkland/Malvinas Islands, the British destroyer HMS *Sheffield* was sunk by a French-built Exocet missile. The missile travelled just above the surface of the water, a distance of some 20 miles and hit the *Sheffield* amidships about six feet above the waterline. The ship's radar reportedly detected the incoming missile, but its computers had been programmed to identify Exocet as a "friendly" missile (apparently because it was French). No alarm was sounded and no attempt was made to take any defensive action. Four of the six Exocets fired by the Argentines during that war hit their targets, destroying two ships and seriously damaging a third.[88]

As for general mobilization, in late 1980, the U.S. military ran a simulation called Proud Spirit to test the nation's readiness to mobilize for war. During the exercise, "A major failure of the computerized Worldwide Military Command and Control System . . . left military commanders without essential information about the readiness of their units for 12 hours at the height of the 'crisis'. . . . [I]t 'just fell flat', one participant said."[89] What happened? With the demands of the major mobilization being simulated, the computer became so overloaded, it temporarily shunted updating readiness information into a buffer memory. Later, the buffer simply refused to discharge it. "The result was that the Army's . . . computer systems went silent for six hours while programmers struggled to find the right code to release the information. That information emerged 12 hours later."[90]

Saboteurs, spies or disgruntled employees might intentionally manipulate software to sabotage computers controlling dangerous technologies. They could also be inadvertently compromised by playful programmers. It could even happen accidentally, as when a "keyboard error" at Rockwell International stripped away software security for *six months,* allowing unauthorized personnel to make changes in the raw flight software code for the space shuttle.[91]

Intentional software manipulation could be used to steal data—that is, to spy. Computers store and manipulate information. Whether spies are industrial, political or military, information is what they are looking for. If they can gain access to the information they want by typing the proper passwords into a computer terminal, they are saved the messy and dangerous business of making their way past armed guards, breaking into locked rooms and rifling filing cabinets. They are also likely to gain access much more quickly to much more information. And especially if they can do it anonymously from a remote location, they are exposed to even less risk.

Virus-like invasions and other programming attacks on computer systems can also be used to steal passwords and other access codes for later use. Access codes could not only allow an unauthorized user to break in to computer files, but to do so in someone else's name. In 1987, the IBM Christmas Tree virus sent itself by e-mail to the entire standard mailing list of every computer user whose station was attacked, in the name of that victim.[92] The Arpanet virus/worm stole passwords and directed messages to a remote monitoring computer to hide the location of the actual perpetrator.

How vulnerable are the computers in dangerous-technology systems to this kind of break-in? Shortly after the November 1988 Arpanet attack, the *New York Times* reported that a few years earlier, without any fanfare, an unspecified number of military computer experts had decided to try to get past the safeguards protecting some vital, highly classified federal government computers to see if it could be done. "One expert familiar with those efforts . . . said that they found those safeguards to be 'like swiss cheese' to enterprising electronic intruders. . . . They went after the big ones and found it incredibly easy to get inside."[93] *Newsweek* reported on a study conducted in 1985 by the office of Robert Brotzman, director of the Department of Defense Computer Security Center at Fort Meade. According to that report, the study "determined that only 30 out of 17,000 DoD computers surveyed met minimum standards for protection. Brotzman's conclusion: 'We don't have anything that isn't vulnerable to attack from a retarded 16-year-old.'"[94]

In 1996, the General Accounting Office issued a report to Congress of its investigation into Department of Defense (DoD) information security. DoD has over 2.1 million computers, and 10,000 local and 100 long-distance networks. According to the study, Defense Information Systems Agency data indicated that there may have been as many as *250,000 attacks* on DoD computers the preceding year, *65 percent of them successful.* GAO reported,

> Attackers have seized control of entire Defense systems, many of which support critical functions, such as weapons research and development, logistics and finance. Attackers have also stolen, modified and destroyed data and software. . . . They have installed unwanted files and 'back doors' which circumvent normal system protection and allow attackers unauthorized access in the future. They have shut down and crashed entire systems and networks, denying service . . . to help meet critical missions. . . . The potential for catastrophic damage is great.[95]

The Department of Defense did not dispute the report's findings.[96]

Spying creates problems. Aside from copying advanced designs, an opponent that better understands how a weapons system works can more easily find ways to defeat it. A terrorist group that can better understand the details of how a nuclear power plant or other dangerous-technology system is laid out is more likely to find its weak points. Yet this may not be the biggest problem.

It is not difficult to imagine that a virus could be used to sabotage a key data processing or communications system during a terrorist attack or a developing international political or military crisis. Planted in advance and designed to hide, the virus could lie dormant until the critical moment. It would then be triggered by an external command, or even by the system itself as it passed some measurable level of crisis activity. Once triggered, the virus would rapidly overload the system, jamming it long enough to render it useless. The IBM Christmas Tree and Arpanet attacks are examples of how large networks of computers can be crippled by programs whose design caused them to multiply out of control. The Christmas Tree attack rendered the 350,000-terminal worldwide IBM computer network unusable for at least several hours. The Arpanet attack reportedly disabled many of the computer systems at the Naval Ocean Systems Command (San Diego) for more than 19 hours by overloading them.[97] Even false alarms or hoaxes, such as that at the Lawrence Livermore (nuclear weapons) Laboratory in 1988 can result in lengthy disruptions of key computer systems.

Juxtaposed against the quick reaction times required in the event of nuclear attack (the *maximum* warning of an ICBM launch is 30 minutes), this kind of jamming may be virtually as effective as destroying the system. Virus-provoked overloading of crucial computer links could also be used in a political or conventional military crisis or terrorist action to temporarily "blind" or "cripple" the opponent. Pre-planted viruses could interfere with computers key to electronic information gathering or communication, making it difficult for military or antiterrorist forces to see what is happening in an area of critical confrontation or to effectively deploy and control their own people. By the time the electronic smoke screen could be cleared, the attacker might have already succeeded or at least gained an important advantage.

Malicious viruses can also be designed to destroy critical data or programs, requiring much more extensive and time-consuming repair.[98] They could interfere directly with critical programming that controls dangerous technological systems. If the programming that launches, targets or otherwise operates individual weapons or weapons systems could be infected by such a virus, it wouldn't be necessary to physically destroy the weapons. Their programming or data could be so scrambled as to render them unusable. In the same way, the computers that control critical functions at nuclear plants, toxic chemical facilities and the like could also be subjected to electronic attack. These computers may not be not accessible from any external network, but that would not prevent disgruntled insiders or agents who had infiltrated the system from launching such an attack.

Given the damage that viruses and related programs can do and how quickly they can do it, restoring an attacked system can be very time consuming. Even when an infection has been quickly detected, trying to prevent the virus from spreading to other systems may require disconnections or "cold shutdowns" that are very disruptive. Likewise, false warnings of virus attack and other credible hoaxes may lead to actions that take the system at least partly out of action for a time. Under the right conditions, such temporary disruptions may be enough to render critical control systems ineffective.

It is possible to design fairly sophisticated electronic security procedures that will make it difficult for outsiders to connect to, or insiders to gain access to, parts of the system they are not authorized to reach. But it has been repeatedly shown, even in the case of military-linked computer systems, that sufficiently creative and/or persistent programmers can find a way to break through these software barriers. In July 1985, the prosecutor involved in the case of seven teenage hackers arrested in New Jersey revealed that they had "penetrated 'direct lines into . . . sensitive sections of the armed forces.'"[99] In 1987, the so-called Hannover hacker broke into the 20,000-computer scientific research and unclassified U.S. military network from his apartment in West Germany. The hacker reportedly "used fairly standard techniques for cracking passwords" but "he showed uncommon persistence . . . and gained access to more than 30" different computers, ranging from "the Naval Coastal Systems Command in Panama City, Florida, to Buckner Army Base in Okinawa."[100] There are no grounds for believing that all of the loopholes can be closed.

If the computers that control military or critical civilian dangerous-technology systems cannot be completely protected by electronic barriers, what about physical barriers? That depends on the system. The networks of satellite-based computers that have become so central to communication, attack warning, espionage and the like cannot be kept under lock and key. They must be accessible from remote locations. Computers that service complex worldwide logistics must be connected to many other computers and terminals in many other places in order to effectively carry out their mission. Their usefulness lies in their ability to receive and process information from many different entry points. They also cannot practically be completely physically isolated.

It might be possible to isolate certain key, smaller, special-purpose dangerous technology computer systems. There is not much information publicly available about the military computers that include nuclear war plans and control nuclear weapons, but they may fit into this category. They are reportedly "completely inaccessible to outsiders by telephone or any other means. . . . Some are guarded by troops. Others are encased in copper-sheathed rooms to discourage electronic eavesdropping or tampering."[101] "Current military protection systems depend to a large degree on isolationism."[102] Won't enough guards, walls, locks, etc. offer virtually perfect protection to such critical systems? The answer is no. Someone, somewhere, sometime must write the programming for all computers. Others must periodically access the programming to modify, update or test it. As long as that is true, and it is always true, those with authorized access would be able to make an unauthorized change. No system of external protection can ever protect perfectly against the actions of insiders.

It is not even theoretically possible to accord critical computers the degree of protection against internal flaws or external mischief that is appropriate to crucial dangerous-technology control systems. Nuclear power plants, toxic chemical–industry facilities and nuclear and toxic chemical waste storage areas are tempting targets for terrorists and other malefactors. So too are arsenals of weapons of mass destruction. In all these cases, the allowable margin of error is far too small and the potential consequence of error far too great.

Deliberate attack by viruses and virus-like programs is a clear threat to the integrity and efficiency of the computer systems on which so many modern technologies—dangerous and otherwise—depend. But in the end, it is not so much this kind of malevolence as it is ordinary human error and the limits imposed by our inherent imperfectability that pose the greatest threat of catastrophic computer failure. We can and should do whatever we are able to do to frustrate deliberate attempts at sabotage, but we should not delude ourselves into thinking that these efforts will solve the problem. It may well be that the faulty component, the misplaced comma, the simple error of logic, the inability to foresee every possibility and the less-than-brilliant job of manufacturing or programming are even more likely to do us in.

PART V

What Can
and Must Be Done

CHAPTER ELEVEN

Understanding and Assessing Risk

Everything we do involves risk. Chance is an unavoidable element in all of life. The world in which we live is not completely predictable, but it is also not completely random.

We have seen that technological systems fail for a great many reasons, rooted in the nature of technology itself and in the fallibility of its human creators. We have also seen that when those failures involve dangerous technologies, the possibility of catastrophe is all too real. If we are to avoid stumbling into technology-induced disaster, we cannot leave the choice of which technological paths we follow and which we forego to chance. We all have a stake in making that choice a conscious and deliberate one in which we all play a role. But in order to make intelligent decisions in a world that combines determinate and random elements, we must first understand the limits imposed by chance itself and by the difficulty of estimating how likely the events that concern us might be.

RISK AND UNCERTAINTY

Although the terms are often used interchangeably, technically "risk" and "uncertainty" are two different concepts. When there is "risk," we do not know exactly what will happen, but we do know all of the possible outcomes, as well as the likelihood that any particular outcome will occur.[1] For example, in rolling a pair of dice, we know that any of six numbers (1-6) can come up on either die. No other number is possible. Furthermore, as long as the dice are not loaded and are rolled in the usual way, each face of each die is equally likely to come up, an average of once every six times the dice are rolled. This is "risk": we don't know what number will come up on any given roll, but we do know every possible outcome and the probability that it will occur.

"Uncertainty" refers to a situation in which we don't know all possible outcomes and/or we don't know the probability of every possible outcome. We are lacking some critical information that we have in a comparable case of risk. That lack of information is important to our ability to predict what will happen. Consider the problem of trying to predict the likelihood of randomly drawing a spade from a standard deck of 52 playing cards. This is risk: we know there are 13 spades; we know there are 52 cards. Therefore, we know that the probability of drawing a spade is 13 out of 52, or one out of four (0.25). If this random drawing were repeated 10,000 times (replacing the card we drew each time before drawing again), we can predict with considerable accuracy that we will come up with a spade about 2,500 (0.25 times 10,000) of those times.

Now suppose we have a deck made up of 52 playing cards that are the randomly assembled odds and ends of other decks. We have absolutely no idea how many cards of each suit there are. This is a situation of uncertainty: we still know all the possible outcomes (any card that we draw will either be a heart, diamond, club or spade), but we don't know the probability of any of them. There is no way to predict the likelihood of randomly drawing a spade from such a deck, no way to accurately predict about how many times a spade will come up in 10,000 random drawings.

The uncertainty would be greater still if an unknown number of the 52 cards was not from an ordinary deck, but instead had an arbitrarily chosen word, symbol or number printed on it. Now it would not only be impossible to predict the likelihood of randomly drawing a spade, it would be impossible to make any useful or sensible prediction as to what would be printed on the face of the card that was chosen. We know neither all the possible outcomes, nor the probability of any of them. Uncertainty is risk minus information. If we repeat the same chance experience over and over again, each time observing the outcome, we may be able to fill in the missing information and so convert uncertainty into risk.

Even in situations of risk, it is impossible to reliably predict what will happen on any particular occurrence. Either heads or tails will come up whenever a fair coin is tossed; each is equally likely. If the coin is tossed 6,000 times, we can predict with real confidence that heads will turn up about 3,000 times, but we have no idea which of the 6,000 tosses will come up heads. When there is risk, repetition and aggregation generally increase predictability. This principle is essential to writing life insurance. Actuarial tables based on past experience allow insurers to estimate every policyholder's probability of living through the year or dying. All possible outcomes are known, and there are reasonably accurate estimates of their likelihood. By insuring a large number of people, companies are able to predict how many claims they will have to pay in any given year and to set premiums accordingly. Though no one knows which particular policyholders will die, aggregation allows accurate prediction of the overall number of deaths that will occur.

The complex and immensely useful mathematical fields of probability and statistics provide vital tools for analysis, prediction and decision making in situations of risk. But their usefulness in situations of uncertainty falls as the degree of uncertainty increases. Greater uncertainty means less information, and information is the essential raw material of analysis and decision making. Because there are much better tools available, analysts and decision makers would rather deal with risk than uncertainty. It is easier to figure out what to do. If this bias encourages careful and systematic data collection, it is a good thing. But if, as is so often the case, it leads analysts and decision makers to assume or to convince themselves that they are dealing with risk when they are actually dealing with uncertainty, it can be misleading and even downright dangerous.

If it were possible, the great majority of us would just as soon eliminate risk entirely in most situations. We'd like to be able to guarantee the outcome we prefer. Of course, we'd rather face risk with some possibility of positive outcomes than the certainty of disaster, but unpredictability means lack of control and lack of control can be very anxiety producing. Not knowing what will happen makes it hard to know what to do, how to adjust our actions to achieve our goals.[2] Though we all like the thrill of taking a chance from time to time, we still want to win. That may partially account for the popularity of amusement park rides like roller coasters, which create the illusion of taking the ultimate life or death risk while virtually guaranteeing that we will survive the experience.

All this adds to the psychological bias toward wanting to believe that there is a higher degree of predictability in situations than actually exists. Events that are risky may be assumed to be deterministic. Cases of uncertainty may be assumed to be cases of risk. And since unpredictability is more anxiety producing when there is more at stake, it is likely that this bias gets worse rather than better in more critical situations.

Then there is the problem of ego. For analysts or decision makers to admit that a situation is one of uncertainty rather than risk is tantamount to admitting that they are unable to see all the relevant outcomes and/or evaluate their probabilities. It is an admission of ignorance or incapability. This is yet another source of psychological pressure to misrepresent what is actually uncertainty as risk. Those we call experts are even more likely to fall into this trap. It is a real boost to the ego to be called an expert. Experts are not supposed to be ordinary human beings who know more than most of us about some things; they are supposed to know everything about everything within their area of expertise. It is therefore more than usually difficult for someone called an expert to admit to ignorance or incapability.

When experts are hired as consultants, clients want to know what will happen, or at least what could happen and just how likely it is. They are not interested in being told that the situation is uncertain and therefore unpredictable. A report like that does not do wonders for the reputation or the future flow of consulting work. The desire to remove the anxiety that unpredictability produces is often stronger than the desire to know the truth.

When experts in risk assessment fail, it is usually not because they have calculated the probability of a possible outcome wrong, but because they have not foreseen all possible outcomes. They either made an error of omission or they were dealing with uncertainty when they believed—or at least pretended—that they were dealing with risk.

Unpredictability interacts with dangerous-technology systems in two important and partly contradictory ways: (1) many dangerous-technology systems are built as an attempt to protect against unpredictability (for example, building nuclear power plants to assure a predictable energy supply), and (2) unpredictability is built into the design, manufacture, operation and maintenance of these technologies (because of the problems of fallibility). An aversion to unpredictability is one of the driving forces behind the buildup of military arsenals, including arsenals of weapons of mass destruction. The belief that armed forces offer protection against unpredictable future attack by external enemies—and therefore against loss of autonomy, loss of property, loss of loved ones and loss of life—is an important source of public support for them. The military may also be seen as a protection against unpredictability because it can be used to coerce others to provide what we need. For example, worry that the invasion and occupation of Kuwait by Iraq would lead to disruption of the predictable and steady flow of cheap oil is clearly one of the reasons why a coalition of nations led by the United States chose to militarily threaten and then attack Iraq in mid January of 1991. Yet as we have seen, every dangerous-technology military system, and every technical and human component of these systems, has some inescapable probability of failure and therefore some embedded unpredictability.

WHEN THE "IMPOSSIBLE"
(OR THE "VERY UNLIKELY") HAPPENS

When faced with the possibility of disaster, it is common to seek refuge in the belief that disaster is very unlikely. Once we are convinced (or convince ourselves) that it is only a remote possibility, we often go one step farther and translate that into "it won't happen." Since there is no point in wasting time worrying about things that will never happen, when we think about them at all, we think about them only in passing, shudder, and then move on.

It is certainly not healthy to become fixated on every remote disaster that might conceivably occur. That is the essence of paranoia. Paranoid thinking is not illogical, but it is grossly distorted. It does not distinguish between possibility and probability. Psychoanalyst Erich Fromm put it this way: "If someone will not touch a doorknob because he might catch a dangerous bacillus, we call this person neurotic or his behavior irrational. But we cannot tell him that what he fears is not *possible*. Even full-fledged paranoia demonstrates that one can be insane and yet that one's capacity for logical thinking is not impaired."[3]

There is a big difference between paranoid fixation and a proper appreciation for the critical distinction between what is highly unlikely and what is completely impossible. We simply cannot afford to ignore events whose consequences are catastrophic, even if their likelihood is very small. Along with its many benefits, the enormous growth of our technological capacity to affect the world in which we live has created the prospect of unparalleled, perhaps terminal disaster. General nuclear war and global ecological catastrophe are among the most extreme forms of such disaster. They may not be very likely, but as long as they are still possible, we can ill afford to ignore them. The fact is, events that are extremely unlikely do happen from time to time.

There is also an important difference between the probability that an event will occur during any one trial of the chance experience and the probability that it will occur *at least once* in many, many repetitions. Even if the probability of a particular event occurring on any *single* repetition is extremely low, as long as it is not zero, the probability that it will happen at least once rises to one (certainty) with continued repetition. Given enough time and enough opportunity, anything that is possible will happen. This is the mathematical principle behind "believe it or not" statements like "a group of monkeys, choosing typewriter keys at random, will eventually type out all the works of Shakespeare."[4]

Our ability to accurately estimate the true probability of a very unlikely event is also questionable, especially if the event is the result of a complex process. It is easy to radically underestimate the likelihood of rare events by looking at the problem the wrong way. And, as the last two chapters make clear, it is extremely difficult, if not impossible, to foresee all the ways things can happen in complicated systems. Unless it is possible to see all the paths to a given outcome, it is hard to be sure that our estimates of its probability are reliable. These problems are worse for very unlikely events because the very fact that they are so unlikely means that there is much less experience with them.

Rare Events

Even exceptionally unlikely events do occur. The great power failure that left New York City without electricity for an entire summer day in 1977 began with a series of lightning strikes of electric supply lines that in the words of the power company's president "just never happens."[5] After-the-fact accounts pointed out that "double lightning hits were a rare event," and went on to outline the extraordinary and very unlikely series of subsequent failures of grounding systems, circuit breakers, timing devices and other technical and human components.[6] This highly improbable blackout led to widespread arson and looting, with property damage estimates running from $350 million to $1 billion.[7]

About 500 meteorites a year survive the fiery passage through the atmosphere and hit the earth's surface. They typically fall into the ocean or strike remote areas. Only a tiny fraction of them are ever recovered. Throughout all of history, it is estimated that only about 2,100 separate meteorites have ever been recovered and scientifically validated worldwide.[8] Yet, on November 7, 1982, a grapefruit-sized meteorite crashed through the roof, second- and first-floor ceilings of a house in Wethersfield, Connecticut, and rolled to a stop under the dining-room table. This event was rare enough by itself. What made it truly remarkable is that 11 years earlier, on April 8, 1971, a meteorite weighing about two-thirds of a pound had penetrated the roof and embedded in the second-floor ceiling of another house in Wethersfield, only about a mile away. In the words of Dr. Ursala Marvin of the Smithsonian Astrological Observatory, "Meteorites are always a dramatic occurrence, but to have two strike the same town is . . . almost incomprehensible."[9]

Natural phenomena are certainly not the only source of rare events. On October 31, 1968, an Australian fighter jet shot itself down when bullets fired from the plane's own cannon ricocheted off the ground.[10] In the late spring of 1989, the Oak Hill Country Club near Pittsford, New York, was playing host to the U.S. Open golf tournament. At 8:15 A.M. on June 16, pro golfer Doug Weaver shot a hole-in-one on the sixth hole. Forty minutes later, Mark Wiebe teed off at the same hole, and also put the ball in the cup in one stroke. At about 9:50 A.M., Jerry Pate stepped up to the tee at the sixth hole, and once again shot a hole-in-one. Then, just after 10:00 A.M., another pro, Nick Price, took on Oak Hill's sixth hole, and carded yet another hole-in-one. In 88 previous Opens, only 17 holes-in-one had ever occurred. *Golf Digest* reportedly estimated the odds of any four golfers having holes-in-one at the same hole on the same day at 332,000 to 1 (0.000003). But, as the *New York Times* put it, "not even a Las

Vegas odds-maker could come up with credible odds on four golfers using the same club to make a hole-in-one at the same hole in the United States Open within two hours."[11]

On February 25, 1988, the *New York Times* reported a story drawn from a British psychiatric journal, yet bizarre enough to sound like a headline story from the *National Inquirer*. Five years earlier, a mentally ill teenager suffering from a disabling obsessive-compulsive disorder that caused him to wash his hands constantly decided to take his own life. He went into the basement of his house, put a .22 caliber rifle in his mouth and pulled the trigger. The bullet lodged in the left frontal lobe of his brain, and, except for a few fragments, was later removed by surgeons. It destroyed only the section of the brain responsible for his compulsive hand washing, without causing any other brain damage. Five years later, there was still no sign of the obsessive-compulsive behavior or any other brain-related problem.[12]

In October 1985, a New Jersey woman won the top prize in that state's lottery. Four months later, she did it again! Her February winnings added $1.5 million to the $3.9 million she had won before. The odds of winning the first prize in the October lottery were about 1 in 3.2 million; the odds of winning February's first prize were about 1 in 5.2 million. After consulting a statistician, state lottery officials announced that the odds of one person winning the top prize in the same state lottery twice in a lifetime were roughly 1 in 17.3 trillion (0.000000000000058).[13]

Neither the world of high technology in general nor the world of the military in particular is immune from such events. In August 1985, the Air Force attempted to launch a Titan 34D rocket, its main vehicle for putting heavy satellites into orbit. The Titan exploded.[14] Then, on January 28, 1986, after 24 successful space shuttle program launches in a row, NASA's attempt to launch the space shuttle *Challenger* ended in a tragic explosion, killing its crew of seven.[15] Two and a half months later, the Air Force's next attempt to launch a Titan 34D ended in another explosion.[16] A week later on April 25, a Nike-Orion research rocket misfired as NASA tried to launch it at White Sands Proving Grounds in New Mexico. It was the first failure in 55 consecutive missions of that rocket.[17] At the end of the next week, a Delta rocket, considered one of the most dependable available, misfired and had to be destroyed in flight by NASA controllers.[18] Within only *nine* months, the United States experienced the failure of *five highly reliable* launch vehicles. This remarkable series of events grounded most of the nation's space launch fleet for a time and called into question America's much vaunted ability to successfully put payloads into orbit.

A few years later, the U.S. Navy experienced an even more unusual series of mishaps. On October 17, 1989, a Navy pilot crashed into the aircraft carrier *Lexington* in the Gulf of Mexico while trying to land. Four sailors and one civilian were killed. The next day, the pilot of an F/A-18 accidentally dropped a 500-pound bomb on the guided-missile cruiser *Reeves* in the Indian Ocean. Five sailors were injured. The following day, a wave washed over the deck of another aircraft carrier, the *Dwight D. Eisenhower,* off the coast of North Carolina. Three sailors were swept overboard (two of whom were rescued), and more than three dozen missiles were lost. The same day, this time in the Pacific, a sailor on the aircraft carrier *Carl Vinson* was lost overboard. The next day, nine sailors suffered smoke inhalation, and four of them were treated for burns, when a fire started in the boiler room of the *Monongahela,* a fleet oiler sailing in the North Atlantic. The next week was relatively quiet. Then, on November 9, a Navy A7E Corsair 2 jet fighter crashed into an apartment complex in Smyrna, Georgia, while trying to land at Dobbins Air Force Base. The plane burst into flames, killing two civilians and injuring four more. Two days later, in the Strait of Malacca (between Sumatra and Malaysia), the Navy destroyer *Kinkaid* collided with a merchant ship, killing one sailor and injuring five. That same day, two Navy A6 attack bombers based in El Centro, California, dropped a dozen bombs in the wrong place. They landed about 300 yards from a campsite, injuring one of the campers. Only three days later, the amphibious assault ship *Inchon* caught fire at Norfolk, Virginia, injuring at least 31 sailors; and a Navy F-14 Tomcat fighter crashed at sea near Key West, Florida.[19] Within only *17 days,* the Navy had had an astonishing series of *ten serious accidents.* They involved a wide variety of equipment and personnel, many different kinds of incidents, and took place all over the world during a calm period on the international scene.

On February 19, 1994, six hospital emergency room workers at Riverside General Hospital in California mysteriously collapsed after drawing blood from a 31 year old woman whose life they were trying to save. Working with samples of the woman's blood and the air from her sealed coffin, scientists at the Lawrence Livermore Laboratory concluded that the workers may have been poisoned when an extremely unusual set of chemical reactions in the woman's body and in the syringe used to draw her blood produced a close relative of nerve gas! They theorized that she had taken in the solvent DMSO (dimethyl sulfoxide), possibly through the use of an analgesic salve, which was then converted in her body—through just the right combination of temperature, the oxygen she was being given and blood chemicals—into dimethyl sulfone. When her blood was drawn, the dimethyl sulfone became dimethyl sulfate gas, a highly poisonous chemical warfare agent.[20]

How Likely Are Rare Events?

As strange as it may seem, the likelihood of an event is greatly affected by exactly how the event is defined. Suppose we want to estimate the probability of getting "heads, tails, heads, tails" when we flip a fair coin four times. If the event is defined as getting two heads and two tails in four flips of a coin, its probability is 6 in 16, or 0.375. If the same event is instead defined as getting alternating heads and tails in four flips of a coin, its probability is only 2 in 16, or 0.125. And if it is defined as getting heads on the first toss, tails on the second, heads on the third and tails on the fourth, its probability drops to 1 in 16, or 0.0625. Now let's reconsider the case of the New Jersey woman who won top prize in that state's lottery twice within four months. What is the probability of *that* happening? Again, the answer depends on how the event is defined. The Rutgers University statistician consulted by the state lottery officials calculated the odds at 1 in 17.3 trillion. That's correct if the event is defined very narrowly: buying one ticket in each of two New Jersey state lotteries, and winning the first prize each time. But suppose the event is defined as someone, somewhere in the United States winning the top lottery prize twice in four months. Stephen Samuels and George McCabe of the Department of Statistics at Purdue University calculated that probability at only about 1 in 30.[21]

How could it be so high? Remember, millions of people buy lottery tickets each month, often buying many tickets in a number of different lotteries. (The woman who won the New Jersey lottery twice had bought multiple tickets each month.) Samuels and McCabe then calculated the odds of a still broader description of the same event. If seven years, rather than four months, were allotted for the double win, the odds grew to better than 50-50 (1 in 2).[22] If the time is extended indefinitely, a double lottery winner becomes a virtual certainty.

In the *Journal of the American Statistical Association*, mathematicians Persi Diaconis and Frederick Mosteller argued for what they call "the law of truly large numbers. . . . With a large enough sample, any outrageous thing is likely to happen." Elaborating further, "truly rare events, say events that occur only once in a million . . . are bound to be plentiful in a population of 250 million people. If a coincidence occurs to one person in a million each day, then we expect 250 occurrences in a day and close to 100,000 such occurrences a year. . . . Going from a year to a lifetime and from the population of the United States to that of the world . . . (5 billion at this writing), we can be absolutely sure that we will see incredibly remarkable events."[23]

Many events that intuitively seem very unlikely are actually much more likely than they seem. Consider what mathematicians call the birthday problem. Suppose 23 people, assembled randomly with respect to their date of birth, are brought into a room. What are the odds that at least two of them have the same birthday? With only 23 people, and 365 days (actually 366) to choose from, it seems very unlikely that there would be any matches. Actually, the odds are 50-50; it is just as likely there will be a match as there won't be. If the birthdays only have to be within one day of each other, the chances are 50-50 with as few as 14 people in the room. If the definition of a match is broadened further to include any pair of people whose birthdays are within one week of each other, the odds are 50-50 with as few as 7 people in the room![24]

It is reasonable to draw three important conclusions from all of this: (1) events that are extremely unlikely do occur from time to time; (2) the probability of at least one occurrence of any event, no matter how unlikely, rises toward certainty given enough time and opportunity (as long as it is not completely impossible); and (3) events that we believe to be very unlikely may not actually be anywhere near as unlikely as they seem. The last of these raises an interesting and important question. Just how accurately are we able to estimate risks in the real world?

RISK ASSESSMENT

In a letter to *Science,* Karl Z. Morgan, professor of nuclear engineering at Georgia Tech, wrote: "I often recall my argument with friends in the Atomic Energy Commission shortly before the United States launched a space rocket which, together with its 17,000 curies of plutonium-238, was incinerated . . . over the Indian Ocean in April 1964 [on the very first launch]. My friends had tried unsuccessfully to convince me before the launch that the probability of something like this happening was of the order of 10^{-7} [one in ten million]; if so we were very unlucky."[25]

On May 25, 1979, one of the engines on an American Airlines DC-10 tore loose as the plane took off from O'Hare Airport in Chicago. Normally, the loss of an engine, even in such a dramatic fashion, does not make the plane impossible to fly. But when the engine tore off, it severed control cables and hydraulic lines that made it very difficult for the pilot to control the plane and at the same time disabled critical warning indicators. With warning, the pilot could have compensated for the loss of control of the wing slats during the takeoff. Without normal control and without proper warning, the plane crashed, taking 273 victims with it.

There had been a similar accident with a DC-10 in Pakistan in 1977. On January 31, 1981, a DC-10 at Washington's Dulles Airport also had an engine failure combined with slat (and flap) control damage that led it to crash during takeoff. Nine months later, a DC-10 blew an engine while taking off from

Miami, and the debris severed the slat control cables. Because it happened early enough in the flight, the crew was able to abort the takeoff successfully.

Experienced aircraft designers like those at McDonnell-Douglas understand the possibility and implications of such a problem. But various company engineering studies had estimated the probability of both loss of engine power and slat control damage during takeoff at less than one in a billion. Yet that exact combination of events had occurred four times in only five years.[26]

How could we be so wrong? Risk assessment can be a difficult and complex process that often runs up against the limits of our technical knowledge and our personal humility. That is especially true when dealing with complicated systems and trying to assess the probabilities of events that occur only infrequently.

The problem of risk assessment has something in common with the problem of computer programming. In programming, as in risk assessment, it is necessary to imagine all the ways things can go wrong. But after programmers write a piece of complex software, they have the enormous advantage of being able to run the program again and again to debug it. Risk assessors can't keep running their "program" to catch their errors. They must correctly see all the ways thing can go wrong *in advance*. For complex systems, especially those that are technologically advanced, this is a monumental—in some cases impossible—task. The problem is that much greater when people are an integral part of the system's operation. We are generally much better at predicting the behavior of technical components than that of human beings.

Formulating Problems and Making Assumptions

While there is a scientific core to risk assessment, it is at least as much an art as a science. Many assumptions must be made, and it is difficult for even the most unbiased and disinterested analysts to avoid the distortions the assumptions may introduce. They may not even know whether on balance the resulting bias tends to overstate or understate the risks being analyzed. Risk assessors are not always as explicit about the assumptions they are making as they should be. Apart from any deliberate attempt to mislead (which is always a possibility), it may be necessary to assume so many things to make risk assessment tractable in complicated situations that it is too tedious and time consuming to state them all. Some may also seem so obvious that they are not worth stating. But it is important not to gloss over them: the underlying assumptions are critical to the ultimate results.

Just how much impact these assumptions can have on the risk assessor's ultimate conclusions is strikingly illustrated by what has become a standard application of risk analysis—judging how toxic a particular chemical is to human beings. There are quite a number of different models available and commonly used for making this assessment. They differ in a variety of ways, including the assumptions they make in translating data from animal testing to human beings. Toxicologist Martyn Smith of the University of California noted that, in assessing the health danger posed by trichloroethylene (a common solvent), the choice among existing mathematical toxicity models could make as much as a *billionfold* difference in the risk estimate. According to Smith, "Since there is no scientific method of determining which mathematical model to use and none of the models have any real biological basis, this level of uncertainty will always be present."[27]

One of the most basic assumptions of risk assessment is that the situation being analyzed is actually one of risk and not uncertainty. Creating a comprehensive list of outcomes in complex systems is often beyond our capabilities. In many real-world situations, our unwillingness to admit how little we actually know (or can reasonably be expected to know) about what will happen can lead us into serious trouble.

One way to deal with the problem of uncertainty in risk assessment is to specify all the known outcomes, and then create a residual category called "all other outcomes." If reasonable estimates of probabilities can be worked out for all the outcomes that can be listed, the probability of the "all other outcomes" category would simply be one minus the sum of those probabilities. The probability attached to this residual category is an estimate of the degree of uncertainty in the situation, of the limits of our knowledge. As long as the residual probability is small enough, the risk assessment should still be useful. Unfortunately, this common strategy does not work nearly as well as one would think. It turns out that the probabilities estimated for specific outcomes tend to be significantly affected by how complete the listing of outcomes and paths to those outcomes is. Making the list less complete should enlarge the probability of the "all other outcomes." But it often also changes the probabilities that assessors attach to those outcomes that haven't been altered. That should not happen.

Consider a common tool of risk assessors, the "fault tree." A fault tree begins by specifying some kind of system failure, then creating branches that represent different paths to that failure. Once the tree has been created, the probabilities of all of the events along all of the paths represented by the branches are estimated. From these, the probabilities of each path to failure, and thus the probability of the failure itself can be estimated. The problem is that the estimated probabilities that risk assessors attach to the main branches of the tree tend to be influenced by how many main branches (paths to failure) they have foreseen.

Consulting authoritative sources, Baruch Fischhoff, Paul Slovic and Sarah Lichtenstein constructed a fault tree for the failure of a car to start.[28] The full tree included seven main branches: (1) battery charge insufficient; (2) starting system defective; (3) fuel system defective; (4) ignition system defective; (5) mischievous acts or vandalism; (6) other engine problems; and (7) all other problems. They then created several versions of the tree that differed from the original in which paths and outcomes were listed and which were left out. First they gave a group of laypeople various "pruned" and complete versions of the fault tree. Each person was asked to estimate the probability that a car would fail to start for each of the reasons listed as branches of the specific version of the tree that they were shown. If the completeness of the fault tree did not distort the subjects' risk estimates, then leaving out a branch should just transfer the probability of that branch to the branch labelled "all other problems." In other words, the probability assigned to "all other problems" by those given *incomplete* fault tree diagrams should be equal to the probability assigned to "all other problems" by those given *complete* trees plus the probabilities attached to the branches that were left out.

Those given incomplete tree diagrams did assign higher probability to "all other problems," but it was nowhere near high enough to account for what had actually been left out of their tree: "Pruned tree subjects clearly failed to appreciate what had been left out. For pruned tree Group 1, 'other' should have increased by a factor of six . . . to reflect the proportion of failures . . . [that were left out]. Instead, 'other' was only doubled, whereas the importance of the three systems that were mentioned was substantially increased".[29] Although risk estimates for the branches that were the same in both pruned and unpruned trees should have been unaffected by whether other branches were present or missing, they were higher when the assessors were looking at incomplete trees.

The experiment was then repeated, with subjects specifically told, "we'd like you to consider . . . [the fault tree's] completeness. That is, what proportion of the possible reasons for a car's not starting are left out, to be included in the category *all other problems?*"[30] While focusing attention on the residual like this did increase the probability that pruned tree subjects assigned to "all other problems," it was still much less than it should have been. In another experiment, the subjects were all given complete trees, but with different branches split or fused in different ways. This too affected the assessment of risks: "In every case, a set of problems was perceived as more important when it was presented as two branches than when presented as one."[31]

When the first version of the experiment (pruned vs. unpruned fault trees) was redone with a group of experienced auto mechanics, those given the complete diagram assigned an average probability of 0.06 to the residual branch. The probabilities they assigned to the branches that were later left out in the pruned version totaled 0.38. Those shown the pruned tree should therefore have assigned a probability of about 0.44 (0.06 + 0.38) to the "all other problems" category. But they actually assigned an average probability of only 0.215. The experts' risk assessment suffered the same basic kind of distortion as the laypersons' assessment had. And among the experts, "neither self-rated degree of knowledge nor actual experience had any systematic relation to ability to detect what was missing from the tree."[32]

These studies strongly suggest the troubling possibility that the way a risk assessment is set up may strongly influence its results. Risk analysis appears to be less solid and objective than its formal structure and mathematical nature leads most of us to believe. Since a proper understanding of risk is critical to many important public policy decisions, the ease with which the results of risk assessments can be manipulated is disturbing.

The history of risk assessment in the civilian nuclear industry is a good case in point. Nuclear power has been heavily subsidized by the federal government since its inception. Early on, nuclear power was pushed especially hard, with relatively little attention to possible public health effects.[33] The mid 1950s saw the first systematic attempt on the part of the Atomic Energy Commission (AEC) to assess the risks of a major nuclear accident. Known as WASH-740, this study used "worst case" analysis, concluding that a large radioactive release could kill 3,400 people, injure 43,000 and cause $7 billion in property damage.[34] Policymakers were not pleased with the results and the AEC did not exactly go out of its way to publicize them. Because neither the nuclear industry nor the insurance companies were willing to assume a financial risk of that magnitude, the federal government rushed to the rescue with the Price-Anderson Act of 1957, which sharply limited private-sector liability for nuclear power accidents. Had it not done so, it is quite possible that nuclear power would have died on the vine.

In 1964, the AEC tried again, conducting a follow-up study to WASH-740. The worst-case results were even more alarming this time, with 45,000 deaths projected as the possible consequence of a major nuclear plant accident. The commission didn't even publish that study.[35] Then in 1972, the AEC set up a group headed by Norman Rasmussen to conduct what became known as the Reactor Safety Study, or WASH-1400. WASH-1400 structured the problem differently. Rasmussen's team was told to calculate the probabilities, as well as the consequences, of a broad array of possible nuclear reactor accidents. By focusing attention on average risk, rather than on the worst case, the AEC hoped that the results would be less alarming. The team used fault tree analysis. This time, the results came out "right," supporting the idea that nuclear power was safe. Among its famous conclusions was that "a person has as much chance of dying from an atomic power accident as of being struck by a meteor."[36] Despite serious objections to the structure and method of this study from inside as well as outside the agency, the AEC released it—all 14 volumes—in the summer of 1974. Five years later, the AEC's successor, the Nuclear Regulatory Commission, finally repudiated the study, though not with as much fanfare as had accompanied its earlier release.[37] This is an effective technique for fixing a distorted message in the public's mind without actually lying, similar to the old journalistic trick of headlining allegations of misconduct on page 1, then later printing a retraction in a small paragraph on the bottom of page 16.

The Dimensions of Risk and Its Consequences

Although "risk" technically refers only to a chance situation in which all outcomes and probabilities are known, in common usage it is loosely applied to any chance situation that involves a dangerous or otherwise undesirable outcome. There is a considerable literature, most of it based in psychology, relating to how people perceive risk in this looser sense, and how their perceptions affect their judgement of the magnitude and acceptability of any particular risk.

Slovic, Fischhoff and Lichtenstein list nine key dimensions of risk in the form of questions: (1) Is the risk one that people expose themselves to voluntarily (such as breaking a leg while skiing) or is it imposed on them (such as getting cancer from polluted air)? (2) Does the risk have immediate consequences (such as getting killed in a high-speed auto accident) or are the consequences delayed (such as becoming infected with the AIDS virus)? (3) Are the people exposed aware of the risk? (4) Are the risks understood by the relevant scientists or other experts? (5) Is it possible to take action to avoid or mitigate the consequences of the risk (such as building levies to protect low-lying areas against flooding)? (6) Is the risk a new one or one that people have known about for a long time? (7) Is the risk chronic, claiming only one or a few victims at a time (such as cirrhosis of the liver in alcoholics), or is it acute and catastrophic, claiming many victims at the same time (such as disastrous earthquakes)? (8) Are people able to think calmly about the risk (such as cancer caused by smoking tobacco), or do they find it terrifying (such as being eaten by a shark)? and (9) When the risk is realized, how likely is it that the consequences will be fatal?[38] We can summarize these dimensions (in the order stated above) as: *voluntariness, immediacy, personal knowledge, scientific knowledge, control, familiarity, catastrophic potential, dread,* and *severity.*

In psychological experiments aimed at understanding how risk is perceived and estimated, an interesting difference has emerged between experts and laypeople asked to judge the riskiness of various possible causes of death: "When experts judge risk, their responses correlate highly with technical estimates of annual fatalities. Laypeople can assess annual fatalities if they are asked to (and produce estimates somewhat like the technical estimates). However, their judgments of 'risk' are related more to other hazard characteristics (for example, catastrophic potential, threat to future generations) and, as a result tend to differ from their own (and experts') estimates of annual fatalities."[39] In other words, the experts in these experiments rely on a narrow, stripped-down, quantitative concept of average risk, while the laypeople have a much broader, richer and more qualitative view of risk.

These differences in concept result in sharp disagreements as to the riskiness of hazards that rate very differently on the nine risk dimensions. When asked to rank 30 activities and technologies in order of risk, well-understood, chronic, voluntary risks with little dread (like smoking) were ranked similarly by experts and laypeople. But those that had considerable catastrophic potential and dread (like nuclear power) were ranked very differently. Smoking was on average ranked third or fourth by laypeople and second by experts; nuclear power was on average ranked first by most laypeople and twentieth by experts.[40]

Who is right? Surely experts have more detailed information than laypeople. But to the extent that the information they have is primarily related to average risk, it may mislead at least as much as it informs. Information, especially "hard" quantitative data, can be seductive as well as informative. It can direct attention away from qualitative issues because they are more vague and subjective than quantitative measures. Yet as "soft" and difficult to integrate into "objective" calculations as they may be, these other dimensions of risk are much too important to ignore.

Experts often seem to believe that if the lay public would only take the time to look at the data, they would come around to the experts' point of view. But these disagreements about risks do not simply come from stubbornness or lack of data. They are the result of fundamentally different ideas about what risk means and how it should be judged. These differences in concept will not evaporate in the presence of more statistical data. Of course, stubbornness and preconception can be a real problem—for experts as well as laypeople. Research shows that when people strongly hold to a particular point of view, they tend to interpret the usefulness and reliability of new evidence on the basis of whether or not it is consistent with their belief. If it is, it is relevant and useful; if not, it is obviously wrong or at least not representative.[41] When Plous gave supporters and opponents of a particular technology identical descriptions of a non-catastrophic disaster involving that technology in three studies at Wesleyan University, "supporters focused on the fact that the safeguards worked, while opponents focused on the fact that the breakdown occurred in the first place." Supporters were therefore reassured by the account while opponents were more convinced than ever that the technology posed a real threat.[42] The tendency to emphasize evidence that reinforces prior beliefs is a common, but not insurmountable problem. After all, we do sometimes change our minds in the face of new information.

There are no scientific grounds for dismissing either the more quantitative expert view or the more qualitative lay view of risk. Quite the contrary, decision theorists have developed strategies for decision making under risk and uncertainty that offer general support for both approaches. The "expected loss" strategy argues that each possible hazardous outcome of an activity or technology should be weighted by the chance that it will occur, then combined in a weighted average estimate of the danger involved. The risk associated with the activities or technologies can then be rated and compared on the basis of this average risk. The possibility of a catastrophic outcome is not ignored, but if that outcome is judged extremely unlikely, it will have a very small weight and thus little effect on the average. Experts are essentially using this approach when they judge the riskiness of hazards on the basis of average annual fatality rates. On the other hand, decision theorists have also developed the "minimax" strategy. This approach focuses attention on the worst possible outcome associated with each activity or technology whose riskiness is being judged. The strategy then calls for ranking the activities or technologies in terms of their worst cases. The riskiest is the one whose worst outcome is the most damaging of all the worst outcomes, the next most risky is the one whose worst outcome is the next most damaging and so on. Assessing the risk of alternative activities or technologies with "minimax" is more conservative than using "expected loss." It minimizes the damage even if you are very unlucky and the worst possible outcome occurs. The minimax approach seems more closely related to how laypeople evaluate risk.

Though it is possible to argue that the "expected loss" strategy is superior to the "minimax" strategy, it is just as possible to argue the opposite. There is no objective way of deciding which approach is better. Which is better depends on which characteristics of risk are considered more important. The lay public's more complex, multidimensional concept of risk is certainly more difficult to work with, but it does seem more realistic. In any case, the attitudes of the public should be taken seriously in matters of public policy—especially when those attitudes are not the result of ignorance (lack of information), but spring from a legitimate difference of opinion.

Overconfidence

Even those who are experts in system design, operation or maintenance tend to overlook some possible pathways to failure. It is easy enough to understand how this could happen, but it is troubling because it leads to systematic underestimates of the chances of failure. If we believe our own risk estimates, we are likely to be overconfident in the system's reliability. When dealing with dangerous-technology systems, overconfidence is hazardous to our health.

There is little evidence that we do—or that we can—completely overcome this problem even when the stakes are extremely high. Among the many criticisms leveled at the Rasmussen nuclear reactor safety study was that it did not consider improper maintenance, failure to obey operating rules when shutting the reactor down, and sabotage as possible causes of system failure. The study did consider human error. In fact, "The results of the Rasmussen report are dominated by single failures, most of which involve identified human error."[43] But it neither enumerated nor evaluated all of the important ways in which human error could lead to potentially disastrous reactor failure.

Fischhoff describes a number of categories of pathways to failure that are particularly prone to being overlooked.[44] First among them are pathways involving human error or intentional misbehavior. It is extremely difficult to think of all the ways people might cause problems. Who would expect that a technician would decide to use a candle to look for air leaks and so begin a fire that would disable the emergency core-cooling system at the Brown's Ferry nuclear power plant? How could it be foreseen that the operators at Three Mile Island would repeatedly misdiagnose the reactor's problems and so take inappropriate actions that made matters much worse? A second common cause of omissions is the failure to anticipate ways that the technological, societal or natural environment in which the system operates may change. Suppose unusually heavy rainfall in a normally dry area causes flooding because of inadequate drainage systems, crippling the transportation system on which the area relies. Transportation designers would have had to foresee both the possibility of excessive rainfall and the fact that designers of the drainage system would not foresee (or would fail to provide for) that possibility in order for flooding to have been included as a pathway to failure of the transportation system.

A third category is failure to foresee all the interactions within the system that affect the way in which the system as a whole functions. The National Academy of Sciences study of the effects of nuclear war, for example, concluded that food supplies would not be endangered for survivors because food plants could tolerate the additional ultraviolet light that would penetrate the partially destroyed ozone layer of the earth.[45] But they failed to consider that the increased flow of ultraviolet (among other things) might make it virtually impossible for people to work in the fields to grow and harvest the crops. Yet another related cause of omissions is overlooking the chances of common mode failure (see Chapter 9). Finally, omissions arise from overconfidence in our scientific and technical capabilities. Most risk analyses assume that the system has been correctly designed, and that it won't fail if all of its components work properly. But as we saw in Chapters 9 and 10, this degree of confidence is not always warranted.

Psychological studies of risk perception and assessment throw some light on just how overconfident we can be about what we know. In one study, participants given paired lists of lethal events were asked to indicate which cause of death in each pair was more frequent. Then they were asked to give the odds that each of their answers was correct. Many people expressed high confidence in the accuracy of their answers: odds of 100 to 1 were given 25 percent of the time. Yet about one out of every eight answers given so confidently were wrong (less than one in a hundred should have been wrong at those odds). About 30 percent gave odds of greater than 50 to 1 that they were correct in their assertion that there were more homicides than suicides in the U.S. each year.[46] Actually, the suicide rate was more than 30 percent higher.[47] In another set of experiments, subjects were again asked to answer a series of questions and then indicate how sure they were that their answers were correct. Answers assigned odds of 50 to 1 were wrong about 25 percent of the time, and therefore should have been given odds of only about 3 to 1. When subjects gave odds of 1,000 to 1 that their answers were correct, they actually should have been giving odds of just 5 to 1.[48] In another experiment, when subjects gave odds of 1,000 to 1, they would have been correct to give odds of only 7 to 1; at 100,000 to 1, they should have given odds of just 9 to 1. There were even a large number of answers for which people gave odds of 1,000,000 to 1 that they were right—yet they were wrong more then 6 percent of the time (true odds of 16 to 1). Subjects gave odds of 1,000 to 1 or greater about 25 percent of the time.[49] They were so sure of their answers that they were willing to bet money on them. Extreme overconfidence was "widely distributed over subjects and items."[50]

Other studies asked diverse groups of people to make many different kinds of quantitative judgements. Instead of giving a single numerical answer, they were asked to give their estimates in the form of upper and lower bounds that they were 98 percent sure included the true answer. Rather than falling outside those "confidence limits" only 2 percent of the time, as it should, the true value fell outside those bounds 20 to 50 percent of the time. People are clearly much more sure of their ability to judge risks than they should be.[51]

There are two unfortunate conclusions to be drawn from this literature. First, overconfidence is not only widespread, it also gets worse as the task at hand becomes more difficult. One would hope and expect that people confronted with more trying tasks would be more tentative about the correctness of their conclusions. Yet, "Overconfidence is most extreme with tasks of great difficulty."[52] Second, and even more troubling, experts seem as likely to be overconfident as laypeople, especially when they must go beyond the bounds of available hard data. For example, seven internationally known geotechnical engineers were asked to predict how high an embankment would cause a clay foundation to fail, and to give confidence limits around this estimate that would have a 50 percent chance of enclosing the correct value. None of them estimated the failure height correctly. Worse, the proper height didn't even fall within *any* of the confidence intervals they specified.[53] The Grand Teton Dam encountered many serious problems during construction. Yet its engineers were absolutely certain that they had solved them—until the dam collapsed in 1976. About 1 in 300 dams fail when the reservoir is first filled, but it still does not seem to be routine practice to calculate failure probabilities for new dams.[54]

According to Slovic, Fischhoff and Lichtenstein, "The psychological basis for this unwarranted certainty seems to be people's insensitivity to the tenuousness of the assumptions upon which their judgments are based. . . . Such overconfidence can keep us from realizing how little we know and how much additional information is needed about the various problems and risks we face."[55] In addition to thinking they know more than they do know, people also tend to assume that the mental processes they use to help them work through complex problems do not lead to systematic bias. That simply does not seem to be true. Some psychologists argue that the "heuristics" (rules of thumb) that people adopt when dealing with difficult problems may easily lead them astray. For example, when asked to give a range of estimate for a particular risk, people often use a heuristic called "anchoring and adjustment." They think first of a specific number (the anchor), then adjust up and down from that number to get confidence limits that seem right to them. It is easy to influence the choice of that initial anchor by the way the question is asked. Yet once the anchor is chosen, it has such a powerful impact on their thinking that there is a strong tendency to be overconfident, to give upper and lower bounds that are much too close together.[56]

It has also been argued that people work through complex decision problems one piece at a time. To make the process less stressful, they ignore the uncertainty in their solutions to earlier parts of the problem when they are working through later parts. Because the solutions of the later parts are built on the solutions of the earlier parts, ignoring the tentativeness of their earlier conclusions makes them overconfident in the final conclusions.[57] Overconfidence can also be a reaction to the anxiety that uncertainty creates: "One way to reduce the anxiety generated by confronting uncertainty is to deny that uncertainty. . . . [P]eople faced with natural hazards . . . often view their world as either perfectly safe or predictable enough to preclude worry. Thus, some flood victims interviewed . . . flatly denied that floods could ever recur in their area. Some thought (incorrectly) that new dams and reservoirs in the area would contain all potential floods, while others attributed previous floods to freak combinations of circumstances unlikely to recur."[58] Since anxiety is greater when the problem faced is more complex or consequential, this might explain why people are often more overconfident in such situations. This does not bode well for our ability to accurately assess the risks posed by dangerous technologies, where the problems we confront are almost always both complex and highly consequential.

A CALCULUS OF CATASTROPHE

Because many critical public policy decisions—especially those involving dangerous technologies—are thoroughly bound up with risk, we have a compelling interest in more accurate risk assessment, as well as in a greater degree of honesty about what we do not, and perhaps cannot, know. As Judge David Bazelon of the federal appeals court in Washington has put it, "[one] kind of uncertainty that infects risk regulation comes from a refusal to face the hard questions created by lack of knowledge. It is uncertainty produced by scientists and regulators who assure the public that there are no risks, but know that the answers are not at hand. Perhaps more important, it is a false sense of security because the hard questions have never been asked in the first place."[59]

Our ways of evaluating risks that involve dangerous technologies are also flawed because we have yet to develop a clear way of thinking about low-probability-high-social-consequence events. As a result, we confuse very unlikely events that have enormous social consequences with very unlikely events that are harmful to those involved but have little effect on society.

In the long debate over nuclear power, it has been common, beginning with the Rasmussen report, to see statements like "the probability of getting killed as a result of a nuclear core meltdown in a power reactor is about the same as the chance of dying from a snakebite in mid-town Manhattan." In one sense, the comparison seems valid. Both possibilities are very unlikely and, after all, dead is dead. But the underlying events are not at all comparable. A snakebite might kill the person bitten, but it has little effect on the wider society. But a core meltdown that breaches a reactor's containment and spews deadly radioactivity all over the landscape is an event of enormous long-term social consequence, with the potential for taking hundreds of thousands of lives and doing billions of dollars worth of property damage. The huge difference in their social impact puts these two kinds of events into entirely different categories. Even if their probabilities were mathematically identical, they are simply not comparable.

We also do not know how to properly aggregate technological risks that have low probability but high social consequence. We look at each risk separately. If its probability is low enough, we call it acceptable, and then often unconsciously take that to mean that it won't happen. Then we develop and use the technology involved. But looking at these risks at a point in time and in isolation from each other is foolish. They are neither static nor independent. They cumulate in three different ways, all of which make catastrophe more likely. First, they grow over time. No matter how unlikely an event is (unless it is completely impossible), the probability that it will happen at least once increases as we consider longer and longer periods of time. Given enough time, it is certain to occur. Second, the larger the number of dangerous technologies we create and the more widespread their use, the more low-probability-high-consequence events there are. The more low-probability-high-consequence events there are, the higher the probability that at least one of them will happen. As we add new toxic chemicals to the environment, the probability of ecological disaster increases; as we add to the size and variety of arsenals of weapons of mass destruction, the probability of catastrophic war or weapons accident increases, and so on. In other words, the more ways there are for technology to generate a catastrophe, the more likely it is that at least one of them will do us in.

Finally, there is the very real possibility of synergies. As the actual and potential damage we do by one route increases, it may increase the probability of disaster by another route. A future Chernobyl, spewing dangerous radioactivity across national borders, might just be the spark that could ignite an already tense international situation and cause a war in which weapons of mass destruction are used. Or chronic ecological damage done by one set of toxics could magnify the impact of a different set of toxics to the point of precipitating a major environmental disaster. With greater numbers of increasingly complex dangerous technologies in use, it becomes impossible to even imagine all the ways in which synergies like these could raise the probability of technology-induced disaster to the point of inevitability.

It would be very helpful if we could develop a "calculus of catastrophe"—a method that focused our attention on low-probability-high-social-consequence events and allowed us to clearly evaluate the risk associated with each of them over time and in combination with each other. We could then monitor the total social risk and take care not to do anything that pushed it above the level of risk we consider acceptable. Such a calculus of catastrophe would be particularly useful in making social decisions about risks involving dangerous technologies. But as with so many other social problems in which technology plays a role, here too we must avoid the illusion of a simple technical fix. No matter how clever, no method of calculation is going to solve the problem of differing judgements as to what dimensions of risk are most important, or save us from the hard work of dealing with the complex social, economic and political problems of making decisions on technological risk. And no method of calculation can completely come to terms with the inherent unpredictability that human fallibility unavoidably introduces.

It may seem strange, even foolish, to be so concerned about the possibility of highly improbable technological disasters, especially those that have never happened. But remember, anything that has ever happened, happened once for the first time. Before that time, it too had never happened. And for some of these catastrophes, we cannot afford even a first time.

In this and in the previous chapters, we have carefully considered the many ways in which disaster might be triggered by the clash of our innate fallibility and the dangerous technologies with which we have filled our world. We have looked at a wide range of often terrifying problems. Now it is time to look for possible solutions. In the next chapter, we consider what might be done to prevent our technological society from succumbing to one or another of the self-induced catastrophes that lie waiting for us along the path we have been following.

CHAPTER TWELVE

Preventing Disaster

The clash between our growing technological power and our unavoidable fallibility has laid us open to grave danger. We have not only gained the knowledge required to trigger any of a variety of global catastrophes, we have constructed the means. There is no longer any corner of the planet so remote that it is safe from the havoc we can now cause, intentionally or by accident.

There is no reason to believe that we will ever forget how to do that which we have learned how to do. And there is every reason to believe that our frailty, our tendency to make mistakes is a permanent part of what makes us human. Are we then doomed to a world of increasing disaster, a world that may someday even see our species die by its own hand?

Ultimately, none of the dangerous technologies we have developed is really beyond our control. We can change them, limit how we use them, even eliminate them entirely. No external force compelled us to create the dangers we now face and no external force will prevent us from getting rid of them.

The technologies that have put us in danger did not come into being by accident. They were developed intentionally to serve purposes important enough to make us want to explore the ideas that lie behind them and then build the devices that gave those ideas physical form. We do not have to give up all the benefits that led us to create dangerous technologies and cause us to continue relying on them. We can find ways to enjoy the advantages of modern technology without exposing ourselves to self-induced catastrophe.

There is no technological fix, simple or complicated, for the problems created by the deadly combination of human fallibility and dangerous technology. But there is a solution. It lies at the intersection of technological capability, social wisdom and political will. There are four steps we must simultaneously take. The first begins with recognizing that there are some technologies whose potential for catastrophe is so great that fallible human beings have no business dealing with them at all. No matter what short-term benefits they seem to provide, in the long run they are simply too dangerous.

STEP I: ABOLISH WEAPONS OF MASS DESTRUCTION

Given the inevitability of eventual technical failure or human-induced disaster, abolition is the only sensible, long-term viable way of dealing with technologies capable of destroying human society and driving our species to extinction. Nuclear, biological and chemical weapons of mass destruction are the clearest present-day examples of such technologies. But is abolition really practical? Can the genie be put back into the bottle?

The Canberra Commission

In 1996, the government-sponsored Canberra Commission presented a comprehensive plan for abolishing nuclear weapons to the United Nations. It was not the first time nuclear abolition was proposed. But this time, the proposal came from a group with very unusual credentials, from four of the five then-declared nuclear weapons states. The commission included General George Lee Butler, former commander-in-chief of the U.S. Strategic Command, the officer in charge of all American strategic nuclear weapons from 1991 to 1994; Michel Rocard, Prime Minister of France from 1988 to 1991; Field Marshal Michael Carver, former chief of the British Defence Staff; Roald Sagdeev, former science advisor to the president of the Soviet Union; and Robert MacNamara, former U.S. secretary of defense and a key figure in developing the policy of security through "mutually assured destruction," which drove much of the nuclear arms race. The lengthy report argues, "The proposition that large numbers of nuclear weapons can be retained in perpetuity and never used—accidentally or by decision—defies credibility. The only complete defense is the elimination of nuclear weapons and assurance that they will never be produced again." [1]

The commission goes farther, arguing that nuclear weapons have little military utility, and do not even provide security, their main raison d'être:

> Possession of nuclear weapons has not prevented wars . . . which directly or indirectly involve the major powers. They were deemed unsuitable for use even

when those powers suffered humiliating military setbacks [as did America in Vietnam and the Soviets in Afghanistan]. . . . Thus, the only apparent military utility that remains for nuclear weapons is in deterring their use by others. That utility . . . would disappear completely if nuclear weapons were eliminated. . . .

The possession of nuclear weapons by any state is a constant stimulus to other states to acquire them. The world faces threats of nuclear proliferation and nuclear terrorism. . . . For these reasons, a central reality is that nuclear weapons diminish the security of all states.[2]

The year before the Canberra Commission report, U.S. Air Force General Charles A. Horner, head of the North American Aerospace Defense Command (NORAD), had become the first active-duty officer to publicly call for the abolition of nuclear weapons, saying, "I want to get rid of them all. . . . Think of the high moral ground we secure by having none. . . . It's kind of hard for us to say . . . 'You are terrible people, you're developing a nuclear weapon' when the United States has thousands of them."[3] On December 8, 1996, 60 retired generals and admirals from the all of the then-declared nuclear-armed nations (the United States, Russia, England, France and China) signed a joint statement at the United Nations endorsing the idea that nuclear weapons can and should be completely eliminated.[4]

Very much the insider, with decades of experience in the U.S. nuclear military, General Butler has continued to forcefully argue the case for abolition. In an article based on his speech to the National Press Club in early 1998, Butler declared,

> I was engaged in the labyrinthine conjecture of the strategist, the exacting routines of the target planner, and the demanding skills of the air crew and the missileer. I was a party to their history . . . witnessed heroic sacrifice and the catastrophic failure of both men and machines. . . . Ultimately, I came to these judgments:
>
> - . . . the risks and consequences of nuclear war have never been properly weighed by those who brandish them [nuclear arms]. . . .
> - The likely consequences of nuclear war have no political, military or moral justification.
> - The threat to use nuclear weapons is indefensible.
>
> . . . We cannot at once keep sacred the miracle of existence and hold sacrosanct the capacity to destroy it. . . . It is time to reassert . . . the voice of reason.[5]

Although most people are not aware of it, for years all the nuclear-armed countries have had a formal treaty commitment to do exactly what the generals, the admirals and the Canberra Commission have urged. Along with 173 other nations, they signed the Treaty on the Nonproliferation of Nuclear Weapons, Article VI of which reads, "Each of the Parties to the Treaty undertakes to pursue negotiations in good faith on effective measures relating to cessation of the nuclear arms race at an early date and to nuclear disarmament. . . ."[6] For the United States, Russia and Britain, that commitment is longstanding: they signed the treaty in 1968. It is long past time for them to make good on their promise.

Inspection and Verification

The elimination of nuclear, chemical and biological weapons of mass destruction is not only practical, it is critical to our security and our survival. Even so, abolition will not and should not be achieved overnight. A careful process of arms reduction requires time to verify that each nation has done what it has agreed to do at each step of the process. It also takes time for everyone to gain confidence that the process is working. Care must be taken to create a world open enough to inspection that any meaningful violation of the treaty will be quickly detected.

For more than seven years after the defeat of Iraq in the Persian Gulf War, UN inspectors combed the country for evidence of clandestine programs aimed at developing biological, chemical and nuclear weapons of mass destruction. As a condition of ending the war, Iraq had agreed to allow free access. Nevertheless, the Iraqi government continually interfered with and harassed the inspectors. From spying on them to find out about surprise inspections to blocking access to so-called presidential sites in late October 1997 (even at the cost of triggering a three-week crisis that nearly led to war), Iraq tried every trick in the book to frustrate the inspection process.[7]

Inspectors from the International Atomic Energy Agency (IAEA, a division of the UN) first tried "environmental monitoring" in Iraq after the war. The technique uses advanced technologies to test for the presence of materials likely to be emitted by forbidden activities. Though IAEA used this approach only to search for nuclear materials, there is no reason why it could not also be used to detect at least some suspicious chemical or biological weapons activities. Environmental monitoring can certainly provide evidence of illicit activities at declared sites and may also be able to detect covert activities at facilities that have never been declared. According to the U.S. Office of Technology Assessment,

> In the month between the end of the war and the start of inspections, Iraq removed much of the most incriminating equipment . . . and concocted stories to explain the remainder. Inspectors took samples of materials within and near facilities, and swipes of dust that had collected on the surfaces of equipment.

These were analyzed at various laboratories . . . [and] played a key part in demolishing Iraq's cover stories and exposing its nuclear weapon program. . . . No industrial process can prevent minute traces of materials from escaping. Even the most sophisticated filtration systems can only reduce, not eliminate, releases. . . .

The IAEA believes that if environmental monitoring had been part of routine safeguards inspections in the 1980's [well before the Persian Gulf War], it would easily have revealed Iraq's weapons activities.[8]

Environmental monitoring can also be used to detect secret nuclear weapons tests carried out in violation of the Comprehensive Test Ban Treaty. An on-site inspection team armed with tubes for boring into the suspected site of the test could find evidence of telltale radioactive trace gases using relatively new, sensitive techniques weeks to more than a year after even a "well-contained" clandestine explosion.[9] Although it would be unwise to depend solely on environmental monitoring (or any other single technique) to verify compliance with abolition agreements, it should be extremely useful as part of an inspection and verification regime. More research aimed at refining and improving the technique would be a very good idea.

Despite all of Iraq's obstructionism, unarmed UN inspectors were able to find and destroy more weapons of mass destruction in Iraq than were destroyed by military action during the Persian Gulf War.[10] Only time will tell if they were ultimately able to find and destroy enough to completely eliminate Iraq's programs. But the lesson of this experience is clear. Effective inspection requires both open access and fierce determination on the part of well-trained inspectors, fully equipped and strongly supported by all other parties to the disarmament agreement.

Ironically, the same problems of error and fallibility that call us to abolish weapons of mass destruction also make it obvious that no system of inspection can be guaranteed to work perfectly all the time. There is simply no way to eliminate risk from the world. Yet the risks involved in carefully and deliberately eliminating weapons of mass destruction are far less threatening than the certainty of eventual catastrophe we face if we do not.

What if some nation someday does manage to slip through a crack in the system and secretly build a nuclear weapon without being detected? It may seem obvious that that would give them a decisive political advantage in a world without weapons of mass destruction. But it is not at all obvious. At some point the rogue government would have to make it known that it had violated the treaty and built such a weapon in order to have any chance of political gain. There is no way to achieve political advantage from a weapon no one knows exists. That revelation would instantly make it a pariah, subject to immediate political isolation, economic boycott or even conventional military attack by every other nation in the world. There is nothing that country could do with one or two nuclear weapons to prevent this from happening. It could not even be sure to devastate the nation it had built the bomb to threaten, let alone prevail against such overwhelming odds.

After all, in the aftermath of the horrors of World War II, the United States was the only nation that had nuclear weapons or the knowledge required to build them. Yet America's nuclear monopoly did not allow it to bend every other nation to its will. If a powerful United States could not even dictate terms to a world of heavily damaged, war-weary nations (the Soviet Union being the most striking example), it is difficult to see how a rogue nation with one or two nuclear weapons would be more able to dictate terms to a strong, healthy, intact world at peace. The leaders of that nation might be able to cause a lot of pain, but there is nothing they could actually gain by having one or two nuclear weapons that would be worth the enormous risks involved in building them.

Inspection and verification systems do not have to be perfect for the abolition of nuclear, chemical and biological weapons of mass destruction to succeed. They only need to be good enough to insure that any attempt to build a significant arsenal will be caught and exposed.[11] It is not easy to design, and even more difficult to implement systems that good, but it is certainly possible. Some interesting, creative and practical ideas have already been suggested.[12] Perfection is simply not required.

Apart from the problems of developing a disarmament process supported by effective systems of verification and inspection, it is not a simple matter to eliminate physically the world's stockpiles of weapons of mass destruction. There are serious technical issues involved in safely destroying the weapons.

Destroying Chemical and Biological Weapons

Chemical weapons can be destroyed by incineration and chemical neutralization.[13] The U.S. Army chose the incineration option, building its first prototype chemical weapons incinerator (JACADS) on Johnston Atoll in the Pacific. JACADS experienced more than eight explosions and at least five releases of deadly nerve gas into the atmosphere between 1990 and 1995.[14] Then, on August 22, 1996, the Army's first chemical weapons incinerator in the continental United States began operating at Tooele Army Depot in Utah. Within six months, it had burned 11,471 M-55 nerve gas rockets and 53 bulk containers filled with almost a ton of sarin nerve gas each. It had also failed repeatedly, forcing the Army to shut it down six times. Three high-ranking former employees dismissed from the incineration project charged that safety at the facility was lax.[15] Despite this, operations at Tooele continued unabated and by April 1998, 14,000 rockets, bombs and bulk containers had been destroyed along with 1,500 tons of sarin.[16]

Fears that this and other chemical weapons incinerators planned for the United States might someday release poisonous gas over a populated area have led to growing public pressure to abandon disposal by incineration. This sentiment was given a boost in the fall of 1996, when a panel of the National Research Council endorsed neutralization over incineration as the technology best able to protect human health and the environment while effectively destroying the chemical weapons by the Army's 2004 deadline.[17] At the end of World War II, the Allies captured nearly 300,000 tons of chemical weapons manufactured by Nazi Germany. More than two-thirds of them were dumped into the oceans by the victorious British, French and American forces. They are still there. The 1992 Chemical Weapons Convention, which requires destruction of all chemical arsenals by 2007, specifically excludes weapons dumped into the sea before 1985. Some scientists believe it is less dangerous to leave them alone than to try to recover and destroy them, especially if they are in deep water. Others argue that they are a ticking time bomb that must be defused, pointing out that some mustard gas bombs have already washed up on beaches in Germany and Poland.[18] This is yet another dimension of the chemical weapons disposal problem we cannot afford to ignore. But we must think long and hard before deciding what to do.

Though there is no public information available on the size or content of any stockpiles of biological weapons that might exist, there has been considerable biological warfare research. In early 1998, Dr. Kanatjan Alibekov ("Ken Alibek"), a medical doctor and former colonel in the Soviet military, began to speak publicly about the size and scope of the Soviet/Russian biological weapons development effort.[19] Alibekov, who left the former Soviet Union for the United States in 1992, had been second-in-command of a branch of that program. He claimed that, by the early 1980s, the Soviets had prepared many tons of anthrax and plague bacteria and smallpox virus that could be mounted for delivery by intercontinental ballistic missile with only a few days notice. Although the offensive program was officially cancelled by Soviet President Gorbachev in 1990 and again by Russian President Yeltsin in 1992, research continues on defense against biological attack (developing vaccines, protective masks, etc.). Alibekov believes some offensive research— including work on "genetically altered antibiotic-resistant strains of plague, anthrax, tularemia and glanders" and work on the Marburg virus—is still being done under the cover of the ongoing defensive program.[20]

The U.S. offensive biological weapons program was officially cancelled in the 1970s, but, as in Russia, "defensive" research continues that could provide cover for otherwise forbidden offensive research. After all, it is hard to develop a good defense, say an effective vaccine, without knowing what form of which organism might be used in the attack, and knowing what organism might be used requires a degree of offensive research.

It should be possible to destroy any biological toxin that has been accumulated as a result of this research (and any clandestine weapons programs) by processes similar to those used to destroy toxic chemicals. But even greater care must be taken to destroy the virulent living organisms that have been studied, especially any that are genetically modified or newly created and do not exist in nature. Biotoxins cannot reproduce, but living organisms can. Not even one of these genetically engineered germs can be allowed to escape into the environment alive.

Dismantling Nuclear Weapons

Nuclear weapons cannot be incinerated or neutralized the way chemical and biological weapons can. They must instead be carefully taken apart. Since the Pantex plant in Amarillo, Texas, was the finally assembly point for all American nuclear weapons during the long Cold War, it logically became the point of disassembly when arms reductions began. In 1998, roughly 10,000 plutonium "pits" (radioactive cores) from dismantled nuclear weapons were in storage at Pantex, a facility with a less-than-pristine safeguard history (see Chapter 3). That amounts to about 30,000 to 40,000 kilograms of plutonium stored as pits in one location. (As little as one-billionth of a kilogram of plutonium can cause fatal cancer if inhaled in particulate form.) According to the GAO, those pits were stored in containers "that both the Department [of Energy] and the Defense Nuclear Facilities Safety Board believe are not suitable for extended storage."[21] Many of the containers were stored in modified World War II–vintage bunkers originally designed and built to hold conventional munitions.[22]

Recognizing that it is far too dangerous to continue to store this much plutonium metal in this form, the Department of Energy (DoE) announced in January 1997 that it had decided to follow a two-track approach to more permanent disposal.[23] Some of the plutonium is to be "vitrified" (mixed with other highly radioactive waste then solidified as a glass) and stored in steel canisters as nuclear waste. The rest is to be converted into oxide form, mixed with oxides of uranium and fabricated into mixed-oxide (MOX) reactor fuel to be "burned" in civilian nuclear power reactors. Weapons-grade uranium from dismantled nuclear weapons can be made less dangerous by mixing with natural uranium and making the "blended down" uranium into standard, low-enriched reactor fuel.

On the surface, the MOX option seems more promising. "Burning" plutonium in reactors sounds like a more complete way to destroy it. Using it to generate electric power also sounds more cost effective than just throwing it away. But a closer look reveals serious technical and economic problems with MOX. For one thing, plutonium does not actually "burn" in reactors the way oil or coal burns in a furnace. Running MOX through reactors does convert weapons-grade plutonium (Pu-239) both into plutonium isotopes less ideally suited for nuclear weapons (Pu-238, 240 and 242) and into other elements, but it also turns some uranium into plutonium.[24] In the end, there is still plenty of plutonium to be disposed of, now contaminated with newly created highly radioactive waste. The plutonium could still be separated and used for weapons, although it would require a fairly sophisticated processing plant.[25] More troubling still, the MOX approach may encourage international commerce in plutonium-based fuels, raising the risk of accidents, theft, nuclear terrorism and nuclear weapons proliferation.

The MOX option is also not cheap. It requires new production facilities as well as changes in existing facilities. According to calculations by William Weida (based on DoE data released in 1996), the total life-cycle cost of fuel fabrication, capital investment and operations involved in using MOX fuel in commercial nuclear power reactors is between $2.5 billion and $11.6 billion.[26] The low-end estimate assumes modifying existing light-water reactors to run on MOX fuel; the high-end estimate assumes building new power reactors specifically designed for this purpose. Either way, the cost of safely storing the extra nuclear waste generated by the MOX approach must also be added.

Yet another problem with the MOX option is that the metallic element gallium is alloyed with the plutonium used in American nuclear weapons. In high concentrations, gallium interferes with fabrication of MOX fuel. It also chemically attacks the zirconium metal used in the tubes that hold reactor fuel, causing potentially serious deterioration problems after the fuel is used. Current processes for extracting gallium are expensive and generate large quantities of liquid radioactive waste. Gallium is not known to pose any particular problem for vitrification.[27]

In March 1996, DoE opened a vitrification plant at Savannah River near Aiken, South Carolina. Since the plant is designed to deal with liquid nuclear waste, using it to vitrify metallic plutonium pits would require first processing them into some form that could be mixed with the waste. The facility mixes radioactive waste with molten glass, pours the mixture into stainless steel cylinders, and solidifies it into glass "logs" ten feet long and two feet in diameter. Operations at the plant are highly automated, since this process only renders the radioactive material easier to handle and harder to extract, not less deadly. (A worker standing by one of the glass-filled steel cylinders would absorb a lethal dose in minutes.) Like the MOX option, this approach is not cheap. The $2.4 billion plant produces only one glass log per day, making the cost over the projected life of the plant about $1.4 million dollars per log. At that rate, it will take 25 years even to vitrify all of the 36 million gallons of liquid waste already stored in 51 underground tanks at the site.[28]

Vitrification is far from ideal, but at the current state of knowledge, it seems to make more sense than MOX. But whatever approach is used, it is important to keep in mind that the end product will still be dangerous radioactive material that has to be safely stored, kept isolated from the environment and protected for very long periods of time.[29] Even so, it will pose much less danger to us than the danger posed by continuing to store pure weapons-grade plutonium metal, let alone by continuing to maintain large stockpiles of intact nuclear weapons.

Talk and Double Talk

In recent years, strategic arms talks have made progress in reducing nuclear arsenals. Under the first Strategic Arms Reduction Treaty (START I), the number of city-destroying Russian and American nuclear warheads mounted on long-range delivery systems (missiles and bombers) is being reduced to 7,000-8,000 each, though both countries will keep thousands more in reserve. Under START II, those limits will be cut in half.[30] In March 1997, Presidents Clinton and Yeltsin agreed in principle that negotiations on START III would aim at reducing arsenals to 2,000-2,500 warheads.[31] The Comprehensive Test Ban Treaty has been signed, prohibiting nuclear test explosions in order to inhibit further nuclear weapons development. The American government insists that it has stopped developing new nuclear weapons. It appears that we are well on our way to ridding the world of nuclear weapons. But appearances are deceiving.

For one thing, that the START agreements constitute a major reduction in strategic nuclear weapons is one measure of how absurdly overbuilt nuclear arsenals had become. Even after START III is negotiated and implemented, each nuclear superpower will still have at least 2,000 nuclear weapons mounted and ready to go, with thousands more doubtless in reserve. Since the 1960s, Pentagon calculations showed that 200-300 warheads delivered on target were enough to destroy, and thus should be enough to deter, any nation on earth. The United States has begun the nuclear "Stockpile Stewardship and Management Program," under which DoE has undertaken a billion-dollar expansion of Los Alamos National Laboratory's capacity to produce new plutonium pits for nuclear weapons.[32] At the same time, DoE is "beating plowshares into swords," breaking a 50-year taboo by using *civilian* nuclear power reactors to manufacture tritium for nuclear weapons use.[33]

Sold as a program for insuring the stability and safety of stockpiled nuclear weapons against deterioration, Stockpile Stewardship is nothing more than a high-tech means of subverting the spirit of the START process and the Comprehensive Test Ban Treaty. The program, developed largely to buy off the nuclear weapons R & D establishment, is lavishly funded at $4.5 billion a year—more than the average annual cost of building up the nuclear arsenal during the Cold War![34] In the Department of Energy's Orwellian language, Stockpile Stewardship will not develop "new weapons," but will instead seek "structural enhancements" to nuclear weapons that might be needed to satisfy "changes in military requirements."[35] What does that mean if not the development of new nuclear weapons? In early 1998, the United States added the newly developed B-61-11 earth-penetrating nuclear warhead to its arsenal. The arsenal did not previously have an earth-penetrating nuclear warhead. But DoE and the Pentagon insist that the B-61-11 is not a new nuclear weapon. We are also manufacturing new, more accurate D-5 missiles to replace the C-4 missiles on Trident ballistic missile submarines. But these too are not "new" nuclear weapons.[36]

This is double talk, plain and simple. It may be technically within the bounds of the treaty commitments we have made, but it is clearly in violation of the spirit of post–Cold War arms reduction. More importantly, it is making the world a more dangerous place. It will not get us where we want and need to go.

We humans are a contentious and quarrelsome lot. It will be a long time, if ever, before we learn to treat each other with the care and respect with which we would all like to be treated. But as cantankerous and conflictual as we may be, there is no reason why we cannot learn to manage and resolve our conflicts without maintaining huge arsenals of weapons that threaten our very existence. In this increasingly interconnected and technologically sophisticated world of fallible human beings, these weapons have become too dangerous for us to continue to rely on them for our security. The very idea that weapons of mass destruction can provide security is an idea whose time has gone. And nothing should be less powerful than an idea whose time has gone.

STEP II: CHOOSE NEW, MORE EFFECTIVE SECURITY STRATEGIES

Abolishing weapons of mass destruction will not by itself make us secure. In a troubled world of nations and people that still have a great deal to learn about getting along with each other, we need some other way of protecting ourselves. But how?

The Canberra Commission's answer is for the major powers to rely on large and powerful conventional military forces instead. While that might be an improvement, there are other more effective and much less expensive ways of achieving security. Among the alternative security strategies worthy of a closer look are nonoffensive defense, the more efficacious use of economic sanctions and using properly structured economic relations to create positive incentives to keep the peace.

Nonoffensive Defense

One form of nonoffensive defense is a military approach known as the porcupine strategy. Its advocates argue that a nation is most secure when, like a porcupine, it has a very strong defense against anyone who tries to attack and yet has no capability to launch an attack against anyone else. Because it has no offensive military capability, there is no reason for other nations to worry that it might attack them. Therefore, there is no reason for other nations to build up their own military forces to counter the threat of attack, and no reason to think about launching a pre-emptive strike to prevent it. At the same time, powerful defensive forces make it clear to any potential aggressor that attacking would be pointless, since the devastating losses they would suffer would overwhelm any possible gains.[37] It is also possible to conceive of nonmilitary or mixed versions of nonoffensive defense.[38]

Nonoffensive defense may be an easier strategy to sell politically to nations that do not have "great power" status, but even the largest and most powerful nations would do well to thoroughly explore the advantages it offers. Unlike more traditional offensive military approaches to security, the more nations decide to adopt it, the more secure all of them become.

Economic Sanctions

A different but compatible approach is to use nonmilitary forms of force against countries that threaten the peace. Most prominent among them are economic sanctions (trade embargoes, financial boycotts and the freezing of assets abroad). The strongest sanction is a total ban on exports to and imports from the target country. Economic sanctions have a reputation for being weak and nearly always ineffective. But they actually work much better than most people think.

In 1990, Gary Hufbauer, Jeffrey Schott and Kimberly Elliott published a comprehensive analysis of the use of economic sanctions in international relations since World War I.[39] They analyzed 116 cases, most of which involved the sanctions applied by the United States. Using a consistent set of criteria to score each case, they judged that economic sanctions were successful about a third of the time. That may not seem terribly impressive, but no security strategy, military or nonmilitary, works all the time. Even if they cannot reliably do the whole job, sanctions clearly do work well enough to play a serious role in providing security.

If we can better understand why sanctions succeed when they do and why they fail when they do, we can surely make them work better. To that end Hufbauer, Schott and Elliott tried to analyze the conditions under which economic sanctions have succeeded in the past. Their conclusions were: (1) the higher the cost sanctions imposed on the target country, the more successful they were. On average, sanctions that cost the target country about 2.5 percent of its gross national product (GNP) succeeded, while sanctions that cost 1 percent or less failed; (2) sanctions do not work quickly. Overall, an average of almost three years was required for success; (3) the success rate was higher when the target country was much smaller economically than those applying the sanctions; (4) applying sanctions quickly and in full force increases the chances that they will succeed.[40] As the world economy becomes more interconnected, it is less likely that economic sanctions can be successfully applied by any one nation acting alone. There are simply too many alternative trading partners and too many ways of circumventing sanctions with the help of nations that do not support them. At the same time, increased international interdependence makes broadly imposed sanctions more likely to work.

The current process for imposing multilateral economic sanctions against countries threatening the peace can and should be improved. One way to do this would be to create a special body within the United Nations to deal exclusively with sanctions and peacekeeping. With a broader membership than the Security Council, this body would decide on what kind of sanctions, if any, should be imposed when accepted norms of international behavior are violated. Certain acts could be defined in advance as so unacceptable as to trigger powerful sanctions automatically. For example, it could be agreed that any nation that has invaded or attacked the territory of another nation would be subject to an immediate and total trade embargo by all UN members. The embargo would be lifted only when the attacks ceased, its military was entirely withdrawn and an international force was installed to monitor compliance.[41]

Economic force can often be a useful and effective substitute for military force. But make no mistake, it is still a strategy of force and it will still cause a lot of pain, often among those most vulnerable. The greatest contribution economic relationships can make to global security lies not as a form of punishment, but as a means for strengthening positive incentives to keep the peace.[42] It allows us to get out of the trap of always reacting to crises and move instead toward preventing conflicts from degenerating into crisis and war.

Economic Peacekeeping

More than 20 years ago, Kenneth Boulding set forth his "chalk theory" of war and peace.[43] A piece of chalk breaks when the strain applied to it is greater than its strength (ability to resist that strain). Similarly, war breaks out when the strain applied to the international system exceeds the ability of that system to withstand the strain. Establishing stable peace requires that strains be reduced and strength be increased. So to realize the potential contribution of the international economic system to global security, we must look for and implement a combination of strain-reducing and strength-enhancing strategies. Strategies that incorporate the four basic principles that follow should help to build the kind of international economic system that will provide nations strong positive incentives to keep the peace:[44]

1. *Balance Relationships.* Mutually beneficial, balanced economic relationships create interdependence that binds the parties together out of mutual self-interest. Everyone has strong incentives to resolve any conflicts that arise to avoid losing or even diminishing the benefits that these relationships provide. The European Community today is a working, real-world example of how effective this approach can be. The people of its 15 member nations have fought many wars with each other over the centuries. But though they still have many disagreements—some of them quite serious—they are now bound together in a web of balanced economic relationships so mutually beneficial that they no longer even think in terms of attacking each other. They debate, they shout, they argue, but they don't shoot.

2. *Balance Independence and Interdependence.* Economic independence reduces vulnerability and therefore increases feelings of security. But economic independence runs counter to the idea of tying nations together in a web of mutually beneficial relationships embodied in the first principle. If independence is emphasized where vulnerability is most frightening and interdependence is emphasized everywhere else, a balance can be struck that resolves this dilemma. Since vulnerability is greatest where critical goods (such as staple foods, water and basic energy supplies) are involved, they should be excepted from the general approach of maximizing interdependence implied by Principle 1.

3. *Emphasize Development.* There have been more than 150 wars, taking the lives of over 23 million people, since the end of World War II. More than 90 percent of them were fought in the less-developed countries.[45] Poverty and frustration can be a fertile breeding ground for conflict. Without

sustained improvement in the material conditions of life of the vast majority of the world's population living in the Third World, there is little hope for a just and lasting peace. It is also much easier to establish balanced relationships among nations at a higher and more equal economic level. They simply have more to offer each other.

4. *Minimize Ecological Stress.* Competition for depletable resources has been a source of conflict over the millennia of human history. Acute environmental disasters that cross borders, such as the nuclear power accident at Chernobyl and the oil fires of the Persian Gulf War, also generate conflict. So do chronic international ecological problems such as acid rain. Developing renewable energy resources, conserving depletable resources by recycling and a whole host of other environmentally sensible strategies are thus not only important for ecological reasons, they also help reduce the strain on our ability to keep the peace.

Dealing with the Terrorist Threat

Neither weapons of mass destruction nor conventional military forces nor economic sanctions can protect us from acts of terrorism perpetrated by subnational groups. Some people are driven to commit acts of terrorism by frustration born of grinding poverty and/or deliberate social and political marginalization, combined with an inability to make themselves heard. The threat they pose will not be effectively countered by any strategy of force. Even the most brutal, unfree police state cannot rid itself of terrorism. Its anti terrorist campaigns will only replace private-sector terrorism with government terrorism. The terrorism that grows from these roots can be countered by providing more civilized, nonviolent avenues for those with serious, legitimate economic, social and political grievances to be heard and to seek remedy for their problems.

The terrorism that arises from unbridled hatred, ethnic prejudice, sadism and other forms of mental illness and social pathology will not yield to this treatment. It must instead be fought with high-quality intelligence work, international cooperation and the full force of legitimate legal authority. Terrorists driven by hate and paranoia are also the most likely to one day see dangerous technologies as weapons of choice or targets of opportunity. Abolishing nuclear, chemical and biological weapons and eliminating stockpiles of weapons-grade materials will take us a long way toward overcoming the real and potentially catastrophic threat to our security of terrorism with dangerous technologies. But the threat will never be completely eliminated until we have found a way to eradicate terrorism itself.

It is important that terrorism, whatever its source, be taken seriously and opposed at every turn. But it is even more important that we keep the threat in perspective. We cannot afford to accept any serious compromise of our civil liberties simply because it is clothed in the rhetoric of antiterrorism. If we do, the terrorists will have done incalculable damage to us. And we will one day wake to find that we have lost both our freedom and our security.

STEP III: REPLACE OTHER DANGEROUS TECHNOLOGIES WITH SAFER ALTERNATIVES

Although all dangerous technologies threaten us with eventual catastrophe, those designed for a beneficial purpose are in a different class from those designed to cause death and destruction on a massive scale. There is no point in looking for alternative technologies capable of doing as much damage as nuclear, chemical and biological weapons of mass destruction. But it makes perfect sense to look for safer alternatives to those technologies whose design purpose was benign. After all, nuclear power plants generate energy, the lifeblood of modern society; many dangerous chemical technologies produce materials or fuels that are enormously useful. There is no doubt that we have the talent and creativity to find alternative technologies that provide similar benefits without imposing the threat of disaster. In more than a few cases, safer alternatives already exist. Why then do we so often marginalize or ignore them?

The search for and choice of less-risky technologies may appear to be a purely technological problem, but it is not. Powerful economic, social and political factors have a lot to do with the research directions we follow and the technological choices we make. Safer alternatives may not be used because they are more expensive, more difficult to operate and maintain, harder to monopolize or control, or because businesses are already heavily invested in the technologies they would replace. They may also be pushed to the side by politically influential vested interests who stand to lose power or money if the alternatives are put into use.

The Role of Government Subsidies

Governments sometimes subsidize the development and use of technologies that turn out to be dangerous even while they withhold subsidy from less-favored but less-risky alternatives. Such subsidies distort the technological choices that would otherwise be made in a free market economy. The nuclear power industry received massive subsidies from the federal government from the very beginning. Not only was the original development of nuclear technology heavily subsidized (in part, in connection with the nuclear weapons program), but the cost of waste disposal and other operating expenses were also partly funded with taxpayer money. Even more striking was the huge under-the-table subsidy that resulted from the Price-Anderson Act of 1957, which limited total liability for damages in the event of a major nuclear power accident to $560 million (a small fraction of the estimated multibillion-dollar damages) and then used taxpayer money to guarantee $500 million of that amount.[46] In effect, taxpayers were underwriting nuclear power through the guarantee, and undercutting their own right to financial compensation in the event of a major accident. Had the federal government simply let the market work, the nuclear power industry might well have died in its infancy.

Not all government subsidies distort choice in the economy. In fact, subsidies can increase economic efficiency when the unaided market does not capture all the benefits to the public generated by the goods or services produced. It makes sense to subsidize public education, for example, because we are all better off living in a more educated society. Everyone gains from everyone else's education as well as from his/her own. If we all had to pay the full cost of our own schooling, many people would get a lot less education, and we would all be deprived of the benefits of their higher skill. This kind of subsidy gives us real value for our money. Subsidizing dangerous technologies, on the other hand, diverts our attention and resources from the search for safer alternatives. This has certainly been true of energy technology.

Alternative Energy Technologies

Oil- and coal-fired electric power plants are safer alternatives to nuclear power, and they are actually much more common than nuclear plants. Unfortunately, they are major sources of environmental pollution, contributors to global warming, and depend on resources that will eventually run out. It would make more sense to look for more environmentally benign alternative technologies based on renewable resources. Fortunately, we do not have to look far. There is a great deal of energy involved in the natural processes that drive the earth's ecological systems. We can learn to tap it much more efficiently with the right kind of research and development. Solar power, geothermal energy, biomass conversion, hydro power, ocean thermal gradients, the tides and the wind can all be used to supply the energy on which all modern societies depend.[47]

Some of these energy sources are also much more evenly distributed around the earth than oil, coal, natural gas or uranium. Their development may be the key to widespread national energy independence, as called for by Principle 2 of international economic peacekeeping (discussed earlier). Still we must take care to use these natural, renewable energy sources wisely. Just because the underlying source of energy is ecologically benign does not mean that the system we use to tap it will be. Suppose we developed a cheap, technically efficient way of generating electric power from sunlight. We must still make sure that manufacturing the energy conversion devices does not either require large quantities of toxic chemicals as raw materials or result in rivers of toxic waste. Similarly, we must be careful not to overuse natural energy resources to the point at which we interfere with critical ecological mechanisms. We must learn to ask whether, for example, locating huge numbers of wind-powered electric generators in one place might affect air flow enough to alter local or regional rainfall patterns.

We must also avoid the trap of taking projections of energy demand as a given and concentrating all of our attention on the supply side of alternative energy. There is no reason why we cannot find technologies that allow us to maintain or improve our quality of life while using a great deal less energy than we do today. Prompted by the energy crisis more than 20 years ago, I explored a wide variety of possibilities in a book called *The Conservation Response: Strategies for the Design and Operation of Energy-Using Systems.*[48] Without even considering the more advanced technological possibilities of the day, I estimated that total U.S. energy consumption could be cut by 30 to 50 percent while maintaining the same standard of living. Many of the energy conservation strategies I recommended were relatively simple, and no single change accounted for more than a small fraction of the total savings.

A serious program of research and development aimed both at exploring key energy conservation technologies and increasing the efficiency of natural energy alternatives to reduce their cost would allow us to move toward a more secure, ecologically sensible energy future. Using public funds to encourage this R & D is easy to justify economically because, like public education, it would generate so many benefits (environmental and otherwise) not captured by the market. It would also help level the playing field by compensating for the enormous subsidies given to fossil fuel and nuclear energy in the past. Ongoing government subsidies that favor nuclear power should also be eliminated.

What is true of energy is also true of many other technologies—there are virtually always alternatives available or within reach. Where the subsidies to dangerous technologies have not been as massive as in the energy business, technologies that are dangerous may still have won out because they were cheaper to make or use, or because they worked better. If that is so, there will be a short-run money and/or performance penalty to pay when safer alternatives are substituted. In nearly all cases, it will be well worth bearing the extra cost. In rare cases, the advantages of certain key technologies may be so overwhelming and so important that it makes sense to keep using them for the time being, despite the danger. But especially in those cases, high priority should be given to research and development programs aimed at finding less risky replacement technologies so that any necessary compromises will only be very temporary.

STEP IV: FACE UP TO THE LEGACY
OF NUCLEAR AND TOXIC CHEMICAL WASTE

Even if we ultimately abandon all dangerous technologies in favor of less-risky alternatives, we will still have to deal with their deadly legacy. Nowhere is this problem more obvious than with nuclear weapons and nuclear power. Long after nuclear weapons have been dismantled, the vexing problem of safely storing the plutonium and enriched uranium in their cores continues. Plutonium and enriched uranium cannot be neutralized in any practical way.[49] They remain lethally radioactive for millennia. Both the vitrification and MOX options for disposing of weapons-grade plutonium and the "blending down" option for disposing of highly enriched uranium will generate large additional amounts of highly radioactive waste. Enormous quantities of high- and low-level radioactive waste have already been generated in the process of producing nuclear weapons and nuclear power. More is being made every day. All of it will have to be safely stored, isolated and protected for a very, very long time.[50]

There is still a great deal of controversy among scientists, not to mention the public at large, over the best ways to treat and store this radioactive garbage. We do know that much of what we are doing now is not a viable solution in the long run. Close to a quarter of the U.S. population is now living within a 50-mile radius of a storage site for military-related nuclear waste.[51] Storing millions of gallons of liquid nuclear waste in above-ground tanks in places like Hanford, Washington, and Savannah River, South Carolina, is far too dangerous to be more than a stopgap measure.

For nearly half a century, the managers of the federal government's Hanford site insisted that radioactive waste leaking from their underground tanks (many of which are the size of the Capitol dome) posed no threat to the environment because the waste would be trapped by the surrounding earth. An estimated 900,000 gallons of radioactive waste leaked from 68 (of 177) tanks and another 1.3 billion cubic meters of liquid radioactive waste and other contaminated fluids were deliberately pumped into the soil. They maintained that none of the waste would reach the ground water for at least 10,000 years. But they were wrong. By November 1997, it was already there.[52]

Every part of the expensive process of dealing with these wastes is fraught with danger and uncertainty. The Department of Energy began "characterizing" (determining the specific contents of) the waste in Hanford's tanks in 1985, as a first step toward cleanup. More than ten years and $260 million later, neither DoE nor its contractor Westinghouse had managed to definitively determine the contents of any of the tanks, and they were unable even to predict when they would finally succeed in characterizing any of the them.[53] Neither DoE nor Westinghouse had even figured out how to draw reliable samples and characterize the waste.[54]

In Russia, the problem is even more severe. The Kola Peninsula has become a graveyard for more than a hundred Soviet-era nuclear-powered submarines, rusting away with their nuclear reactors still on board. At present rates, it will take decades to transport the more than 50,000 fuel assemblies from those reactors for reprocessing or permanent storage. Meanwhile, they sit in storage tanks, some of which are leaking, and open-air bins on military bases and shipyards. In its present depleted economic condition, it is hard to know how Russia will be able to find the resources necessary to do anything serious about this disaster waiting to happen.[55]

There is no cheap or clever end run around these problems. But the longer we put off the day when we abandon dangerous nuclear technologies, the worse the problems become.

Storing Nuclear Waste

Despite decades of effort and billions of dollars of research, there is as yet no generally persuasive solution to the problem of storing nuclear waste. Unless we manage to make a spectacular technological breakthrough soon, we are stuck with choosing the best of a lot of unappealing options. The sooner we phase out the technologies that generate this radioactive refuse the better. We do not need to keep adding to the problem.

Among the options that have been proposed are space disposal, burial by tectonic subduction, disposal under the seabed and retrievable storage underground on land. The most interesting space disposal option involves shooting rockets loaded with radioactive waste into the surface of the sun. It is appealing because the sun is already a gigantic nuclear (fusion) reactor, so additional nuclear waste would fit right in. But the number of launches required makes both the cost and the risk of this approach much too great. By one calculation, on the order of 10,000 spacecraft would have to be launched to carry accumulated U.S. military waste alone.[56] This would be extraordinarily expensive and it would divert available launch facilities from other uses for quite a long time. More important, the chances of at least one of these rockets exploding, failing to achieve escape velocity or turning itself around and heading back toward earth is far too high, given the potential consequences.

Burial by tectonic subduction involves placing canisters containing the nuclear waste in deep ocean trenches, in the subduction zone where one tectonic plate is sliding beneath another. The idea is that the canisters would be carried deep into the earth by the "diving" plate and thus would be permanently buried far out of reach. The problem is that these subduction zones are geologically unpredictable. At our present state of knowledge, we could not know with confidence where the wastes would actually go.[57] In addition, although making the wastes impossible to retrieve seems like a good idea, there are at least two reasons why preserving some access makes sense. First, if and when we succeed in developing breakthrough technology that will somehow allow radioactive waste to be neutralized, it would be a great advantage to be able to recover and treat it. Second, if something goes wrong and the waste begins to leach into the environment or otherwise go out of control, being able to get to it would allow us do what we can to mitigate the problem. Preserving access does increase the risk that some day the waste might be recovered and reprocessed for nefarious purposes. On balance, though, storing the radioactive waste so that it is difficult but possible to recover seems to be the least-dangerous option.

Disposal under the seabed seems less risky. It could be done at stable locations far from the edges of the tectonic plates, using standard deep sea drilling techniques. Drills could bore cylindrical shafts 700 to 1000 feet deep into the abyssal muds and clays of the seabed (which are already some three miles below the surface of the sea). Canisters loaded with nuclear waste could then be lowered into the shafts, separated from each other by 70 to 100 feet of mud pumped into the bore hole. Then the shafts could be backfilled and sealed in such a way that the wastes could be retrieved should that prove necessary or desirable.[58] The seabed disposal option is attractive, though the chances for catastrophic accident at sea, the environmental effects of this proposal and the implications of storing waste in locations in which surveillance and protection are difficult require further investigation.

Land-based deep underground disposal in specially designed facilities also isolates the waste, while at the same time simplifying surveillance and (if done properly) protection and retrieval. Although very far from ideal, it appears that either this approach or deep under-seabed disposal are the most viable near term options.

At present, two land-based sites have received the most attention as candidates for the highly dubious honor of becoming the nation's first long-term nuclear garbage dumps: Yucca Mountain, adjacent to the Nevada Nuclear Test Site (about a hundred miles from Las Vegas), and the Waste Isolation Pilot Project (WIPP) site, in the southeastern corner of New Mexico.

Yucca Mountain is a ridge made of rock formed from volcanic ash. The plan is to create a series of tunnels under the ridge large enough to store 63,000 metric tons of spent fuel from commercial nuclear power plants and 7,000 metric tons of military waste. This would be a huge and expensive undertaking, and yet the planned capacity for military waste storage there would not even accommodate all the waste from Hanford alone. If no new commercial nuclear reactors were built and all of those currently in operation in the United States were retired at the end of their licensed operating life, the inventory of spent commercial nuclear fuel (84,000 metric tons) would still be about 30 percent greater than planned commercial waste storage capacity.[59] The repository was supposed to be ready by 1998. It is now slated to be available no earlier than 2010.[60]

Yucca Mountain was chosen because it was believed to be extremely geologically stable. It had been quiet for thousands of years, so geologists considered the risk of earthquake or volcanic intrusion to be very low. The nature of the formation and its desert location also led them to believe that rainwater, which could ultimately disturb the waste, would take centuries or millennia to penetrate the rock and reach the site deep inside the mountain. But in mid 1997, researchers reported that rainwater was infiltrating the site much faster than had been predicted—rather than taking thousands of years, it had seeped 800 feet into the rock in only 40 years.[61] In March 1998, a team of geophysicists and geologists reported that the earth's crust at the site was stretching ten times faster than it had been on average in the past, raising the risk of volcanic or other disruption of the repository. The estimated risk was still low—roughly a 1 in 1,000 chance of a volcano piercing the storage area in the next 10,000 years.[62] But of course, we know that expert risk assessments are often inaccurate and very rare events do sometimes occur. And in February 1999, another group of experts hired by DoE to assess its report on the site was highly critical of the agency's sweeping, confident claim that Yucca Mountain "would protect public health and the environment for thousands of years."[63]

Unlike Yucca Mountain, WIPP is being developed to store low-level nuclear waste. Located in the same county as the world famous Carlsbad Caverns, WIPP consists of a huge network of vaults carved out of deep underground salt deposits. The storage area is nearly a half mile below the surface and covers almost a square mile. After 20 years of development effort at a cost of $2 billion, WIPP will still only be able to handle about half of the low-level waste that has been generated by nuclear weapons production over the past 50 years. Furthermore, because they are carved out of salt rather than less plastic rock, government scientists predict that the walls and ceilings of the vaults will slowly engulf and entomb the waste.[64] If so, unlike Yucca Mountain, the buried waste will be extremely difficult if not impossible to retrieve.

Perhaps even more than at Yucca Mountain, WIPP has given rise to serious public opposition. Significant issues have been raised as to the geologic suitability of the site, even as to whether the site meets DoE's own selection criteria. It is an open question whether the choice of the location may have been driven more by politics than by scientific criteria and just plain common sense.[65] In March 1999, WIPP began receiving the first of an estimated 37,000 shipments of nuclear waste.[66]

Meanwhile, since 1995, Congress has allocated only $50 million a year to scientific research aimed at finding cheaper and safer ways to treat and store nuclear waste.[67] That is two-thirds the cost of one single F/A-18 Super Hornet Navy fighter plane.[68] Funding for this vital research should be sharply increased so that a wider range of alternatives can be thoroughly explored.[69] It is inconceivable that the impact on national security of having two or three fewer fighter planes is at all comparable to the contribution that would be made by real progress in treating nuclear waste.

Treating Chemical Waste

Like chemical weapons, highly toxic chemical waste can be converted into much less dangerous compounds. That's the good news. The bad news is that it is not particularly easy to do, and it can be extremely expensive. There are many different types of toxic chemicals, found in many different forms and combinations. No single approach can deal with them all. Although treatment and chemical detoxification is virtually always possible, it can be so expensive or difficult that it is technically or economically impractical. There are times when, as with nuclear wastes, we are left with simply trying to store dangerous chemicals safely and keep them isolated from the environment.

In addition to its complexity, the problem of chemical toxics is extremely widespread. There are highly toxic chemical waste dumps all over the world. In the United States alone, in mid 1997 there were nearly 1,200 so-called "operable units" requiring preferential cleanup on the Environmental Protection Agency's National Priorities List (the "Superfund").[70] As of April 1997, an estimated $6.4 billion was needed just to clean up the 60 percent of sites already on the list that were at nonfederal facilities—an additional $18 billion would be required if only 300 more such sites were added.[71] Yet an average of only slightly over $1 billion a year was spent on Superfund cleanup from 1981 to 1997. The Superfund has been considerably less than super.

The chemical pollution situation at federal facilities monitored by the Environmental Protection Agency looks even worse. According to EPA, it will cost in excess of $26 billion to clean up the more than 1,700 facilities the

Department of Defense operates that are "potentially contaminated" with fuels, solvents, industrial waste and unexploded ordnance. An estimated $4-8 billion more will be required to treat the mining, municipal and industrial waste that is contaminating facilities operated by the Department of the Interior, and another $4-5 billion to clean up the various chemicals polluting Department of Agriculture and NASA sites. As of 1997, the total estimated bill for all of these agencies combined was thus $34-39 billion, yet as a group they have been allocating only slightly more than $2 billion a year to this effort.[72]

The whole process of cleanup is also taking longer. For sites added to the National Priorities List between 1986 and 1990, it had taken an average of less than six years from discovery to listing. For sites added in 1996, it had taken an average of almost ten years. From 1986 to 1989, the average Superfund site cleanup had been completed in just under four years; by 1996, it was taking more than a decade.[73] Furthermore, EPA was not using inspectors from its hazardous waste program to evaluate the operation of all its Superfund incinerators, in violation of its own management directives. In a textbook example of bureaucratic fallibility (see Chapter 8), "EPA regional staff responsible for hazardous waste incinerator inspections were unaware that the Superfund incinerators were supposed to be inspected; and EPA headquarters officials were unaware that the inspections were not occurring."[74]

In sum, we have only begun to seriously address the problem of cleaning up the mess. Meanwhile, we continue to produce lethal chemicals at an alarming rate.

In the mid 1990s, I served under contract as a consultant to the Industrial Partnership Office of the Los Alamos National Laboratory on expanding civilian-oriented research at the lab. Among other things, we began to explore the possibilities for cooperative research between the chemicals industry and the lab aimed at dealing with the toxic waste problem by finding alternative chemical processes that would neither require nor produce toxics in the first place. The project was in its infancy when, for all practical purposes, it was killed by Congress, which lavishly increased funding for nuclear weapons and related military research, while directing the lab to stay away from more civilian-oriented projects. Yet it is exactly that kind of joint research enterprise with the chemicals industry that has the greatest promise for finding a long-term solution to the toxic chemical threat to our health and security.

Even when we have found substitutes for the more dangerous chemicals we currently use, it is very likely that there will still be some chemical toxics in the waste stream produced by some manufacturing processes. A long-term viable solution to this problem and many other pollution problems ultimately lies in taking a more ecological view of modern industrial life. If we can develop technologies that allow us to use the waste produced by one industry as the feedstock of another, we can design an industrial system that mimics the closed cycles so common in nature. That will allow us to prevent the buildup of toxics at the same time we conserve our natural resources. It is the ultimate recycling.

WHAT HAVE WE LEARNED?

All human/technical systems are unavoidably subject to failure. Design errors, flaws in manufacturing, mistakes in maintenance and the complexity of modern technological systems conspire to make them less than perfectly reliable. Substance abuse, mental illness, physical disease and even the ordinary processes of aging render the people who interact with them unreliable and prone to error as well. So too do normal human reactions to what are often stressful, boring and isolating work environments, and to the traumas and transitions that are an ordinary part of everyone's life.

The problems of human fallibility cannot be circumvented by putting critical decisions and actions in the hands of groups rather than individuals. Groups are subject to a whole set of fallibility problems of their own, problems that sometimes make them even less reliable than individuals. There is the distortion of information and directives in bureaucracies and the bravado of groupthink. Given the right circumstances, ordinary people can even become so wrapped up in the delusions of a crazy but charismatic leader that they ultimately do terrible things together that none of them would ever have dreamed of doing alone.

The spectacular advance of computer technology has made it seem more feasible than ever to do an end run around human-reliability problems by creating automated systems that leave people "out of the loop." But this too is a false hope. Computers are not only subject to all of the design and hardware failure problems that afflict other equipment, they are subject to an even more perplexing set of problems related to their most important advantage—their programmability. There is plenty of empirical evidence to back up theoretical arguments that flaws are ultimately unavoidable in computer software of any real complexity. Despite the seductive names of such techniques as automatic programming, expert systems and artificial intelligence, there is no real prospect that they will completely eliminate these problems in the foreseeable future, if ever. Besides, fallible people are always "in the loop" anyway—they are the designers, manufacturers and programmers behind these impressive machines.

Our imperfectability and the limitations of the devices that embed the technology we have developed create boundaries inside which we must learn to live. These boundaries only challenge us to channel our creativity. They need not stifle it.

Boundaries have certainly not stifled human creativity in art, music or literature. Many of the world's most compelling works have been created within the bounds of a structured form and style. The poetry of Wordsworth, the music of Beethoven, the art of Michaelangelo seem so powerful and unconstrained— and yet they are all contained within such limits. All of our advances in science and engineering have been achieved within the boundaries imposed by the biological, chemical and physical mechanisms of nature. There has been no other choice. But if we can respond with such creativity and inventiveness to the external boundaries set by nature, there is no reason we cannot respond with similar ingenuity to the internal boundaries that come from the inherent limitations of being human.

We can no more avoid the boundaries imposed by our fallibility than we can revoke the laws of nature. We cannot allow our fascination with the power of what we can do to blind us to what we cannot. It is no longer a matter of humility. It is a matter of survival.

NOTES

PROLOGUE

1. Wald, Matthew L., "Nuclear Waste Tanks at Hanford Could Explode," *New York Times* (July 31, 1990).
2. Bukharin, Oleg and Potter, William C., "Potatoes Were Guarded Better," *Bulletin of the Atomic Scientists* (May-June 1995), p. 48.
3. Abrams, Herbert, "Human Reliability and Safety in the Handling of Nuclear Weapons," Center for International Security and Arms Control, Stanford University (unpublished, 1990), pp. 2-4.

CHAPTER ONE

1. The pioneering work of Amartya Sen is of special interest here. See, for example, "Starvation and Exchange Entitlement: A General Approach and Its Application to the Great Bengal Famine," *Cambridge Journal of Economics* (1, 1977); and *Poverty and Famines: An Essay on Entitlement and Deprivation* (Oxford: Clarendon Press, 1981).
2. Sivard, R. L. et al., *World Military and Social Expenditures, 1993* (Washington, D.C.: World Priorities, 1993), p. 21. Overall, the twentieth century has seen more than 250 wars that have taken the lives of more than 100 million people (Sivard, R. L. et al., *World Military and Social Expenditures, 1996,* Washington, D.C.: World Priorities, 1996, p. 7).
3. A number of interesting scientific analyses of some of the likely ecological consequences of nuclear war have been published since the mid 1970s. These include: National Research Council, *Long-Term Worldwide Effects of Multiple Nuclear-Weapons Detonations* (Washington, D.C.: National Academy of Sciences, 1975); Ehrlich, P. R., Sagan, C., Kennedy, D. and Roberts, W.O., *The Cold and the Dark: The World After Nuclear War* (New York: W.W. Norton, 1984); Harwell, M.A., *Nuclear Winter: The Human and Environmental Consequences of Nuclear War* (New York: Springer-Verlag, 1984); and National Research Research Council, *The Effects on the Atmosphere of a Major Nuclear Exchange* (Washington, D.C.: National Academy of Sciences, 1985).
4. In 1998, sugar cubes containing anthrax bacilli confiscated from a German courier in 1917 (apparently intended for germ warfare use in World War I) were found in a museum in Norway. Stored without any special precautions, the anthrax was revived, still viable and virulent, after 80 years. "Norway's 1918 Lump of Sugar Yields Clues on Anthrax in War," *New York Times* (June 25, 1998).
5. Acquired Immune Deficiency Syndrome (AIDS) is caused by a number of related viruses of an unusual type. They are called "retroviruses" because they work backwards. Whereas most viruses interfere with the DNA (deoxyribonucleic acid) of an infected cell directly and indirectly affect the cell's RNA (ribonucleic acid), retroviruses affect the RNA directly. The first human retrovirus implicated as a cause of AIDS was HIV-1 (human immunodeficiency virus-1). Since that time an HIV-2 and possibly HIV-3 have been discovered. The AIDS viruses have a number of peculiarities that make them particularly resistant to attack by the body's immune system, and make the search for an effective treatment or vaccine extremely difficult. See, for example: Marx, J.L., "The AIDS Virus—Well-Known But a Mystery," *Science* (April 24, 1987), pp. 390-392; Edwards, D.D., "New Virus, Growth Factor Found for AIDS," *Science News* (June 6, 1987), p. 356; and Barnes, D.M., "AIDS: Statistics But Few Answers," *Science* (June 12, 1987), pp. 1423-1425.
6. In a sense, biologists are today where physicists were in the early part of this century. Their research having led them to the discovery of the basic forces of the physical universe locked within the nucleus of the atom, physicists collaborated to use that knowledge to create the most destructive weapon in

human history. It was not the imperative of science that led them to use this remarkable knowledge in that way, but a political decision taken under the urgings of what was thought to be military necessity. Biologists today have uncovered the basic forces of the biological universe locked within the nucleus of the living cell. It is certain that some will again argue for the military necessity of using that knowledge to create horrendous weapons. Let us hope the biologists have learned something from the experience of the physicists, that those whose work is centered on the physical understanding of life will not allow themselves to be coerced or convinced to create instead the means of mass destruction.

While it is clear that stockpiles of chemical weapons of mass destruction exist, it is not clear whether there are comparable stockpiles of biologicals.

7.   Barnaby, F., "Environmental Warfare," *Bulletin of the Atomic Scientists* (May 1976), pp. 37-43; also Westing, A.H., *Weapons of Mass Destruction and the Environment* (London: Taylor & Francis and the Stockholm International Peace Research Institute, 1977), chap. 3, pp. 49-63.

8.   Ehrlich, et. al., op. cit.

9.   Westing, A.H., *Ecological Consequences of the Second Indochina War* (Stockholm: Almqvist & Wiksell and the Stockholm International Peace Research Institute, 1976).

10. Van Dorn, W.G., LeMehaute, B. and Hwang, L., *Handbook of Explosion-Generated Water Waves,* Tetra Tech Report No. TC-130 (Pasadena, California: Tetra Tech, Inc., October 1968), prepared for the Office of Naval Research (Contract No. N00014-68-C-0227).

11. Larsen, C.A., "Ethnic Weapons," *Military Review* (published monthly by the U.S. Army Command and General Staff College, Fort Leavenworth, Kansas) (November 1970), pp. 3-11. An unexplained, oblique reference to the possibility of this type of weapon appeared almost 30 years later in an article in the *New York Times,* with the alarming title "Iranians, Bioweapons in Mind, Lure Needy Ex-Soviet Scientists" (December 8, 1998; by Judith Miller and William J. Broad). Such developments, we are reassured are "years away."

12. What is not widely appreciated is that the majority of high-level nuclear waste generated since the beginning of the nuclear era is from weapons production, not from civilian nuclear power plants.

13. Diamond, S., "Union Carbide's Inquiry Indicates Errors Led to India Plant Disaster," *New York Times* (March 21, 1985).

14. Sullivan, W., "Health Crisis Could Last Years, Experts Say," *New York Times* (December 7, 1984).

15. Burns, John F., "India Sets Three Nuclear Blasts, Defying Worldwide Ban; Tests Bring Sharp Outcry," *New York Times* (May 12, 1998) and "Indians Conduct 2 More Atom Tests Despite Sanctions," *New York Times* (May 14, 1998). For more details, see Albright, David, "The Shots Heard 'Round the World," *Bulletin of the Atomic Scientists* (July-August 1998).

16. Burns, John F., "Pakistan, Answering India, Carries Out Nuclear Tests; Clinton's Appeal Rejected," *New York Times* (May 29, 1998). See also, Albright, David, "Pakistan: The Other Shoe Drops," *Bulletin of the Atomic Scientists* (July-August 1998).

17. Gilmartin, Patricia A., "U.S. Officials Assess Status of Former Soviet Weapons Programs," *Aviation Week and Space Technology* (January 20, 1992), p. 27.

18. Miller and Broad, op. cit.

19. Taylor, Theodore B., "Worldwide Nuclear Abolition," in Nuclear Age Peace Foundation, *Waging Peace Bulletin* (Summer 1996), p. 3.

20. These include Bulgaria, Burma (Myanmar), China, Czech Republic, Egypt, France, India, Iran, Iraq, Israel, Libya, North Korea, Pakistan, Romania, Russia, Saudi Arabia, Slovakia, South Africa, South Korea, Syria, Taiwan, United States, Vietnam and Yugoslavia (Serbia). Hogendoorn, E.J., "A Chemical Weapons Atlas," *Bulletin of the Atomic Scientists* (September-October 1997).

21. Tucker, Jonathan B., "The Future of Biological Warfare," in Wander, W. Thomas and Arnett, Eric H., *The Proliferation of Advanced Weaponry: Technology, Motivations and Responses* (Washington, D.C.: American Association for the Advancement of Science, 1992), p. 57.

22. Ibid., p. 53.

23. Ibid., p. 54, citing Fairhall, David, "Eleven Countries 'Defying Ban on Germ Weapons,'" *London Guardian* (September 5, 1991).

24. Wander, W. Thomas, "The Proliferation of Ballistic Missiles: Motives, Technologies and Threats," in Wander and and Arnett, op. cit., p. 77; see also Spector, Leonard S., McDonough, Mark G. and Medeiros, Evan S., *Tracking Nuclear Proliferation* (Washington, D.C.: Carnegie Endowment for International Peace, 1995), pp. 185-187.

25. WuDunn, Sheryl, "North Korea Fires Missile Over Japanese Territory," *New York Times* (September 1, 1998).

26. Lorber, Azriel, "Tactical Missiles: Anyone Can Play," *Bulletin of the Atomic Scientists* (March 1992), pp. 39-40.

CHAPTER TWO

1. McFadden, Robert D., "Blast Hits Trade Center, Bomb Suspected, 7 Killed, Thousands Flee Smoke in Towers: Many Are Trapped for Hours in Darkness and Confusion," *New York Times* (February 27, 1993); and Bernstein, Richard, "Trade Center Bombers Get Prison Terms of 240 Years," *New York Times* (May 25, 1994).

2. Kifner, John, "At Least 21 Are Dead, Scores Are Missing After Car Bomb Attack in Oklahoma City Wrecks 9-Story Federal Office Building: 17 Victims Were Children in 2d-Floor Day Care Center," *New York Times* (April 20, 1995); and Treaster, Joseph B., "The Tools of a Terrorist: Everywhere for Anyone," *New York Times* (April 20, 1995).

3. A similar, simple definition of terrorism can be found in the Rand Corporation's body of work on terrorism. See, for example, Cordes, Bonnie, Hoffman, Bruce, Jenkins, Brian M., Kellen, Konrad, Moran, Sue and Sater, William, *Trends in International Terrorism, 1982 and 1983* (Santa Monica, California: Rand Corporation, August 1984), p. 1.

4. "Some Poisoned Israeli Oranges Discovered in Europe," *New York Times* (February 2, 1978). In a letter sent to the West German Ministry of Health, a group calling itself "Arab Liberation Army—Palestine Commando" claimed responsibility for poisoning the oranges.

5. "Medical Sciences," *1984 Yearbook of Science and the Future* (Chicago: Encyclopaedia Britannica, Inc., 1984), p. 353. It is unclear whether the Tylenol poisonings were the work of someone with a grudge against the company or someone intending to extort money.

6. U.S. Department of Defense, *Terrorist Group Profiles* (Washington, D.C.: U.S. Government Printing Office, November 1988), p. 25.

7. Klanwatch Project, *False Patriots: The Threat of Anti-Government Extremists* (Montgomery, Alabama: Southern Poverty Law Center, 1996).

8. The so-called Reign of Terror was carried out by the infamous Maximilien Robespierre's Jacobin government between the summer of 1793 and the summer of 1794.

9. Darnton, John, "Ethiopia Uses Terror to Control Capital," *New York Times* (February 10, 1978).

10. According to the testimony, the campaign included cigarettes infected with anthrax, candies spiked with botulism, sugar containing salmonella and whiskey laced with the herbicide paraquat. See Daley, Suzanne, "In Support of Apartheid: Poison Whisky and Sterilization," *New York Times* (June 11, 1998).

11. Burns, John F., "A Network of Islamic Terrorism Traced to a Pakistan University," *New York Times* (March 20, 1995).

12. Ibid. See also Wren, Christopher S., "U.S. Jury Convicts 3 in a Conspiracy to Bomb Airliners," *New York Times* (September 6, 1996) and Weisler, Benjamin, "Mastermind Gets Life for Bombing of Trade Center," *New York Times* (January 9, 1998).

13. McFadden, Robert D., "FBI Seizes 8, Citing a Plot to Bomb New York Targets and Kill Political Figures," *New York Times* (June 25, 1993).

14. Burns, op. cit.

15. Hoffman, Bruce, deLeon, Peter, Cordes, Bonnie, Moran, Sue Ellen and Tompkins, Thomas C., *A Reassessment of Potential Adversaries to [sic] U.S. Nuclear Programs* (Santa Monica, California: Rand Corporation, March 1986), p. 5.

16. "Aum Paid ¥4 Billion to Foreign Companies," *Japan Times* (July 2, 1995). Members of the Aum Shinrikyo sect also reportedly visited the Russian nuclear research center at Obninsk in 1994 seeking nuclear materials. Obninsk is the same nuclear research institute that U.S. investigators believed to be the source of the small amount of weapons-grade nuclear material smuggled into Germany in 1994; see Katz, Gregory, "Uranium Smuggling Case Confirms Security Fears," *Dallas Morning News* (May 26, 1996).

17. This money is to be spread across all ten "closed cities" in the Russian nuclear weapons complex to help start commercial ventures. Gordon, Michael R., "Russia and U.S. Plan to Guard Atom Secrets," *New York Times* (September 23, 1998).

18. Kilborn, Peter T., "British Group's Sale of Bacteria Stirs Fear of Use by Terrorists," *New York Times* (February 11, 1977).

19. WuDunn, Sheryl, "Elusive Germ Sows Panic in Japan," *New York Times* (August 23, 1996).

20. Klanwatch Project, op. cit., p. 31.

21. Ibid., p. 25. See also Purdum, Todd S., "Two Charged with Possessing Toxin Thought to Be Anthrax," *New York Times* (February 20, 1998).

22. Ambrosino, Michael J., executive producer, "The Plutonium Connection" (Boston: WGBH-TV, NOVA #214, first transmission PBS, March 9, 1975), transcript, p. 8; see also Burnham, David, "Bill Asks Curb on Plutonium Use to Prevent Building of Homemade Bomb," *New York Times* (February 27, 1975).

23. *Newsday* (October 10, 1976).

24. Ostrow, Ronald J., "Backpack A-Bomb Possible, FBI Says," *Los Angeles Times* (April 26, 1979).

25. Morland, Howard, "The H-Bomb Secret: To Know How Is to Ask Why," *The Progressive* (November 1979), pp. 14-23.

26. Carmody, Dierdre, "U.S. Hunting for Copies of Classified Data on Bomb," *New York Times* (May 18, 1979).

27. Willrich, Mason and Taylor, Theodore B., *Nuclear Theft: Risks and Safeguards* (Cambridge, Massachusetts: Ballinger Publishing Company, 1974), pp. 20-21.

28. Burnham, op. cit.

29. Harper, George, "A Do-It-Yourself A-Bomb," *New Scientist* (March 27, 1980), p. 998.

30. Taylor, Theodore B., "Nuclear Power and Nuclear Weapons," Global Security Study No. 22 (Santa Barbara, California: Nuclear Age Peace Foundation, July 1996), p. 2.

31. Gillette, Robert, "Impure Plutonium Used in '62 A-Test," *Los Angeles Times* (September 16, 1977).

32. Gillette, Robert, "Proves Civilian Plants Can Be Atomic Arsenals," *Los Angeles Times* (September 14, 1977).

33. "The critical mass of metallic uranium at 10% enrichment, with a good neutron reflector [like beryllium] is about 1000 kilograms . . . a sphere . . . about a foot and a half in diameter. At 20% enrichment, the critical mass drops to about 250 kilograms . . . and at 50% enrichment, it is fifty kilograms." Willrich and Taylor, op. cit., p. 17.

34. In 1993, 4.5 kilograms of enriched reactor fuel was stolen from a Russian Navy base at Kola Bay. See Bukharin, Oleg and Potter, William, "'Potatoes Were Guarded Better,'" *Bulletin of the Atomic Scientists* (May-June 1995), p. 46.

35. Waller, Douglas, "Nuclear Ninjas: A New Kind of SWAT Team Hunts Atomic Terrorists," *Time* (January 8, 1996), pp. 39-41.

36. "Germ Warfare: New Threat from Terrorists," *Science News* (May 18, 1996), p. 311.

37. Clifford, Frank, "U.S. Drops Anti-Terrorist Tests at Nuclear Plants: Shrinking Budget Is Cited," *Los Angeles Times* (November 3, 1998).

38. Gordon, Michael R., "Russia Struggles in Long Race to Prevent an Atomic Theft," *New York Times* (April 20, 1996).

39. Ibid.

40. Spector, Leonard S., "Clandestine Nuclear Trade and the Threat of Nuclear Terrorism" in Leventhal, Paul and Alexander, Yonah, editors, *Preventing Nuclear Terrorism* (Lexington, Massachusetts: Lexington Books, 1987), p. 80.

41. Waller, op. cit. pp. 39-41.

42. Cockburn, Andrew, "Can We Stop A Bomb Smuggler?" *Parade* (November 3, 1985), p. 12.

43. Associated Press, "FBI Arrests Two in Plot to Steal Nuclear Sub," *Washington Post* (October 5, 1978).

44. From the computerized chronology of terrorist incidents at Rand Corporation, as cited in "Appendix: Nuclear-Related Terrorist Activities by Political Terrorists," in Leventhal and Alexander, op. cit., p. 125.

45. Associated Press, "Plutonium in NYC Water Probed," *Dallas Morning News* (July 28, 1985).

46. Katz, Gregory, "Uranium Smuggling Case Confirms Security Fears," *Dallas Morning News* (May 26, 1996).

47. Gates, Mahlon, E., "The Nuclear Emergency Search Team," Levanthal and Alexander, op. cit., pp. 397-402; also, Waller, op. cit., pp. 39-42.

48. Emshwiller, John R., "In Atom Bomb Scare, Federal NEST Team Flies to the Rescue," *Wall Street Journal* (October 21, 1980).

49. Operations Branch, Division of Fuel Safety and Safeguards, Office of Nuclear Material Safety and Safeguards, *Safeguards Summary Event List (SSEL)* (Washington D.C.: U.S. Nuclear Regulatory Commission, NUREG-0525, vol. 2, Rev. 3; July 1995).

50. Wald, Matthew L., "Doubts Raised on New Rules to Protect Nuclear Reactors," *New York Times* (February 11, 1994).

51. Shenon, Philip, "Saudi Truck Bomb at a U.S. Complex Kills 11, Hurts 150," *New York Times* (June 26, 1996).

52. "Weekly Information Report to the NRC Commissioners" (April 20, 1984), enclosure E, p. 3, as cited in Hirsch, Daniel, "The Truck Bomb and Insider Threats to Nuclear Facilities," in Leventhal and Alexander, op. cit., p. 209.

53. Hirsch, Daniel, "The Truck Bomb and Insider Threats to Nuclear Facilities," in Leventhal and Alexander, op. cit., p. 210.

54. Schlessinger, James R., quoted from transcript of "Meet the Press" in a *Senator Mike Gravel Newsletter* (October 31, 1973) by Krieger, David, "Terrorists and Nuclear Technology," *Bulletin of the Atomic Scientists* (June 1975), p. 32.

55. Associated Press, "Information on How to Sabotage Nuclear Plant Available to Public," *Atlanta Journal and Constitution* (November 25, 1979).

56. Fainberg, A. and Bieber, A.M., Brookhaven National Laboratory *Barrier Penetration Database,* prepared under Contract No. E-1-76-C-02-0016 to the Division of Operating Reactors, Office of Nuclear Reactor Regulation, U.S. Nuclear Regulatory Commission (NUREG/CR-0181, published July 1978).

57. Meyer, Christina, Duncan, Jennifer and Hoffman, Bruce, *Force-on-Force Attacks: Their Implications for the Defense of U.S. Nuclear Facilities* (Santa Monica, California: Rand Corporation, 1993).

58. Hoffman, Bruce, Meyer, Christina, Schwarz, Benjamin and Duncan, Jennifer, *Insider Crime: The Threat to Nuclear Facilities and Programs* (Santa Monica, California: Rand Corporation, February 1990), p. 33.

59. Ibid., p. 41.

60. Kellen, Konrad, "The Potential for Nuclear Terrorism: A Discussion" in Leventhal and Alexander, op. cit., p. 109. Note that some of these groups continue to be active, while the activities of others have waned.

61. Ibid.

62. Kristof, Nicholas D., "At Trial in Tokyo, Guru Says Aim Was to Give 'Ultimate Joy,'" *New York Times* (April 25, 1996).

63. Klanwatch Project, op. cit., pp. 58-68. See also Intelligence Project, Southern Poverty Law Center, *Intelligence Report* (Spring 1998), pp. 6-28.

64. Ibid., p. 8.

65. Ibid., p. 17.

66. Hoffman, Bruce, *Terrorism in the United States* (Santa Monica, California: Rand Corporation, January 1986), p. 51.

CHAPTER THREE

1. U.S. Department of Energy, *Plutonium: The First 50 Years* (Washington, D.C.: February 1996), p. 52.

2. Marshall, Eliot, "Nuclear Fuel Accounts Books in Bad Shape," *Science* (January 9, 1981), pp. 147-148.

3. U.S. Department of Energy, op. cit., p. 52.

4. Special AEC Safeguards Study Group report as excerpted in U.S. Senate, *The Congressional Record: Proceedings and Debates of the 93rd Congress, Second Session* (Washington, D.C.: April 30. 1974), p. S 6625.

5. Bolt, Richard, "Plutonium for All: Leaks in Global Safeguards," *Bulletin of the Atomic Scientists* (December 1988), pp. 14-17.

6. Total cumulative acquisitions of plutonium from all sources amounted to 111,400 kilograms. Total cumulative inventory difference (ID) was given as 2,800 kilograms. The ID's for the seven major DoE facilities were: Hanford, Washington = +1,265.6 kg; Rocky Flats, Colorado = +1,191.8 kg; Savannah River, South Carolina = + 232.0 kg; Los Alamos National Laboratory, New Mexico = +47.5 kg; Lawrence Livermore National Laboratory, California = +5.5 kg; Idaho National Engineering Laboratory = -5.6 kg; and Argonne National Laboratory West, Idaho = -3.4 kg. (a positive ID means that there was more shown on the books than a physical inventory could locate, while a negative ID means that there was less on the books than a physical inventory revealed.) U.S. Department of Energy, op. cit., pp. 22 and 54.

7. The total size of the combined plutonium inventories of the Departments of Energy and Defense (before adjustment for removals) is given as 111.4 metric tons (1 metric ton = 1,000 kilograms). Of this, some 2.8 metric tons was in the "inventory differences" category. Therefore, the ID was 2.5 percent of the inventory (2.8/111.4). Ibid., pp. 22 and 52-54.

8. Ibid., p. 52.

9. Holden, Constance, "NRC Shuts Down Submarine Fuel Plant," *Science* (October 5, 1979), p. 30.

10. U.S. Department of Energy, op. cit., p. 22.

11. Some estimates of world plutonium stocks are as high as 1,100 metric tons, nearly five times the size of world inventories used in these calculations. (See Williams, Phil and Woessner, Paul, "The Real Threat of Nuclear Smuggling," *Bulletin of the Atomic Scientists* (January 1996), p. 41.)

12. In Langham, W.H., "Physiology and Toxicology of Plutonium-239 and Its Industrial Medical Control," *Health Physics* 2 (1959), p. 179, it is assumed that humans would respond to plutonium as experimental animals do. On a weight-adjusted basis, this implies that a dose of 20 mg in systemic circulation would be lethal to half the exposed humans within 30 days. Using Langham's 10 percent lung absorption rate, 200 mg would have to be inhaled.

13. Halloran, Richard, "Defense Department Fights to Curb Big Losses in Arms and Supplies," *New York Times* (February 12, 1987).

14. Mohr, Charles, "Military Faulted as Unaware of Equipment Losses," *New York Times* (March 14, 1986).

15. Halloran, op. cit.

16. Army Audit Agency, *Training Ammunition Management* (January 16, 1986).

17. U.S. General Accounting Office, *Inventory Management: Problems in Accountability and Security of DOD Supply Inventories,* (Washington, D.C.: May 1986), pp. 16-18.

18. Ibid.

19. Ibid., pp. 30-32.

20. Ibid., p. 37.

21. Ibid., p. 43.

22. Ibid., pp. 53-56.

23. Ibid., p. 1.

24. Halloran, op. cit.

25. U.S. General Accounting Office, op. cit., p. 16.

26. News story broadcast on WNBC-TV (New York City: September 28, 1978, 7:25 P.M.).

27. U.S. General Accounting Office, *Inventory Management: Vulnerability of Sensitive Defense Material to Theft* (September 1997), pp. 1, 2 and 5.

28. Ibid., p. 59. These "inventory adjustments" involve a wide variety of ordinary goods and materials, not just weapons and explosives. Even the weapons and explosives they do include are conventional, mostly small arms and ammunition.

29. Arkin, William M., Norris, Robert S. and Handler, Joshua, *Taking Stock: Worldwide Nuclear Deployments 1998* (Washington, D.C.: Natural Resources Defense Council, 1998), p. 1. See also Center for Defense Information (Washington, D.C.), *Defense Monitor* (September-October 1995), p. 1.

30. Vartabedian, Ralph, "Nuclear Bomb Records Are Lost, U.S. Says," *New York Times* (October 24, 1997).

31. Martin, David, "Mystery of Israel's Bomb," *Newsweek* (January 9, 1978), pp. 26-27.

32. Burnham, David, "Computer Safety of U.S. Is Faulted," *New York Times* (February 26, 1986).

33. Burns, John F., "China Reveals Major Scandal Among Top Island Officials," *New York Times* (August 1, 1985).

34. Ingram, Timothy H., "Nuclear Hijacking: Now Within the Grasp of Any Bright Lunatic," *Washington Monthly* (January 1973), p. 22.

35. Sawicki, Michael, "Fort Bliss Sends Rockets to Mexico in Procedural Error," *Dallas Times-Herald* (January 30, 1987).

36. The following discussion of security conditions at Tooele Army Depot is based on the written and oral congressional testimony of Dale Van Atta, including a series of articles he did for the *Deseret News* of Salt Lake City that were reprinted along with his testimony. The reprinted articles by Dale Van Atta in the *Deseret News* included "World's Death Stored at TAD, and Security is Substandard" (1978); "Charges Loom in Nerve Gas Inventory Quirk" (January 16, 1978); "'Don't Talk to Deseret News'— Keep Mum TAD Workers Told" (February 6, 1978); "Anyone Free to Fly Over TAD Cache" (February 14, 1978); "Security at TAD Broken 30 Times in Decade" (February 17, 1978); and "FAA Restricts Flights Over TAD" (February 20, 1978). See *Military Construction Appropriations for 1979,* Hearings before the Subcommittee on Military Construction Appropriations of the Committee on Appropriations, House of Representatives, 95th Congress, 2nd Session (Washington, D.C.: U.S. Government Printing Office, 1978), pp. 179-209.

37. Ibid., p. 177.

38. Ibid., pp. 177 and 188.
39. Ibid., pp. 176 and 187-188.
40. Ibid., pp. 177-178 and 182.
41. Ibid., pp. 177-178 and 182-184.
42. Ibid., p. 182.
43. Ibid., p. 176.
44. Ibid., pp. 176-177 and 189.
45. Zorpette, Glenn, "Hanford's Nuclear Wasteland," *Scientific American* (May 1996), p. 91.
46. Ibid.
47. Long, Clarence, "Views of the Honorable Clarence D. Long, Submitted to Accompany Fiscal 1975 Military Construction Appropriations" (September 24, 1974), pp. 2-3.
48. Ibid., p. 4.
49. Pastore, John, "Security Review of Certain NATO Installations," *Congressional Record—Senate* (April 30, 1975), pp. S.7184-S.7188.
50. United Press International (September 27, 1974).
51. Albright, Joseph, "Prepared Statement," *Military Construction Appropriations for 1979,* Hearings Before the Subcommittee on Military Construction Appropriations of the Committee on Appropriations, U.S. House of Representatives, 95th Congress, Second Session (Washington, D.C.: U.S. Government Printing Office, 1978), pp. 139-140.
52. Ibid., p. 142.
53. "The Nation," *Los Angeles Times* (December 6, 1978).
54. "Uranium-Laden Truck Deserted," *Dallas Morning News* (September 14, 1980).
55. Cox News Service, "Mock Raiders Break Security at Bomb Plant," *The West Palm Beach Post* (September 15, 1982).
56. Lubasch, Arnold, "8 Charged With Arms Plots Including Plan for $2 Billion Sale to Iran," *New York Times* (July 28, 1983).
57. Hirsch, Daniel, "The Truck Bomb and Insider Threats to Nuclear Facilities," in Leventhal, Paul and Alexander, Yonah, editors, *Preventing Nuclear Terrorism* (Lexington, Massachusetts: Lexington Books, 1987), p. 218.
58. Keller, Bill, "Security at U.S. Bases in Philippines Is Called Lax," *New York Times* (September 12, 1985).
59. Drummond, Bob, "Weapons Plant Fails 'Terrorist Test,'" *Dallas Times-Herald* (February 12, 1987).
60. Ibid.
61. Frantz, Douglas and O'Shea, James, "Arms Smugglers Corral Navy Parts," *Chicago Tribune* (August 17, 1986).
62. Associated Press, "Civilian Crews Transport N-Missiles Across Atlantic," *Dallas Morning News* (September 22, 1986).
63. U.S. General Accounting Office, *Army Inventory Management: Inventory and Physical Security Problems Continue* (Washington, D.C.: October 1987), pp. 36-37.
64. Lightbody, Andy and Poyer, Joe, *The Complete Book of U.S. Fighting Power* (Lincolnwood, Illinois: Beekman House, 1990), p. 290.
65. "2 in Security Police Accused of Stealing Jet Fighter Engines," *New York Times* (July 12, 1989).
66. Reuters News Service, "Seven Indicted for Thefts from Army Base" (October 3, 1996, 5:47 A.M.), Yahoo! Reuters Headlines, http:///www.yahoo.com/headlines/961003/news/stories/military_1.html.
67. Johnston, David, "Marines Accused of Weapons Trafficking," *New York Times* (October 17, 1997).
68. "Guards Charge Security Is Lax at Indian Point," *New York Times* (October 17, 1979).
69. "Reporter Cracks Three Mile Island's 'Security,'" *West Palm Beach Post* (February 3, 1980).
70. Fialka, John J., "Internal Threat: Soviets Begin Moving Nuclear Warheads Out of Volatile Republics," *Wall Street Journal* (June 22, 1990).
71. Bukharin, Oleg and Potter, William, "'Potatoes Were Guarded Better,'" *Bulletin of the Atomic Scientists* (May-June 1995), p. 49.
72. Ibid.
73. Ibid., p. 46.
74. Ibid., p. 48.
75. Ibid.
76. Associated Press, "A Power Struggle in Russia's Military," *Dallas Morning News* (September 23, 1995).
77. Ibid.
78. Specter, Michael, "Occupation of a Nuclear Plant Signals Russian Labor's Anger," *New York Times* (December 7, 1996).

79. Cockburn, Andrew and Leslie, producers, Steve Croft, reporting, *60 Minutes,* CBS Television News (September 7, 1997).

80. Berkhout, Frans, Diakov, Anatoly, Feiveson, Harold, Miller, Marvin and von Hippel, Frank, "Plutonium: True Separation Anxiety," *Bulletin of the Atomic Scientists* (November 1992), p. 29.

81. Wald, Matthew L., "U.S. Privatization Move Threatens Agreement to Buy Enriched Uranium from Russia," *New York Times* (August 5, 1998). See also Passell, Peter, "Profit Motive Clouding Effort to Buy Up A-Bomb Material," *New York Times* (August 28, 1996).

82. Wald, op. cit.

83. Panofsky, Wolfgang K.H., "No Quick Fix for Plutonium Threat," *Bulletin of the Atomic Scientists* (January-February 1996), p. 3; and Berkhout etal., op. cit., p. 29.

84. Timothy Ingram, op. cit., p. 24. Though he would not give any details, the former chief of safeguards at the AEC at the time confirmed the general character of these reports.

85. "Mystery of Israel's Bomb," *Newsweek* (January 9, 1978), p. 26; Burnham, David, "Nuclear Plant Got U.S. Contracts Despite Many Security Violations," *New York Times* (July 4, 1977); and Burnham, David, "U.S. Documents Support Belief Israel Got Missing Uranium for Arms," *New York Times* (November 6, 1977).

86. "The Uranium Trade: All Roads Lead to Khartoum's Black Market," *Africa Report* (January-February 1988), p. 6. (The title of the British television documentary cited in the *Africa Report* article is "Dispatches: The Plutonium Black Market.")

87. Ibid.

88. Maceda, Jim, *NBC Nightly News,* NBC-TV, (June 30, 1992), transcript, p. 6.

89. Zimmermann, Tim and Cooperman, Alan, "Special Report: Its Regional Mafias Are Strong, Its Nuclear Wealth Vast; In Russia, the Former Are Vigorously Pursuing the Latter, and that Means Trouble," *U.S. News and World Report* (October 23, 1995), pp. 56-67.

90. Hibbs, Mark, "Plutonium, Politics and Panic," *Bulletin of the Atomic Scientists* (November-December 1994), p. 25-26; and Williams and Woessner, op. cit., p. 43.

91. Perlez, Jane, "Tracing a Nuclear Risk: Stolen Enriched Uranium," *New York Times* (February 15, 1995); Gordon, Michael R., "Czech Cache of Nuclear Material Being Tested for Bomb Potential," *New York Times* (December 21, 1994); and Williams and Woessner, op. cit., p. 43.

92. Perlez, Jane, "Radioactive Material Seized in Slovakia; 9 Under Arrest," *New York Times* (April 22, 1995).

93. Lee, Rensselaer, "Smuggling Update," *Bulletin of the Atomic Scientists* (May-June 1997), p. 55.

94. Ibid.

95. Goldstein, Steve, "A Nightmare Scenario Catches a Second Wind," *Philadelphia Inquirer* (January 10, 1999).

96. Ibid.

97. Williams and Woessner, op. cit., p. 42.

98. The report further claimed that Russian civilian nuclear research institutes did not even do physical checks of their inventories of nuclear materials, they only kept book records. Whitney, Craig R., "Smuggling of Radioactive Material Said to Double in Year," *New York Times* (February 18, 1995).

99. Lee, op. cit., p. 52.

CHAPTER FOUR

1. For an interesting sociological analysis of this phenomenon, see Perrow, Charles, *Normal Accidents: Living with High-Risk Technologies* (New York: Basic Books, 1984).

2. For a theoretical macroeconomic analysis of the long-term costs of using economic resources for military purposes, see Dumas, Lloyd J., *The Overburdened Economy: Uncovering the Causes of Chronic Unemployment, Inflation and National Decline* (Berkeley, California: University of California Press, 1986).

3. The most common design for a nuclear weapon, the so-called implosion design, wraps wedges of conventional explosive around a spherical nuclear core in which the atoms are not densely enough packed at normal pressure to sustain a runaway chain reaction. The atomic nuclei are too far apart for the neutrons released from the splitting of one nucleus to have a high enough probability of striking another nucleus and causing it to split apart, releasing energy and more neutrons. Sometimes the neutrons from the splitting of one nucleus strike and split another, but too many miss to cause an

explosive chain reaction. By directing the force of the conventional explosives inward, the implosion design causes a strong enough shock wave to pass through the nuclear core to temporarily overcome the weak electron forces that normally determine the spacing of the atoms. The nuclei are thus forced much closer together, raising the probability that the neutrons flying out of a splitting nucleus will strike and split another nucleus. A runaway nuclear fission chain reaction occurs, producing a devastating nuclear explosion. If the shock wave is not precisely uniform from every direction or the nuclear core is not spherical, the device will not work.

4. The uranium isotope U-235 is much more readily fissionable and thus much more efficient for nuclear weapons purposes than the isotope U-238. However, naturally occuring uranium consists mainly of the isotope U-238 (about 99.3 percent) with a very small percentage (about 0.7 percent) of the isotope U-235. Thus the core of uranium-based nuclear weapons is made of "highly enriched uranium," which has been processed to greatly increase the ratio of U-235 to U-238. See Glasstone, Samuel and Dolan, Philip J., editors, *The Effects of Nuclear Weapons* (Washington, D.C.: U.S. Department of Defense and U.S. Department of Energy, 1977), p. 5.

5. International Labor Office, "Toxicity Classification of Radionuclides," *Guidelines for the Radiation Protection of Workers in Industry (Ionizing Radiations),* Occupational Safety and Health Series 62 (Geneva: International Labor Office, 1989).

6. Sax, N. Irving et al., *Dangerous Properties of Industrial Materials* (New York: Van Nostrand Reinhold Company, 1984), p. 2247.

7. The following discussion of the physical and biophysical characteristics of alpha, beta and gamma radiation draws heavily on detailed correspondence and conversation between the author and Dr. Warren F. Davis, physicist and founder of Davis Associates, Inc., West Newton, Massachusetts (summer 1995). Some additional information was provided in conversation with Dr. Georgia C. Davis, biological physicist, Francis Bitter Magnetic Laboratory, MIT (November 12, 1988).

8. The slowing alpha (or beta) particle may even be "captured" by the atom, which then gains all of the particle's remaining energy. This can cause enormous disruption.

9. Alpha particles are helium nuclei and beta particles are electrons. Alpha particles are thus much more massive.

10. Because gamma rays have no mass, they must always travel at the speed of light. This is a consequence of Einstein's special theory of relativity. Einstein showed that it would take infinite energy to accelerate a particle with mass to the speed of light. Consequently, any particle that is already moving at the speed of light must have zero mass. On the other hand, a particle of zero mass cannot be decelerated to a speed even infinitesimally less than the speed of light, since that would require extracting an infinite amount of energy from the particle. Therefore, any particle of zero mass must always be moving precisely at the speed of light. Gamma rays therefore exhibit their loss of energy not by slowing down, but by reduced frequency (color or wavelength).

11. This difference is particularly relevant in cancer radiation therapy. Beams of massive particles (such as protons or pions) can be tuned to have the correct initial energy so that they release most of their energy at the depth of a tumor, thereby destroying the tumor with minimal damage to surrounding tissues. Because massless particles (such as cobalt gamma radiation and X-rays) deposit energy uniformly along the beam, a different approach is necessary to use them to preferentially damage tumors. They are used to irradiate the tumor from many different angles, producing most of the damage at their geometric point of intersection.

12. See Richmond, C.R., "Current Status of Information Obtained from Plutonium-Contaminated People," *Radiation Research* (New York: Academic Press, 1974) and Morgan, Karl Z., "Suggested Reduction of Permissible Exposure to Plutonium and other Transuranic Elements," *American Industrial Hygiene Association Journal* (1975). Note that the physical half-life of plutonium is 24,000 years.

13. See Langham, W.H., "Physiology and Toxicology of Plutonium-239 and its Industrial Medical Control," *Health Physics* 2 (1959), p. 179.

14. Ibid. It is assumed that humans would respond to plutonium as experimental animals do. On a weight-adjusted basis, this implies that a dose of as little as 20 mg in systemic circulation would be lethal to half the exposed humans within 30 days. Using the lung absorption rate of plutonium of 10 percent given by Langham, this would mean that a total dose of 200 mg would have to be inhaled in order for 20 mg to get into systemic circulation.

15. The following calculation is based on the data presented in Edsal, John T., "Toxicity of Plutonium and Some Other Actinides," *Bulletin of the Atomic Scientists* (September 1976), pp. 28-30. In experiments carried out on dogs, at doses of as little as 100-1000 nanocuries per gram of lung tissue, most died from pulmonary fibrosis within 1-2 years. (1 nanocurie = 37 radioactive disintegrations per

second; for plutonium, this means that 37 alpha particles are shot off per second). Given an average weight of a human lung of 300-380 grams (source: Dallas Medical Examiner's Office), there is about 760 grams of lung tissue in an average person. Assuming that every gram of lung tissue is dosed at 100-1000 nanocuries, a total dosage of 76,000-760,000 nanocuries should certainly be lethal. For plutonium-239, one nanocurie equals 16.3 nanograms (billionths of a gram). Therefore, 76,000 nanocuries equal 1.2 milligrams (76,000 nc X 16.3 ng/nc = 1,238,800 ng = 1.2 mg); and 760,000 nanocuries equal 12.4 milligrams (760,000 nc X 16.3 ng/nc = 12,388,000 ng = 12.4 mg).

16. Craig, Paul P. and Jungerman, John A., *Nuclear Arms Race: Technology and Society* (New York: McGraw-Hill, 1986), p. 304.

17. Depending on the particular type and design of a nuclear weapon, there will be varying amounts of plutonium (or highly enriched uranium) in its nuclear core. The more sophisticated the design, the smaller the amount of nuclear material required to produce a blast of a given size. Less sophisticated weapons and/or weapons designed to produce larger explosions will contain more nuclear material.

18. Harris, Robert and Paxman, Jeremy, *A Higher Form of Killing: The Secret Story of Chemical and Biological Warfare* (New York: Hill and Wang, 1982), p. 54; see also Bodin, F. and Cheinisse, F., *Poisons* (New York: McGraw-Hill, 1970), pp. 129-130; and Harris, William H. and Levey, Judith S., editors, *The New Columbia Encyclopedia* (New York and London: Columbia University Press, 1975), pp. 1910 and 2426.

19. Shenon, Philip, *New York Times,* "Gulf War Illness May Be Related to Gas Exposure" (June 22, 1996); "Czechs Say They Warned U.S. of Chemical Weapons in Gulf" (October 19,1996); "U.S. Jets Pounded Iraqi Arms Depot Storing Nerve Gas" (October 3, 1996); and "Pentagon Says It Knew of Chemical Arms Risk" (February 26, 1997).

20. Shenon, Philip, *New York Times,* "U.S. Jets Pounded Iraqi Arms Depot Storing Nerve Gas" (October 3, 1996); and "Pentagon Says It Knew of Chemical Arms Risk" (February 26, 1997).

21. Shenon, Philip, "New Study Raises Estimate of Troops Exposed to Gas," *New York Times* (July 24, 1997).

22. The catapult they used, called a *trebuchet,* was an even older technology. The Chinese invented it more than 1,500 years earlier.

23. Chevedden, Paul E., Eigenbrod, Les, Foley, Vernard and Sodel, Werner, "The Trebuchet," *Scientific American* (July 1995), p. 68.

24. Harris, William H. and Levey, Judith S., editors, *The New Columbia Encyclopedia,* (New York and London: Columbia University Press, 1975), p. 2161.

25. Kristof, Nicholas D., "Japan Confronting Gruesome War Atrocity: Unmasking Horror," *New York Times* (March 17, 1995). In one of the more bizarre episodes of World War II, the Japanese began to launch huge balloons containing conventional explosives at the American heartland in 1944. Carried by prevailing winds, 200 or so landed in western states. Some Japanese generals wanted to load the balloons with plague- or anthrax-carrying germ warfare agents. Ironically, that plan was vetoed by the infamous general Hideki Tojo, whom the United States later executed as a war criminal.

26. Ibid. See also Tyler, Patrick E., "China Villagers Recall Horrors of Germ Attack," *New York Times* (February 4, 1997).

27. Ibid.

28. Cole, Leonard A., "The Specter of Biological Weapons," *Scientific American* (December 1996), p. 62.

29. "'Emerging Viruses' in Films and Best Sellers," *New York Times* (May 12, 1995).

30. Cohen, Jon, "AIDS Virus Traced to Chimp Subspecies," *Science* (February 5, 1999), pp. 772-773. See also Le Guenno, Bernard, "Emerging Viruses," *Scientific American* (October 1995), pp. 58-59.

31. Preston, R., *The Hot Zone* (New York: Random House, 1994); and *Outbreak* (Los Angeles: Warner Brothers Studios, 1995).

32. Altman, Lawrence K., "Deadly Virus Is Identified in the Outbreak in Zaire," and contiguous boxed story "A Closer Look: The Ebola Virus," *New York Times* (May 11, 1995).

33. Altman, Lawrence K., "Deadly Disease Attributed to New Virus," *New York Times* (April 7, 1995).

34. Nowak, Rachel, "Cause of Fatal Outbreak in Horses and Humans Traced," *Science* (April 7, 1995), p. 32.

35. Altman, op. cit. (April 7, 1995).

36. Reuters News Service, "List Reveals Pentagon Hid 14 N-Accidents," *Dallas Morning News* (December 22, 1980).

37. Associated Press, "U.S.: 32 Nuclear Weapons Accidents Since 1952," *West Palm Beach Post* (December 23, 1980).

38. Hansen, Chuck, "1,000 More Accidents Declassified," *Bulletin of the Atomic Scientists* (June 1990), p. 9.

39. Weiner, Tim, "Safety Officer Says Military Hid Crashes," *New York Times* (June 24, 1995).

40. The sources of those accidents cited in the discussion that follows that are not specifically noted here can be found in Tables 4-1 and 4-2.

41. Admiral Gene La Rocque, "Proliferation of Nuclear Weapons," Hearings, Subcommittee on Military Applications, Joint Committee on Atomic Energy, 1974.

42. More than 40 years after this terrible accident, new details continue to emerge. See, for example, Stone, Richard, "Retracing Mayak's Radioactive Cloud," *Science* (January 9, 1999).

43. Lapp, Ralph, *Kill and Overkill* (New York, Basic Books, 1962), p. 127; and Larus, J., *Nuclear Weapons Safety and the Common Defense* (Columbus, Ohio: Ohio State University Press, 1967), pp. 93-99.

44. Decontamination required removal of 1,750 tons of radioactive soil and vegetation. By some reports, the material was returned to the United States and buried at sea off the coast of the Carolinas.

45. Sanger, David E., *New York Times*, "U.S.-Japan Ties Worsen on News that Warhead Was Lost in 1965" (May 9, 1989); and "U.S. Says the Bomb Lost Close to Japan Scattered Radiation" (May 16, 1989).

46. Broad, William J., "Russians Fear Leak from Sunken Nuclear Sub, but U.S. Disagrees," *New York Times* (May 21, 1993). See also Spaasky, Igor, "Apathy Above, Terror Below," *New York Times* (June 3, 1993).

47. Broad, William J., "Soviet Sub Sunk in '86 Leaks Radioactivity Into the Atlantic," *New York Times* (February 8, 1994).

48. Ibid.

49. Arkin, William M. and Handler, Joshua, *Neptune Papers No. 3: Naval Accidents 1945-1988* (Washington, D.C.: Greenpeace/Institute for Policy Studies, June 1989), p. 3; Table 5, "Nuclear Reactors Lost in the Oceans," p. 98; and Table 6, "Accidents Involving Nuclear Powered Ships and Submarines," p. 99. See also Eriksen, Viking O., *Sunken Nuclear Submarines* (Oslo: Norwegian University Press, 1990).

50. In August 1989, the Chinese government finally admitted that nuclear accidents in China between 1980 and 1985 had killed 20 people and injured 1200 more. They did not specify the number of accidents and gave little information about the type, claiming only that they resulted from the careless handling of nuclear waste and other radioactive materials. However, since there were no nuclear power plants operating in China at that time, the main source of nuclear waste must have been the military production of their nuclear weapons. See Associated Press, "China Admits Nuclear Accident Deaths," *Dallas Morning News* (August 6, 1989).

51. The Soviets did not admit that the epidemic was the result of an accident until 1991. Questions remain as to whether the biological warfare program that gave rise to this accident remains in operation despite Boris Yeltsin's order to abandon it in 1992. Miller, Judith and Broad, William J., "Germ Weapons in the Soviet Past or in the New Russia's Future," *New York Times* (December 28, 1998). Tucker, Jonathan B., "The Future of Biological Warfare," in Wander, W.T. and Arnett, E.H., editors, *The Proliferation of Advanced Weaponry* (Washington, D.C., American Association for the Advancement of Science, 1992), pp. 59-60. See also Leitenberg, Milton, "Anthrax in Sverdlovsk: New Pieces to the Puzzle," *Arms Control Today* (April 1992). Early reports of this accident include, "1000 Are Said to Die in Soviet Accident," *New York Times* (July 16, 1980); Gwertzman, Bernard, "Soviet Mishap Tied to Germ-War Plant," *New York Times* (March 19,1980) and Gwertzman, Bernard, "Moscow Tells U.S. Infected Meat Caused Outbreak of Anthrax," *New York Times* (March 21, 1980).

52. Clines, Francis X., "Disaster Zone Urged in Soviet Nuclear Explosion," *New York Times* (September 29, 1990); see also Associated Press, "Soviet Region Seeks Kremlin Aid after Nuclear Fuel Blast," *Dallas Morning News* (September 29, 1990).

53. Associated Press, "Suit Discloses '57 Denver Fire Emitted Radioactive Material," *New York Times* (September 9, 1981).

54. Treaster, Joseph B., "Army Discloses Three New Cover-Ups," *New York Times* (September 20, 1975).

55. These tests typically involved bacteria such as *Bacillus subtilis* and *Serratia marcescens*. See Cole, Leonard A., *Clouds of Secrecy: The Army's Germ Warfare Tests over Populated Areas* (Totowa, New Jersey: Rowman and Littlefield, 1988). See also Schmeck, Harold M., "Army Tells of U.S. Germ War Tests; Safety of Simulated Agents at Issue," *New York Times* (March 9, 1977); Altman, Lawrence K., "Type of Germ Army Used in Test Caused Infection Four Days Later," *New York Times* (March 13, 1977); Horrock, Nicholas M., "Senators Are Told of Test of a Gas Attack in Subway," *New York Times* (September 19, 1975).

56. Schneider, Keith, "U.S. Secretly Fixes Dangerous Defects in Atomic Warheads," *New York Times* (May 24, 1990).

57. Ibid.

58.   Schneider, Keith, "Interest Rises in Studies of Atomic Shells," *New York Times* (May 28, 1990).

59.   Schneider, Keith, "Nuclear Missiles on Some Bombers To Be Withdrawn," *New York Times* (June 9, 1990); see also Schneider, Keith, "Interest Rises in Studies of Atomic Shells," *New York Times* (May 28,1990).

60.   "Plutonium Fallout Theory Revealed Here," *Japan Times* (29 October 1969); and Associated Press, "Satellite Fallout Said Not Harmful," *Japan Times* (12 April 1970).

61.   Aftergood, Steven, "Nuclear Space Mishaps and Star Wars," *Bulletin of the Atomic Scientists* (October 1986), p. 40.

62.   Ibid.

63.   Teltsch, Kathleen, "Canada to Continue Search for Fragments of Satellite that Fell," *New York Times* (July 6, 1978); and "Follow-Up on the News: 'Hot' Spy in the Cold," *New York Times* (April 26, 1981).

64.   Broad, William J., "Despite Danger, Superpowers Favor Use of Nuclear Satellites," *New York Times* (January 25, 1983).

65.   Broad, William J., "Soviet Nuclear-Powered Satellite Expected to Hit Earth in Summer," *New York Times* (May 14, 1988); and Broad, William J., "New Plans on Reactors in Space Raise Fears of Debris," *New York Times* (October 18, 1988). See also Leslie, Connie and Hager, Mary, "Lost Among the Stars: A Falling Satellite Raises Global Fears," *Newsweek* (August 29, 1988).

66.   Purdum, Tod S., "Russian Mars Craft Falls Short and Crashes Back to Earth," *New York Times* (November 18, 1996); and Gordon, Michael R., "Mystery of Russian Spacecraft: Where Did It Fall to Earth?" *New York Times* (November 19, 1996).

67.   "Challenger Disaster," *Dallas Observer* (August 27, 1987); see also *Nation* (February 22, 1986).

68.   Ibid.

69.   Major Mike Morgan, deputy director of the Space Control Center, U.S. Space Command, Cheyenne, Wyoming, in interview by Derek McGinty, *All Things Considered,* National Public Radio (August 13, 1997).

70.   Broad, William J., "New Plans on Reactors in Space Raise Fears of Debris," *New York Times* (October 18, 1988).

71.   Wald, Matthew L., "Nuclear Waste Tanks at Hanford Could Explode," *New York Times* (July 31, 1990).

72.   Wald, Matthew L., "4 Decades of Bungling at Bomb Plant," *New York Times* (January 25, 1992).

73.   *Morning Report,* National Public Radio (May 15, 1997: 9:00 A.M.), broadcast on KERA-90.1FM in Dallas.

74.   Pollack, Andrew, "After Accident, Japan Rethinks Its Nuclear Hopes," *New York Times* (March 25, 1997).

75.   By contrast, the terrorist bombings of the World Trade Center in New York and the Murrah Federal Building in Oklahoma City each used less than three tons of ammonium nitrate.

76.   "Fifty Years Later, Texas Town Recalls Horrors of Deadly Blast," *New York Times* (April 16, 1997).

77.   Cushman, John H., "Chemicals on Rails: A Growing Peril," *New York Times* (August 2, 1989).

CHAPTER FIVE

1.   Anderson, Harry with Mazmazumdar, Sudip and Foote, Donna, "Rajiv Gandhi Stumbles: Has India's Young Leader Become 'Directionless'?" *Newsweek* (February 23, 1987), p. 38.

2.   Ford, Daniel, *The Button: America's Nuclear Warning System—Does It Work?* (New York: Simon and Shuster, 1985), pp. 52-53.

3.   Ibid., pp. 25 and 52.

4.   Editorial, *New York Times* (April 22, 1958), p. 10.

5.   Von Hippel, Frank, "De-Alerting," *Bulletin of the Atomic Scientists* (May/June 1997), p. 35.

6.   Ibid. Von Hippel was assistant director for national security in the White House Office of Science and Technology Policy from 1993 to 1994. See also Hall, Brian, "Overkill Is Not Dead," *New York Times Magazine* (March 15, 1998); and Arkin, William M. and Kristensen, Hans, "Dangerous Directions," *Bulletin of the Atomic Scientists* (March-April 1998).

7.   Blair, Bruce, *The Logic of Accidental Nuclear War* (Washington, D.C.: Brookings Institution, 1993), p. 255.

8.   Unless specifically noted, the incidents discussed in this section are discussed in Chapter 4 or its Appendix.

9. Keith Schneider wrote a series of articles on this issue in the *New York Times,* including "U.S. Secretly Fixes Dangerous Defects in Atomic Warheads" (May 24, 1990); "Interest Rises in Studies of Atomic Shells" (May 28, 1990); and "Nuclear Missiles on Some Bombers To Be Withdrawn" (June 9, 1990).

10. Sagan, Scott, *The Limits of Safety: Organizations, Accidents and Nuclear Weapons* (Princeton, New Jersey: Princeton University Press, 1993), pp. 181-183.

11. What makes this scenario particularly interesting is that it does not assume that any of the warning or communications systems fail to function properly.

12. Wall, Don, "The Secret Arsenal," *Evening News,* WFAA-TV, Dallas, Texas (November 2, 1995).

13. For a more detailed description, see Ford, op. cit., pp. 118-119.

14. The following brief discussion is drawn from Blair, op. cit., pp. 188-191. For a detailed description of the warning systems, see also Ford, op. cit., Bracken, Paul, *The Command and Control of Nuclear Forces* (New Haven, Connecticut: Yale University Press, 1983); and Blair, Bruce, *Strategic Command and Control: Redefining the Nuclear Threat* (Washington, D.C.: Brookings Institution, 1985).

15. Nelan, Bruce W., "Nuclear Disarray," *Time Magazine* (May 19, 1997), pp. 46-47; and Blair, Bruce G., Feiveson, Harold A. and von Hippel, Frank N., "Taking Nuclear Weapons Off Hair-Trigger Alert," *Scientific American* (November 1997), pp. 74-76.

16. Ibid.

17. Center for Defense Information, *Defense Monitor* 15 (no. 7, 1986), p. 6. NORAD officials have said that the figures for the years prior to 1977 are "unavailable". These data were classified beginning in 1985 by the Reagan administration, though it is hard to imagine any legitimate security reason why these data could not be made public. As far as I have been able to determine, they are still classified.

18. Major Mike Morgan, deputy director of the Space Control Center, U.S. Space Command, in interview by Derek McGinty, *All Things Considered,* National Public Radio (August 13, 1997).

19. Blair, op. cit. (1993), pp. 192-193.

20. Ibid., p. 193.

21. Public reports of this incident did not appear until late November, and surfaced first in the British press (Frayn, Michael, "Miscellany," *The Manchester Guardian,* November 28, 1960). It was not until the first week in December that reports surfaced in the United States, after which the story was quickly swept under the rug. Various versions appeared: "Moon Stirs Scare of Missile Attack," *New York Times* (December 8, 1960, where the story was relegated to p. 71); "Canadian Is Praised Over Missile Scare," *New York Times* (December 23, 1960); and "A Nuclear Warning Disregarded on Pearl Harbor Day" (referring to the date of the first Associated Press dispatch of the story in the United States) *I.F. Stone's Weekly* (December 19, 1960). Additional accounts were published in the *Washington Star* (December 7, 1960); *Washington Post* (December 8, 1960); and *Boston Traveller* (December 12, 1960).

22. Sagan, op. cit., pp. 99-100.

23. Ibid., pp. 146-148.

24. This explanation is less likely if Penkovsky was also the source of information about the location of the drop. If he were trying to trick the Soviets into triggering a nuclear attack, the odds of succeeding would have been better if they knew nothing about the drop, or at least did not know its true location. On the other hand, the KGB had been on to Penkovsky for some time before his arrest and may have learned about the drop location another way.

25. Malnak, Lewis D., sworn affidavit (Washington, D.C.: January 10, 1974). Malnak was the president of a company that undertook a project for the Special Communications Project Office of the Navy, which interfaced with the SECT program. (It is unclear whether this second incident involved another SECT buoy from the same sub or a buoy from a different sub.)

26. Sagan, op. cit., p. 238.

27. Sulzberger, A.O., "Error Alerts U.S. Forces to a False Missile Attack," *New York Times* (November 11, 1979); Halloran, Richard, "U.S. Aides Recount Moments of False Missile Alert," *New York Times* (December 16, 1979).

28. Hart, Gary and Goldwater, Barry, "Recent False Alerts from the Nation's Missile Attack Warning System," *Report to the Committee on Armed Services of the United States Senate: October 9, 1980* (Washington, D.C.: U.S. Government Printing Office, 1980), pp. 5-6.

29. Halloran, Richard, "Computer Error Falsely Indicates A Soviet Attack," *New York Times* (June 6, 1980); Burt, Richard, "False Nuclear Alarms Spur Urgent Effort to Find Flaws," *New York Times* (June 13, 1980). See also Sagan, op. cit., p. 231.

30. Ford, op. cit., p. 78. See also footnotes 23 and 24.

31. Ford, op. cit., p. 79; Sagan, op. cit., p. 232.

32. Hart and Goldwater, op. cit., p. 7.

33. Ford, op. cit., p. 79.

34. Asakura, Toshio, "Soviet War Message Reported: Sent Shortly After Reagan Bomb Joke," *Sacramento Bee* (October 3, 1984).

35. Bracken, op. cit., pp. 65-66.

36. Halloran, Richard, "Computer Error Falsely Indicates a Soviet Attack," *New York Times* (June 6, 1980).

37. Beecher, William, "Cuban MiG Flight to Florida Spurs Review of Air Defense Procedures," *New York Times* (October 7, 1969).

38. "Soviet Plane From Cuba Flies Over a U.S. Force," World News Briefs, *New York Times* (April 23, 1977).

39. Kifner, John, "U.S. Officer Says Frigate Defenses Were Turned Off," *New York Times* (May 20, 1987); Kifner, John, "Captain of Stark Says Ship Failed to Detect Missiles," *New York Times* (May 21, 1987); Cushman, John, "Blind Spot Left Stark Vulnerable, U.S. Officials Say," *New York Times* (June 1, 1987).

40. Barringer, Felicity, "Lone West German Flies Unhindered to the Kremlin," and Reuters News Service, "Pilot Said to Have Passion for Flying," *New York Times* (May 30, 1987); Taubman, Philip, "Moscow Frees Young German Pilot," *New York Times* (August 4, 1988).

41. That might not be the case, though, if the nations tricked into war were themselves minor nuclear powers. Suppose, for example, a nuclear-capable country outside the Middle East triggered a catalytic war between, say, Iran and Israel in order to promote its own agenda.

42. The goal of designers is to make PALs difficult and time consuming to defeat, but it is generally understood that they cannot be made infallible. As one analyst put it, "Although no PAL can be certified as tamper-proof, a designer can hope to achieve a degree of tamper resistance, which is generally measured by the period of time it would take a skilled but unauthorized person with some knowledge of the PAL's design to defeat it." Zimmerman, Peter D., "Navy Says No PALs for Us," *Bulletin of the Atomic Scientists* (November 1989), p. 38.

43. Blair, op. cit. (1993), p. 50. (note that the order of some of this quote has been rearranged).

44. Gillert, Doug, "Through the Looking Glass," *Airman* (October 1987), p. 13.

45. For a simple discussion of these controls, see Ford, op. cit., p. 151-152.

46. The source of this information is Daniel Ellsberg, who was closely involved with American nuclear war planning at the Pentagon in the 1960s, as noted in Bracken, op. cit., pp. 198-199.

47. Blair, op. cit., (1993) p. 50.

48. Halloran, Richard, "U.S. Downs Iran Airliner Mistaken for F-14; 290 Reported Dead; A Tragedy, Reagan Says," *New York Times* (July 4, 1988).

49. Engleberg Stephen, "Failures Seen in Safeguards on Erroneous Attacks," *New York Times* (July 4, 1988); Gordon, Michael, "Questions Persist on Airbus Disaster," *New York Times* (July 5, 1988); "U.S. Pushes Inquiry on Downing of Jet: Questions Mount"; Halloran, Richard, "Team Goes to Gulf," and Johnson, Julie, "No Shift in Policy," *New York Times* (July 5, 1988); Hess, John, "Iranian Airliner," *New York Observer* (July 18, 1988); Aspin, Les, "Witness to Iran Flight 655," *New York Times* (November 18, 1988).

50. Shribman, David, "Korean Jetliner: What Is Known and What Isn't," *New York Times* (September 8, 1983).

51. Shribman, David, "U.S. Experts Say Soviet Didn't See Jet Was Civilian," *New York Times* (October 7, 1983).

52. Ibid., and "Korean Jetliner: What Is Known and What Isn't," *New York Times* (September 8, 1983); Pearson, David, "KAL 007: What the U.S. Knew and When We Knew It," *Nation* (April 18-25, 1984), pp. 105-124; Weisman, Steven R., "U.S. Says Spy Plane Was in the Area of Korean Airliner," *New York Times* (September 5, 1983); Polman, Dick, "Challenging the U.S. on Flight 007," *Philadelphia Inquirer* (October 18, 1984); Gwertzman, Bernard, "A New U.S. Transcript Indicates Soviet Pilot Fired 'Canon Bursts,'" *New York Times* (September 12, 1983). See also Hersh, Seymour M., *The Target Is Destroyed: What Really Happened to Flight 007 & What America Knew About It* (New York: Vintage, 1987).

53. Weisman, op. cit.; Erlanger, Steven, "Similarities with KAL Flight Are Rejected by U.S. Admiral," *New York Times* (July 4, 1988).

54. Gordon, Michael R., "U.S. Helicopters Given No Warning by Attacking Jets," *New York Times* (April 16, 1994); see also "F-15 Pilot Says Radar Plane Did Not Warn Him," *New York Times* (June 6, 1995).

55. Lietenberg, Milton, "The Case of the Stranded Sub," *Bulletin of the Atomic Scientists* (March 1982), pp. 10-13.

56. Stevenson, Richard, "Russian Submarines, or 'Ghosts' of the Cold War," *New York Times* (March 8, 1993).

57. Simon, Scott and Haydon, Simon reporting, "Swedish Government Makes an Embarrassing Discovery," *Weekend Edition/Saturday,* National Public Radio (Washington, D.C.: February 18, 1995).

58. Ford, op. cit., p. 87.

59. Ibid., p. 88.

60. Ibid., pp. 20-21.

61. Kaplan, Richard, executive producer and Koppel, Ted, anchor, "Surprise Attack: Choosing a Response," *Nightline* (New York: ABC News, Show #1169, November 14, 1985), pp. 5-6.

62. Ford, op. cit., p. 136.

63. Ibid., pp. 91-92.

64. "Pentagon Warning System Can't Handle Pressure, Says GAO," *Electronic Design* (April 12, 1980), p. 59.

65. "Warning System 'Fragile': Bad Computer Military Jinx," *West Palm Beach Post* (March 10, 1980).

66. Ford, op. cit., pp. 188-189.

67. The exercise was run in late November 1980. "Simulated Mobilization for War Produces Major Problems," *New York Times* (December 22, 1980).

68. Hudson, Richard, "Molink Is Always Ready," *New York Times* Magazine (August 26, 1973).

69. "National Warning Plan Fails," *Sacromento Bee* (February 18, 1984).

70. Keller, Bill, "Pentagon Test on News Coverage Hurt by Communications Lapses," *New York Times* (April 27, 1985).

71. *Sacramento Bee* (November 19, 1989).

72. "Norway Military Mail Sent to East Germans," *New York Times* (July 30, 1975).

73. "Coordination Plans of Military Services Attacked," *New York Times* (March 22, 1984).

74. "Army Phones Need $30 Million Repair," *New York Times* (March 2, 1985).

75. U.S. House of Representatives, Armed Services Investigating Subcommittee, Committee on Armed Services, *Review of Department of Defense World-Wide Communications* (Washington, D.C.: U.S. Government Printing Office, 1971).

76. U.S. House, Armed Services Investigating Subcommittee, op. cit., p. 6.

77. Ibid., pp. 6-11.

78. "Israeli Payment to Close the Book on '67 Attack on U.S. Navy Vessel," *New York Times* (December 19, 1980); See also "C.I.A. Papers Cite Israeli's in Attack on U.S. Navy Ship," *New York Times* (September 19, 1977).

79. U.S. House, Armed Services Investigating Subcommittee, op. cit., p. 14.

80. Ford, op. cit., pp. 173-174, and Ball, Desmond, "Can Nuclear War Be Controlled?" *Adelphi Papers* (London: International Institute for Strategic Studies, No. 169), pp. 12-13; see also "Resistance Met: All 39 From Vessel Safe—U.S. Jets Hit Cambodian Base," *New York Times* (May 15, 1975).

81. Center for Defense Information, *Defense Monitor* 15 (no. 7, 1986), p. 3. See also Sagan, op. cit., p. 64.

82. Blair, op. cit. (1993), p. 225. Bayesian analysis is a probabilistic technique for calculating the extent to which additional information causes one to revise earlier estimates of the likelihood of an event.

83. Rusten, Lynn and Stern, Paul C., *Crisis Management in the Nuclear Age,* Committee on International Security and Arms Control of the National Academy of Sciences, and Committee on Contributions of Behavioral and Social Science to the Prevention of Nuclear War of the National Research Council (Washington, D.C.: National Academy Press, 1987), p. 26.

84. Bernstein, B.J., "The Week We Almost Went to War," *Bulletin of the Atomic Scientists* (February 1976), p. 17.

85. Kennedy, John F., "Presidential Address to the Nation" (October 22, 1962), reprinted in Kennedy, Robert F., *Thirteen Days: A Memoir of the Cuban Missile Crisis* (New York: W.W. Norton, 1969), p. 168.

86. Ellsberg, Daniel, "The Day Castro Almost Started World War III," *New York Times* (October 31, 1987).

87. Ibid.

88. Paul, Derek, Intrilligator, Michael and Smoker, Paul, editors, *Accidental Nuclear War: Proceedings of the 18th Pugwash Workshop on Nuclear Forces* (Toronto: Samuel Stevens and Company and University of Toronto Press, 1990), pp. 43-51.

89. Sagan, op. cit., p. 95.

90. Ibid., p. 96.

91. Ibid., pp. 79-80.

92. Ibid., pp. 136-137.

93. Kennedy, Robert, op. cit., p. 111.

94. Hall, Brian, "Overkill Is Not Dead," *New York Times Magazine* (March 15, 1998), p. 84.

95. Stein, Janice G., "The Challenge of the Persian Gulf Crisis," *Peace and Security* (Canadian Institute for Peace and Security: Winter 1990-91), pp. 4-5.

96. Blair, Feiveson and von Hippel, op. cit., p. 80.

97. Blair, Bruce, "Russia's Doomsday Machine," and Broad, William J., "Russia Has a Nuclear 'Dooms-day' Machine," *New York Times* (October 8, 1993).

CHAPTER SIX

1. U.S. General Accounting Office, "Human Factors: FAA's Guidance and Oversight of Pilot Crew Resource Management Training Can Be Improved" (November 1997: GAO/RCED-98-7), p. 2.

2. Wald, Matthew L., "Crew of Airliner Received Warning Just Before Guam Crash," *New York Times* (March 24, 1998).

3. U.S. General Accounting Office, "Military Aircraft Safety: Serious Accidents Remain at Historically Low Levels" (March 1998: GAO/NSIAD-98-95BR), p. 2.

4. Lochbaum, David, *The Good, The Bad and the Ugly: A Report on Safety in America's Nuclear Power Industry,* Union of Concerned Scientists (June 1998), p. v.

5. Bureau of the Census, U.S. Department of Commerce, *Statistical Abstract of the United States, 1998,* Table 249, p. 157. These data were converted assuming alcohol content of 3.5 percent for beer, 12 percent for wine and 40 percent for distilled spirits.

6. Cushman, J. H., "Three Pilots Dismissed in Alcohol Abuse," *New York Times* (March 17, 1990).

7. Weiner, E., "Drunken Flying Persists Despite Treatment Effort," *New York Times* (July 14, 1990).

8. Newhall, D., acting assistant secretary of defense for health affairs, "Memorandum to the Assistant Secretaries of the Army, Navy and Air Force (M&RA), Subject: Error in 'Heavy Drinkers' Category in the 1982, 1985, and 1988 Worldwide Survey of Substance Abuse and Health Behaviors Among Military Personnel" (15 August 1989).

9. Bray, R.M., et al., *1995 Department of Defense Survey of Health-Related Behaviors Among Military Personnel* (Research Triangle Institute, December 1995). Note that because the corrections were distributed some time after the 1988 report was issued, the figures cited here may differ from those given in the text of the original reports of the 1982, 1985 and 1988 surveys.

10. Bray, op. cit., Table 3, "Substance Abuse Summary for Total DoD, 1980-1995."

11. Ibid.

12. Sociologists use the term "total institution" to refer to an organizational setting in which the institution encompasses all aspects of the "inmate's" life. Examples include prisons, mental institutions and the military.

13. Bray, R.M. et.al., *1988 Worldwide Survey of Substance Abuse and Health Behaviors Among Military Personnel* (Research Triangle Institute: December 1988), p. 17.

14. To their credit, the surveyors do, in fact, mention this problem. "Many individuals question the validity of self-reported data on alcohol and drug use, claiming that survey respondents will give socially desirable rather than truthful answers." They go on to argue that their method provides "useful and meaningful data," while at the same time recognizing that "self-reports may sometimes underestimate the extent of substance abuse." They cite a 1979 study of alcohol problems among Air Force personnel which tried to look at this question by cross-checking available records. That study found some categories of answers seemed to be accurate, but it also found, "Air Force beverage sales data . . . suggested that self-reports underestimate actual prevalence of alcohol use by as much as 20 percent." Ibid., pp. 21-22.

15. Treml, Vladimir G., "Drinking and Alcohol Abuse in the USSR in the 1980's," in Jones, A., Connor, W.D. and Powell, D.E., editors, *Soviet Social Problems* (Boulder, Colorado: Westview Press, 1991), p. 121.

16. Treml, Vladimir G., "Soviet and Russian Statistics on Alcohol Consumption and Abuse," in Bobadilla, J.L., Costello, C.A. and Mitchell, F., editors, *Premature Death in the New Independent States* (Washington, D.C.: National Academy Press, 1997), pp. 222-224.

17. Treml, op. cit. (1991), p. 121.

18. This pattern is all the more striking considering the cost of alcohol in the USSR at the time, particularly after the price increases that accompanied the antidrinking campaign. Treml estimates that the average industrial worker in Western Europe or the U.S. had to work roughly 2-3 hours to earn enough money

to buy a liter of vodka, while the average Soviet industrial worker had to work 19 hours. Ibid., pp. 125 and 131.

19. 19.Gabriel, R.A., *The New Red Legions: An Attitudinal Portrait of the Soviet Soldier* (Westport, Connecticut: Greenwood Press, 1980), p. 153.

20. Ibid., p. 154.

21. Ibid., pp. 156-157.

22. "The Red Army," *World,* WGBH-TV (Boston) (May 6, 1981), transcript, p. 21.

23. For a discussion of the problems of conversion in the former Soviet Union, see Oden, Michael, "Turning Swords into Washing Machines: Converting Russian Military Firms to Consumer Durable Production" and Ullmann, John, "Conversion Problems in the Changing Economies," in Dumas, L.J., editor, *The Socio-Economics of Conversion: From War to Peace* (Armonk, New York: M.E. Sharpe, 1995).

24. Brinkley, J., "Vast Undreamed of Drug Use Feared," *New York Times* (November 23, 1984).

25. Crowe, A., "Teen Cocaine Use Triples Since 1975," *New York Times* (March 21, 1985).

26. Treaster, J.B., "Cocaine Use Found on the Way Down Among U.S. Youths," *New York Times* (January 25, 1991).

27. Goodwin, M. and Chass, M., "Baseball and Cocaine: A Deepening Problem," *New York Times* (August 19, 1985).

28. Kerr, P. , "17 Employees of Wall Street Firms Are Arrested on Cocaine Charges," *New York Times* (April 17, 1987).

29. LSD (d-lysergic acid diethylamide) was first synthesized at the research laboratories of the Sandoz drug company in Basel, Switzerland.

30. Despite the strength of the drug, it is extremely difficult if not impossible to become physically addicted to LSD. The body builds up resistance to its effects in 2-3 days. Resistance fades just as quickly. See Grinspoon, L. and Bakalar, J.B., *Psychedelic Drugs Reconsidered* (New York: Basic Books, 1979), p. 11.

31. Ibid., pp. 12-13 and 158.

32. Ibid.

33. Ibid., p. 13-14.

34. Ibid., p. 159.

35. Ibid., p. 160.

36. Ibid., pp. 14-35.

37. Of course, illegal drugs also have a variety of interaction and side effects that can magnify their already troubling effects on reliability and performance. Smoking marijuana is probably the single most common trigger of LSD flashbacks. The stimulant methamphetamine, the use of which had penetrated deep into the American heartland by 1996, has dangerous and relatively frequent side effects that include paranoia and extremely violent behavior. See Johnson, Dirk, "Good People Go Bad in Iowa," *New York Times* (February 22, 1996).

38. Angel, J.E., publisher, *Physicians Desk Reference* (Oradell, New Jersey: Medical Economics Company, 1983), pp. 629-630, 1622, 1702, 1771 and 2075-2076.

39. Griffith, H.W., *Complete Guide to Prescription and Non-Prescription Drugs* (The Body Press, 1990), pp. 116, 236, 336 and 790.

40. Ibid., pp. 156, 514 and 856.

41. Strange as it seems, in Chapter 11 we will see that even extremely rare events occur more often than one might think.

42. U.S. Department of Commerce, Bureau of the Census, *Statistical Abstract of the United States, 1998* (Washington, D.C.: U.S. Government Printing Office, 1998), Table 571, p. 359.

43. Bray, op. cit., Table 10.4, p. 209.

44. If it were suspected that a particular individual had recently taken LSD, a sophisticated gas chromatography mass spectroscopy test could be done on an isolate of his/her urine. That test, costing perhaps a few hundred dollars, would probably pick up evidence of the LSD use, though it is not likely to yield indisputable proof. Trying to screen tens of thousands of people this way each day would rapidly become prohibitively expensive. And the test would have to be repeated every day, because it would turn up negative if the person had used LSD even as recently as the day before. Source: Shulgin, A.T., private correspondence with the author (27 April and 10 May 1991). Dr. Shulgin is a biochemist who has authored more than 150 scientific works in chemistry, pharmacology and botany, and is recognized as an expert on the chemistry and pharmacology of psychedelics.

45. Ibid.

46. Barinaga, M., "Drug Use a Problem at U.S. Weapons Laboratory?" *Nature* (June 23, 1988), p. 696.

47. Kramer, J. M., "Drug Abuse in the USSR," in Jones, A., Connor, W.D. and Powell, D.E., editors, *Soviet Social Problems* (Boulder, Colorado: Westview Press, 1991), p. 94.

48. Ibid., pp. 94-96.
49. Ibid., p. 99.
50. Ibid., pp. 100-101.
51. Ibid., p. 103.
52. Ibid., p. 104.
53. In 1981, for example, in a television interview a U.S. State Department spokesperson reported on problems in the Soviet army: "We didn't find instances of drug abuse being what they really worried about, but now in Afghanistan the troops seem to have been able to get the drugs that they couldn't get before and this is becoming a problem that concerns Soviet military leaders." "The Red Army," op. cit., p. 22.
54. Schmeck, H.M., "Almost One in Five May Have Mental Disorder," *New York Times* (October 3, 1984), p. 1.
55. Bower, B., "Mental Disorders Outpace Treatment," *Science News* (February 27, 1993).
56. Hinsie, L.E. and Campbell, R.J., *Psychiatric Dictionary* (New York: Oxford University Press, 1970), pp. 200 and 221.
57. Ross, C.A., Joshi, S. and Currie, R., "Dissociative Experiences in the General Population," *American Journal of Psychiatry* (November 11, 1990), pp. 1550 and 1552.
58. Goleman, Daniel, "Severe Trauma May Damage the Brain as Well as the Psyche," *New York Times* (August 1, 1995).
59. Bower, B., "Trauma Disorder High, New Survey Finds," *Science News* (December 23 and 30, 1995).
60. Plag, J.A., Arthur, R.J. and Goffman, J.M., "Dimensions of Psychiatric Illness Among First-Term Enlistees in the United States Navy," *Military Medicine* 135 (1970), pp. 665-673.
61. Department of the Army, "Disposition and Incidence Rates, Active Duty Army Personnel, Psychiatric Cases, Worldwide, CY1980-1984 and CY1985-1989," U.S. Army Patient Administration Systems and Biostatistics Activity (Washington, D.C.: 1985, 1990).
62. Department of the Navy, "Distribution of Psychiatric Diagnoses in the U.S. Navy (1980-1989)," Naval Medical Data Services Center (Bethesda, Maryland: 1990).
63. Abrams, H. L., "Human Reliability and Safety in the Handling of Nuclear Weapons" (Stanford, California: Center for International Security and Arms Control, Stanford University, 1990, unpublished), p. 5.
64. Bray et al., op. cit., 1995, Table 29.
65. "3 Atom Guards Called Unstable: Major Suspended," *New York Times* (August 18, 1969).
66. "Three at Key SAC Post Are Arrested on Drug Charges," *International Herald Tribune* (March 29, 1971).
67. Dobish, A.P. , "U.S. Missile Unit Used Drugs Regularly, GI Says Eased Boredom of Missile Base," *Milwaukee Journal* (December 16, 1974).
68. Abrams, H.L., "Human Instability and Nuclear Weapons," *Bulletin of the Atomic Scientists* (January-February 1987), p. 34.
69. Earley, P. , *Family of Spies: Inside the John Walker Spy Ring* (New York: Bantam, 1988), p. 358. See also Abrams, H. L., "Sources of Human Instability in the Handling of Nuclear Weapons," in *The Medical Implications of Nuclear War,* Institute of Medicine, National Academy of Sciences (Washington, D.C.: National Academy Press, 1986), p. 512.
70. Abrams, Herbert L., "Human Reliability and Safety in the Handling of Nuclear Weapons," *Science and Global Security* 2 (1991), pp. 325-327.
71. Ibid. See also Associated Press, "Man Pleads Guilty to Two Murders," *Seattle Post-Intelligencer* (June 20, 1990).
72. Abrams, op. cit. See also "Courts," *Seattle Post-Intelligencer* (October 5, 1993).
73. Luongo, Kenneth L. and Bunn, Matthew, "Some Horror Stories Since July," *Boston Globe* (December 29, 1998).
74. Ibid.

CHAPTER SEVEN

1. Cornell, Charles E., "Minimizing Human Errors," *Space/Aeronautics* (March 1968), p. 72.
2. "Stray Flights Often Human Error," *Dallas Times-Herald* (September 20, 1983).
3. Flournoy, Craig, "2 of 3 Air Crashes Tied to Pilot Error," *Dallas Morning News* (August 25, 1985).

4. Wald, Matthew L., "American Airlines Ruled Guilty in '95 Cali Crash," *New York Times* (September 12, 1997); Mercer, Pamela, "Inquiry into Colombia Air Crash Points Strongly to Error by Pilot," *New York Times* (December 29, 1995).

5. Wald, Matthew L., "Peru Crash Is Attributed to Maintenance Error," *New York Times* (November 16, 1996).

6. Gordon, Michael R., "Astronaut Error Adds New Anxiety on Space Station," *New York Times* (July 18, 1997).

7. Eichenwald, Kurt, "Death and Deficiency in Kidney Treatment," *New York Times* (December 4, 1995).

8. Altman, Lawrence K., "State Issues Scathing Report on Error at Sloan-Kettering," *New York Times* (November 16, 1995).

9. Norris, Floyd, "Salomon's Error Went Right to the Floor," *New York Times* (March 27, 1992).

10. "Report Criticizes Scientific Testing at FBI Crime Lab," *New York Times* (April 16, 1997); see also "FBI Practices Faulted in Oklahoma Bomb Inquiry," *New York Times* (January 31, 1997).

11. "Bounty Hunters Kill Couple in Case of Mistaken Identity," *New York Times* (September 2, 1997).

12. Gruson, Lindsey, "Reactor Shows Industry's People Problem," *New York Times* (April 3, 1987).

13. Ingraham, Larry H., "'The Nam' and 'The World': Heroin Use by U.S. Army Enlisted Men Serving in Vietnam," *Psychiatry* 37 (May 1974), p. 121.

14. Private communication.

15. Frankenhaeuser, Marianne, "To Err is Human: Psychological and Biological Effects of Human Functioning," in *Nuclear War by Mistake: Inevitable or Preventable?*, report from an international conference in Stockholm, Sweden, February 15-16, 1985 (Stockholm: Spangbergs Tryckerier AB, 1985), p. 45.

16. "Why Planes Crash," NOVA #1403, WGBH-TV (Boston: WGBH Transcripts, 1987), pp. 10-13.

17. Flournoy, op. cit.

18. Reason, James, "The Psychopathology of Everyday Slips: Accidents Happen When Habit Goes Haywire," *The Sciences* (October 1984), p. 48.

19. The pilot escaped unharmed. Ibid., pp. 45 and 49.

20. Ibid.

21. Bower, B., "Emotional Stress Linked to Common Cold," *Science News* (August 31, 1991), p. 132.

22. Fackelmann, K. A., "Stress Puts Squeeze on Clogged Vessels," *Science News* (November 16, 1991), p. 309.

23. Vaughan, Christopher, "The Depression-Stress Link," *Science News* (September 3, 1988), p. 155.

24. Bower, B., "Stress Hormone May Speed up Brain Aging," *Science News* (April 25, 1998); Goleman, Daniel, "Severe Trauma May Damage the Brain as Well as the Psyche," *New York Times* (August 1, 1995).

25. Slaby, Andrew E., *Aftershock: Surviving the Delayed Effects of Trauma, Crisis and Loss* (New York: Villard Books, 1989), p. 33.

26. Heron, Kim, "The Long Road Back," *New York Times Magazine* (March 6, 1988).

27. Slaby, op. cit., p. 77.

28. Grady, Denise, "War Memories May Harm Health," *New York Times* (December 16, 1997).

29. Official figures at the time indicated a much lower rate of psychiatric casualties in the 1973 war because those who were both physically and psychologically wounded were counted only as physical casualties. More important, official figures also excluded those whose psychological symptoms did not appear until sometime later. Shipler, David K., "Other Israeli Wounded: Mentally Scarred in War," *New York Times* (January 8, 1983).

30. The study by Naomi Breslau and her colleagues, reported in *Archives of General Psychiatry* (March 1991), is discussed in Bower, B., "Trauma Disorder Strikes Many Young Adults," *Science News* (March 30, 1991), p. 198.

31. Bower, op. cit. The first three ranked disorders are phobias, severe depression and dependency on drugs or alcohol.

32. Bower, B., "Emotional Trauma Haunts Korean POWs," *Science News* (February 2, 1991), p. 68.

33. Slaby, op. cit., p. xiv.

34. Ibid., p. 71.

35. Since the lower military ranks are heavily populated with young adults from urban areas, there may be a real stress-related reliability problem in the lower ranks. Since traumatic stress can have a very long reach and many career officers have experienced the trauma of battle and/or captivity, there may also be a real stress-related reliability problem in the higher ranks.

36.　Moore-Ede, Martin C., Sulzman, Frank M. and Fuller, Charles A., *The Clocks that Time Us: Physiology of the Circadian Timing System* (Cambridge: Harvard University Press, 1982), p. 1.

37.　In the absence of environmental time cues, the free-running sleep-wake cycle for humans is typically about 25 hours. When people are isolated for long periods, the cycle tends to change, drifting to as long as 30-50 hours. There appear to be two major distinct circadian pacemakers in human beings that are coupled to each other. Together they coordinate the various daily cycles. In the absence of external information, the pacemakers eventually seem to uncouple, desynchronizing the body's biological clock. The two pacemakers each seem to control the patterns of different stages of sleep: for example, one controls REM (dream) sleep, while the other controls slow-wave sleep (deep sleep). Thus, decoupling the pacemakers can cause disruptions in the normal sequence of sleep stages and so can degrade the quality of sleep (that is, the degree of rest and refreshment it produces). Ibid., pp. 207 and 298-301.

38.　Monk, T.H. and Aplin, L.C., "Spring and Autumn Daylight Savings Time Changes: Studies in Adjustment in Sleep Timings, Mood and Efficiency," *Ergonomics* 23 (1980), pp. 167-178.

39.　Given enough time, typically a few days to a week, an out-of-phase biological clock will reset itself, as environmental *zeitgebers* (literally "time givers"; signals such as light and temperature variations) get the circadian rhythms back in phase with the external world. Resynchronization can be accelerated by intentionally manipulating *zeitgebers*. Certain aspects of diet, such as specific patterns of carbohydrate versus protein intake, proper timing of meals and variation of meal size can speed readjustment. So can controlled exposure to bright light. See Ehret, C.F., "New Approaches to Chronohygiene for the Shiftworker in the Nuclear Power Industry," in Reinberg et al., editors, *Advances in the Biological Sciences, Volume 30: Night and Shift Work—Biological and Social Aspects* (Oxford: Pergamon Press, 1981), pp. 267-268.

　　　　　The hormone melatonin, secreted by the pineal gland, plays such a key role in controlling circadian rhythms that it has become popular for travelers to take melatonin pills to beat jet lag. Recent research suggests, however, that melatonin supplements may not be as safe as originally thought. They might "numb" the brain's suprachiasmic nucleus, which contains the circadian clock. Barinaga, Marcia, "How Jet-Lag Hormone Does Double Duty in the Brain," *Science* (July 25, 1997), p. 480. It is also possible that excess melatonin could damage the retina of the eye, which may also contain another key circadian clock. Morell, Virginia, "A 24-Hour Circadian Clock Is Found in the Mamallian Retina," *Science* (April 19, 1995), p. 349.

40.　Tilley, A.J., Wilkinson, R.T., Warren, P. S.G., Watson, B. and Drud, M., "The Sleep and Performance of Shift Workers," *Human Factors* 24 (no. 6, 1982), p. 630.

41.　Moore-Ede, Martin C., Czeisler, Charles A. and Richardson, Gary S., "Circadian Timekeeping in Health and Disease: Part 2: Clinical Implications of Circadian Rhythmicity," *The New England Journal of Medicine* 309 (no. 9, September 1, 1983), p. 534.

42.　Tilley, op. cit., pp. 629-631.

43.　Ibid., p. 634.

44.　Ibid., p. 638-639.

45.　Moore-Ede, Czeisler and Richardson, op. cit., p. 531.

46.　Ehret, C.F., op. cit., pp. 263-264.

47.　Wald, Matthew L., "Sleepy Truckers Linked to Many Deaths," *New York Times* (January 19, 1995); and Wald, Matthew L., "Truckers Need More Sleep, a Study Shows," *New York Times* (September 11, 1997).

48.　Moore-Ede, Sulzman, and Fuller, op. cit., pp. 332-333.

49.　Ibid., pp. 333-334.

50.　Ibid., p. 336.

51.　By itself, this association does not imply either that disruptions of the biological clock necessarily trigger mental disorders or that those who suffer from mental illness always experience circadian disruptions.

52.　Ibid., p. 369.

53.　Ibid., p. 372.

54.　Post, Jerrold M., "The Seasons of a Leader's Life: Influences of the Life Cycle on Political Behavior," *Journal of the International Society of Political Psychology* 2 (Fall-Winter 1980, no. 314), p. 42.

55.　Abrams, Herbert L., "The Age Factor in the Election of 1996 and Beyond," Stanford University School of Medicine and Center for International Security and Arms Control (unpublished).

56.　Post, op. cit., pp. 44-47.

57.　Ibid., p. 45.

58.　Ibid., p. 46.

59. L'Etang, Hugh, *The Pathology of Leadership* (London: William Heineman Medical Books Ltd., 1969), p. 84.

60. Ibid., pp. 87 and 90.

61. Ibid., p. 149.

62. Ibid., p. 150.

63. Ibid., p. 155.

64. Ibid., p. 179.

65. Ibid., pp. 181-183.

66. Kunz, Jeffrey R.M. and Finkel, Asher J., editors, *American Medical Association: Family Medical Guide* (New York: Random House, 1987), p. 535.

67. L'Etang, op. cit., pp. 184-188.

68. Woodward, Bob and Bernstein, Carl, *The Final Days* (New York: Simon and Shuster, 1976), pp. 404 and 423.

69. Ibid., p. 424.

70. Abrams, Herbert L., "The Vulnerable President and the Twenty-Fifth Amendment, with Observations on Guidelines, a Health Commission, and the Role of the President's Physician," *Wake Forest Law Review* (Fall 1995).

71. For a thorough documentation of the medical events and political implications of the disabling attempted assassination of Ronald Reagan, see Abrams, Herbert L., *The President Has Been Shot: Confusion, Disability and the 25th Amendment in the Aftermath of the Attempted Assassination of Ronald Reagan* (New York: Norton, 1992).

72. Abrams, Herbert L., "The Power Vacuum," *Stanford Magazine* (March 1992), pp. 51-52.

73. Ibid., p. 53.

74. Ibid.

75. "Memory and Aging," "Is It Alzheimer's: Warning Signs You Should Know" and "If You Think Someone You Know Has Alzheimer's," Alzheimer's Association (919 North Michigan Ave., Chicago, Illinois 60611).

76. Abrams, Herbert L., "Disabled Leaders, Cognition and Crisis Decision Making," in Paul, Derek, Intrilligator, Michael D. and Smoker, Paul, editors, *Accidental Nuclear War: Proceedings of the Eighteenth Pugwash Workshop on Nuclear Forces* (Toronto: Samuel Stevens, 1990), pp. 137-138.

77. Abrams, op. cit., "Age Factor," p. 3.

78. Ibid., pp. 139-140.

79. Ibid.

CHAPTER EIGHT

1. Azar, Beth, "Teams That Wear Blinders Are Often the Cause of Tragic Errors," *American Psychological Association Monitor* (September 1994), p. 23.

2. Holsti, Ole R. and George, Alexander L., "The Effects of Stress on the Performance of Foreign Policy-Makers," in Cotter, Cornelius P. , editor, *Political Science Annual: An International Review* 6 (1975), p. 294.

3. Halperin, Morton H., *Bureaucratic Politics and Foreign Policy* (Washington, D.C.: Brookings Institution, 1974), pp. 235-279.

4. Thomson, James, "How Could Vietnam Happen?" *Atlantic* (1968), pp. 47-53; and Halberstam, David, *The Best and the Brightest* (New York: Random House, 1972).

5. Schmitt, Eric, "How U.S. Immigration Officials Deceived a Group of Lawmakers," *New York Times* (June 29, 1996).

6. Argyris, Chris, "Single-Loop and Double-Loop Models in Research on Decisionmaking," *Administrative Science Quarterly* (September 1976), pp. 366-367.

7. Holsti and George, op. cit., pp. 293-295.

8. Schlesinger, Arthur Jr., *The Imperial Presidency* (Boston: Houghton-Mifflin, 1973).

9. Eberhardt, J., "Challenger Disaster: Rooted in History," *Science News* (June 14, 1986), p. 372.

10. Wald, Mathew L., "Energy Chief Says Top Aides Lack Skills to Run U.S. Bomb Complex," *New York Times* (June 28, 1989).

11. Ibid.

12. Schneider, Keith, "U.S. Takes Blame in Atom Plant Abuses," *New York Times* (March 27, 1992).

13. Ibid.

14. Marshall, Eliot, "Savannah River Blues," *Science* (October 21, 1988), p. 363.

15. Ibid., p. 364.

16. Ibid., pp. 363-364.

17. Spotts, Peter, "Nuclear Watchdog Agency Shows Signs of Losing Bite," *Christian Science Monitor* (June 6, 1996).

18. U.S. General Accounting Office, *Nuclear Regulation: Preventing Problem Plants Requires More Effective NRC Action* (May 1997: GAO/RCED-97-145).

19. Wayne, Leslie, "Getting Bad News at the Top," *New York Times* (February 28, 1986).

20. Ibid.

21. Shenon, Philip, "Pentagon Reveals It Lost Most Logs on Chemical Arms," *New York Times* (February 28, 1997).

22. Wye, Ted (a pseudonym), "Will They Fire in the Hole?" *Family,* supplement of *Air Force Magazine* (November 17, 1971).

23. Violation of strict rules against fraternization between the ranks were part of the legal charges leveled against drill instructors and other military officers during the flood of sexual harrassment and rape complaints lodged by women in all branches of the armed forces during the 1990s.

24. Kilborn, Peter T., "Sex Abuse Cases Stun Pentagon, But the Problem Has Deep Roots," *New York Times* (February 10, 1997).

25. A majority of the servicewomen responding to a 1996 Pentagon poll said that they had been sexually harassed. Many said their attempts to complain had resulted in "ridicule, retaliation or indifference". Later that year, the Army set up a telephone "hot line" to facilitate and destigmatize reporting sexual harassment complaints. In the first 12 weeks alone, they logged almost 7,000 calls, nearly 1,100 of which they considered legitimate enough to follow up. See Kilborn, Peter T., "Sex Abuse Cases Stun Pentagon, But the Problem Has Deep Roots," *New York Times* (February 10, 1997).

26. Holsti and George, op. cit., p. 295.

27. MacNamara, Robert S., "One Minute to Doomsday," *New York Times* (October 14, 1992).

28. Kennedy, Robert F., *Thirteen Days: A Memoir of the Cuban Missile Crisis* (New York: Norton, 1969), pp. 94-95.

29. Burlatsky, Fedor, "Castro Wanted a Nuclear Strike," *New York Times* (October 23, 1992).

30. Wilson, Richard, "A Visit to Chernobyl," *Science* (June 26, 1987), pp. 1636-1640.

31. Ibid., p. 1639.

32. Daniels, Arlene K., "The Captive Professional: Bureaucratic Limitations in the Practice of Military Psychiatry," *Journal of Health and Social Behavior* 10 (1969), pp. 256-257.

33. Ibid., pp. 255-265.

34. Ibid., p. 263.

35. Bem, D.J., "The Concept of Risk in the Study of Human Behavior," in Carney, R.E., editor, *Risk Taking Behavior* (Springfield, Illinois: Thomas, 1971).

36. Janis, Irving L., *Victims of Groupthink: A Psychological Study of Foreign-Policy Decisions and Fiascoes* (Boston: Houghton-Mifflin, 1972), p. 9.

37. Johnston, David and Revkin, Andrew C., "Report Finds FBI Lab Slipping From Pinnacle of Crime Fighting," *New York Times* (January 29, 1997).

38. Janis, op.cit., pp. 197-198.

39. Ibid., p. 67.

40. Ibid., p. 78.

41. Ibid.

42. Ibid., p. 88. It is interesting to note that this is yet another example of the common tendency discussed earlier to jump from the conclusion that an event has a low probability to the operating assumption that it will not happen, that is, that it is impossible.

43. Ibid., p. 89.

44. Ibid., p. 91.

45. Ibid., pp. 138-166.

46. Most of the following discussion of the Manson case draws heavily on the most comprehensive account of this murder case available, written by Vincent Bugliosi, chief prosecutor at the subsequent trial. See Bugliosi, Vincent with Gentry, Curt, *Helter Skelter: The True Story of the Manson Murders* (New York: W.W. Norton, 1974).

47. Ibid., p. 245.

48. Much of this discussion is drawn from two primary sources: "The Death of Representative Leo J. Ryan, People's Temple, and Jonestown: Understanding a Tragedy," *Hearing before the Committee on*

Foreign Affairs, House of Representatives (96th Congress, 1st Session, May 15, 1979); and "The Cult of Death," special report, *Newsweek* (December 4, 1978), pp. 38-81.

49. Axthelm, Pete et al., "The Emperor Jones," *Newsweek* (December 4, 1978), p. 55.

50. Committee on Foreign Affairs, op. cit., p. 15.

51. Everbach, Tracy, "Agents Had Trained for Months for Raid" and Hancock, Lee, "Question's Arise on Media's Role in Raid," *Dallas Morning News* (March 1, 1993).

52. Loe, Victoria, "Howell's Revelation: Wounded Sect Leader, in a Rambling Interview, Says that He Is Christ," and "A Look at Vernon Howell," *Dallas Morning News* (March 1, 1993).

53. Ibid.

54. Ibid.

55. Ibid.

56. England, Mark and McCormick, Darlene, "Waco's Paper's Report Called Cult 'Dangerous,'" *Dallas Morning News* (March 1, 1993).

57. Ibid.

58. The discussion of the Heaven's Gate cult in this section is based largely on information drawn from the following articles published in the *New York Times:* Purdum, Todd S., "Tapes Left by 39 in Cult Suicide Suggest Comet Was Sign to Die" (March 28, 1997); Niebuhr, Gustav, "On the Furthest Fringes of Millennialism" (March 28, 1997); Bruni, Frank, "Leader Believed in Space Aliens and Apocalypse" (March 28, 1997); Heaven's Gate, "Our Position Against Suicide," reprinted as "Looking Forward to Trip Going to the Next Level" (March 28, 1997); Ayres, B. Drummond, "Families Learning of 39 Cultists Who Died Willingly" (March 29, 1997); Bruni, Frank, "A Cult's 2-Decade Odyssey of Regimentation" (March 29, 1997); Steinberg, Jacques, "The Leader: From Religious Family to Reins of U.F.O. Cult" (March 29, 1997); Goldberg, Carey, "The Compound: Heaven's Gate Fit In With New Mexico's Offbeat Style" (March 31, 1997); Ayres, B. Drummond, "Cult Members Wrote a Script to Put Their Life and Times on the Big Screen" (April 1, 1997); Purdum, Todd, S. "Last 2 Names Released in Mass Suicide of Cult Members" (April 1, 1997); Brooke, James, "Former Cultists Tell of Believers Now Adrift" (April 2, 1997); Bearak, Barry, "Eyes on Glory: Pied Pipers of Heaven's Gate" (April 28, 1997); and Purdum, Todd S., "Ex-Cultist Kills Himself, and 2nd Is Hospitalized" (May 7, 1997).

59. Steinberg, Jacques, "The Leader: From Religious Family to Reins of U.F.O. Cult," *New York Times* (March 29, 1997).

60. Niebuhr, Gustav, "On the Furthest Fringes of Millennialism," *New York Times* (March 28, 1997).

61. Heaven's Gate, "Our Position Against Suicide," reprinted as "Looking Forward to Trip Going to the Next Level," *New York Times* (March 28, 1997).

62. Lifton, Robert Jay, *Thought Reform and the Psychology of Totalism: A Study of "Brainwashing" in China* (New York: W.W.Norton, 1961).

63. This and much of the discussion on cults that follows draws heavily from Conway, Flo and Siegelmen, Jim, *Snapping:America's Epidemic of Sudden Personality Change* (New York: Dell Publishing, 1979).

64. Conway and Siegelman, op. cit., p. 225.

65. Ibid., pp. 22 and 27.

66. On October 30, 1980, and again on November 6, 1980, ABC-TV News aired a two-part story focused on Lifespring on *20/20*. The discussion that follows draws heavily from both parts of that program.

67. Ibid., October 30, 1980, transcript, p. 15.

68. Ibid., transcript, pp. 9 and 6, respectively.

69. Ibid., transcript, pp. 16-17.

70. Ibid., November 6, 1980, transcript, pp. 14-15.

71. Ibid., transcript, p. 16.

72. Ibid.

73. Poe, James, *The Bedford Incident,* (based on the novel by Mark Rascovich), produced by James B. Harris and Richard Widmark (Columbia Pictures: Bedford Productions Limited, 1965).

CHAPTER NINE

1. Suppose Machine A has three key parts and all of them must work right for the machine to do its job. For simplicity, assume the probability of failure of any part is independent of the failure of any other part. If each part is 95 percent reliable (that is, has only a 5 percent chance of failure), Machine A will

be 86 percent reliable (95 percent x 95 percent x 95 percent)—it will have a 14 percent chance of failure. Now suppose we change the design so that it now has five key parts, each still 95 percent reliable. Even though every part is just as reliable as before, the new Machine A would now be only 77 percent reliable (95 percent x 95 percent x 95 percent x 95 percent x 95 percent)—its probability of failure would *increase* from 14 percent to 23 percent. The added complexity has caused the machine to become less reliable.

2. Suppose Machine A is redesigned. Machine B is much more complex, having 20 key parts. Assume each part must still work properly for the machine to work. Even if each part of the new machine is only 1/5 as likely to fail as before (that is, 99 percent reliable), the machine as a whole will still be less reliable. The more complex Machine B has an 18 percent chance of failure (82 percent reliability), compared to the 14 percent chance of failure (86 percent reliability) of the original, simpler Machine A. Even though each part of the original machine was *less* reliable, its simpler design made the whole machine *more* reliable.

3. Suppose Machine C has three parts, each 95 percent reliable. Unlike Machine A, it has two failure modes: if one part fails, the machine's performance degrades, but it still keeps working; if two or three parts fail, the machine stops working. Machine C might be an electrical generator designed for hospital use so it can still provide enough power for minimal lighting, key life-support equipment and other critical functions, even though one part fails. But if more than one part fails, the generator stops working. If the probability of failure of any part does not influence the probability of failure of any other part, there will be a 13.5 percent chance that Machine C will have a partial failure, and less than a 1 percent chance that it will fail completely.

    Suppose Machine D is just like C, except that the failure of one part results in a much higher strain on the remaining parts, reducing their reliability. This might result from designing the generator to provide a higher-than-emergency level of power when one part fails by having the others bear a greater load. Say that as soon as any part fails, the reliability of the remaining parts drops from 95 percent to 75 percent because of the increased load. There will still be a 13.5 percent chance of partial failure, but the chance of complete failure will soar from less than 1 percent to 6 percent.

4. Broad, William J., "Fragmenting Space Debris Could Put Satellites at Risk," *New York Times* (May 17, 1994).

5. Peterson, Ivars, "A Digital Matter of Life and Death," *Science News* (March 12, 1988), p. 171.

6. Spinney, Franklin C., *Defense Facts of Life,* (Washington, D.C.: December 5, 1980), unpublished Department of Defense staff paper, unreviewed by the Department and therefore not an official document.

7. Wilford, John N., "Air Control Technology Refined to Meet Hazards," *New York Times* (January 22, 1980).

8. Sims, Calvin, "Disruption of Phone Service Is Laid to Computer Program," *New York Times* (January 17, 1990).

9. Ibid.

10. Markoff, John, "Superhuman Failure: AT&T's Trouble Shows Computers Defy Understanding Even in Collapse," *New York Times* (January 17, 1990).

11. For an interesting if somewhat overlong discussion of these so-called "revenge effects" of technology, see Tenner, Edward, *Why Things Bite Back* (New York: Vintage Books, 1997).

12. Markoff, op.cit.

13. Ibid.

14. Ibid.

15. Perrow, Charles, *Normal Accidents: Living with High-Risk Technologies* (New York: Basic Books, 1984), p. 83.

16. Waldrop, M. Mitchell, "Hubble Managers Start to Survey the Damage," *Science* (July 6, 1990), and "Astronomers Survey Hubble Damage," *Science* (July 13, 1990).

17. Cowen, Ron, "The Big Fix: NASA Attempts to Repair the Hubble Space Telescope," *Science News* (November 6, 1993), pp. 296-298.

18. Cowen, Ron, "Hubble Finally Gets a Heavenly View," *Science News* (January 22, 1994), p. 52.

19. Cowen, Ron, "Trying to Avoid Hubble Trouble," *Science News* (March 27, 1999), p. 203.

20. Petroski, Henry, *To Engineer Is Human: The Role of Failure in Successful Design* (New York: St. Martin's Press, 1985), pp. 176-179.

21. Randell, B., Lee, P. A. and Treleaven, P. C., "Reliability Issues in Computing System Design," Association for Computing Machinery, *Computing Surveys* 10 (no. 2, June 1978), p. 135. In the case of crucial decisions with great consequences, there are two ways to strengthen TMR. One can require that all three modules agree in order for the system to take action, or the modules can be designed to

receive inputs from different, isolated systems. Unfortunately, insisting on unanimity not only destroys fault tolerance, it also biases the system against reacting. Requiring the modules to agree while receiving inputs from different sources reduces the likelihood of failure through improper action, but may increase the likelihood of failure through inaction.

22. For example, suppose Machine E has two key parts, each of which is 95 percent reliable. Assume that the probability of failure of either part does not influence the probability of failure of the other. If both parts must work for the machine to work, Machine E will then be 90 percent reliable (95 percent x 95 percent)—much less reliable than either of its parts. But, if it were designed with a third 95 percent reliable part which could serve as a backup system for either of the other two, Machine E would be 99 percent reliable—much more reliable than any of its parts.

23. Wooten, James T., "Sound of Debate Is Off Air for 27 Minutes," *New York Times* (September 24, 1976).

24. Diamond, Stuart, "The Bhopal Disaster: How It Happened," *New York Times* (January 28,1985).

25. Perrow, op. cit., p. 45.

26. The airline later reported that only two of the three inertial navigation systems had failed, but the crew could not determine which of the devices was working properly. Weiner, Eric, "Jetliner, in Flight to Tokyo, Lands in Alaska After a 600-Mile Error," *New York Times* (December 23, 1989).

27. Perrow, op. cit., p. 260.

28. Ford, Daniel, "Three Mile Island: II—The Paper Trail," *The New Yorker* (April 13, 1981), pp. 52-53.

29. *New York Times* (September 26, 1981).

30. According to John Galipault, president of the Aviation Safety Institute, as quoted in Barron, James, "Tail Fan May Have Come Apart, Crippling DC-10," *New York Times,* (July 21, 1989).

31. Weiner, Eric, "DC-10 Jets to Get Hydraulic Device," *New York Times* (September 16, 1989).

32. See Hanley, Robert, "Newark Airport Is Closed as Crew Cuts Power Lines" and "Blackout at Airport Shows Need for a Backup System," *New York Times* (January 10 and 11, 1995, respectively).

33. Waldrop, M. Mitchell, "Computers Amplify Black Monday," *Science* (October 30, 1987), p. 604.

34. Weiner, Tim, "B-2, After 14 Years, Is Still Failing Basic Tests," *New York Times* (July 15, 1995).

35. Shenon, Philip, "B-2 Gets a Bath To Prove That It 'Does Not Melt,'" *New York Times* (September 13, 1997).

36. Clark, Christopher, "B-2 Bomber Makes Combat Debut," WashingtonPost.com (March 25, 1999: 2:39 A.M. EST).

37. Wilford, John N., "NASA Considered Shuttle Boosters Immune to Failure," *New York Times* (Febraury 3, 1986).

38. Shcherbak, Yuri M., "Ten Years of the Chornobyl Era," *Scientific American* (April 1996), p. 45.

39. Wilford, John N., "Space Shuttle Is Grounded By Lovesick Woodpeckers," *New York Times* (June 3, 1995).

40. Bishop, Katherine, "Experts Ask If California Sowed Seeds of Road Collapse in Quake," *New York Times* (October 21, 1990).

41. Weiner, Eric, "Jet Lands After an Engine Drops Off," *New York Times* (January 5, 1990).

42. Petroski, op. cit., pp. 85-90.

43. Ibid., p. 197.

44. Waldrop, op. cit.,1990.

45. Eberhart, J., "Solving Hubble's Double Trouble," *Science News* (July 4, 1990).

46. Petroski, op. cit., p. 198-199.

47. Rosenthal, Andrew, "Design Flaw Seen as Failure Cause in Trident II Tests," *New York Times* (August 17, 1989).

48. Petroski, op. cit., p. 179.

49. Perrow, op. cit., p. 34.

50. Eisman, Dale, "Super Hornet Critics Charge that Navy Hid Flaws to Get Funding," *The Virginian-Pilot* (Norfolk, Virginia) (February 2, 1998).

51. Petroski, op. cit., pp. 94-95.

52. Pollack, Andrew, "Japan Quake Toll Moves Past 4,000, Highest Since 1923," *New York Times* (January 20, 1995).

53. Blakeslee, Sandra, "Brute Strength Couldn't Save Kobe Structures," *New York Times* (January 25, 1995).

54. Browne, Malcolm W., "Doomed Highway May Have Pulsed to the Rhythm of the Deadly Quake," *New York Times* (October 23, 1989).

55. "Investigators Agree N.Y. Blackout of 1977 Could Have Been Avoided," *Science* (September 15, 1978).

56. Stevens, William K., "Eastern Quakes: Real Risks, Few Precautions," *New York Times* (October 24, 1989).

57. Hancock, Lee, "Chances Are 50-50 for Major Quake by Decade's End," *Dallas Morning News* (July 22, 1990).

58. Ibid.

59. Stevens, op. cit.

60. Goldberger, Paul, "Why the Skyscrapers Just Swayed," *New York Times* (October 19, 1989).

61. Albright, David, Paine, Christopher and von Hippel, Frank, "The Danger of Military Reactors," *Bulletin of the Atomic Scientists* (October 1986), p. 44.

62. Fuller, John G., *We Almost Lost Detroit* (New York: Reader's Digest Books, 1975), pp. 102 and 215-220.

63. Perrow, op. cit., p. 36.

64. Ibid., p. 37.

65. Shcherbak, op. cit., p. 48.

66. Wartzman, Rick and Pasztor, Andy, "Doubts Are Raised About the Reliability of the Cruise Missile," *Wall Street Journal* (July 27, 1990.

67. Ibid.

68. Waldrop, M. Mitchell, "Hubble: the Case of the Single-Point Failure," *Science* (August 17, 1990), p. 735.

69. Ibid., p. 736.

70. Smith, R. Jeffrey, "Pentagon Hit by New Microchip Troubles," *Science* (November 23, 1984).

71. Sanger, David E., "Chip Testing Problems Abound, Pentagon Says," *New York Times* (April 16, 1985).

72. Hart, Gary and Goldwater, Barry, "Recent False Alerts from the Nation's Missile Attack Warning System," Report to the Committee on Armed Services, United States Senate (October 9, 1980).

73. "Pentium Flaw Has Some Scientists Steamed," *Wall Street Journal* (November 25, 1994),

74. For example, if X is divided by Y, then the result is multiplied by Y, the answer should again be X. If this answer is then subtracted from X, the result should be zero. The particular values of X and Y used do not matter. But performing this sequence of operations with the flawed Intel Pentium when X = 4195835.0 and Y = 3145727.0 gives an answer of 256.

75. Rosenthal, Andrew, "Defects in $12 Circuit Board Stall Deployment of New Fighter Missile," *New York Times* (September 15, 1989).

76. Cushman, John Jr., "Findings Suggest Jet's Flaw Was Visible Before Crash," *New York Times* (November 1, 1989).

77. Cushman, John Jr., "Metallurgical Mystery in Crash that Killed 112," *New York Times* (November 3, 1989).

78. "Built to Break," ABC Television News, *20/20* #917, (April 28, 1989).

79. Ibid.

80. Ibid.

81. Broad, William J., "Major Flaw Found in Space Station Planned by NASA," *New York Times* (March 19, 1990). See also Broad, William J., "NASA Is Faulted Over Space Station," *New York Times* (March 21, 1990).

82. "Space Base Heads Back to Drawing Board," *Science News* (July 28 1990), p. 53.

83. Broad, op. cit., March 21, 1990.

84. Broad, op. cit., March 19,1990.

85. Gordon, Michael R., "Russian Space Station Damaged in Collision With a Cargo Vessel," *New York Times* (June 26, 1997). See also in the *New York Times:* Specter, Michael, "Mir Computer Failure Sends Space Station Out of Control" (August 19, 1997); Gordon, Michael R., "Mir Shadow Falls on Russian Role in Space" (August 2, 1997) and "Russian and American Astronauts Start Hazardous Space Walk to Inspect Mir Damage" (September 6, 1997).

86. Nather, David, "Safety Issues Plague Aging Airline Fleets," *Dallas Morning News* (July 22, 1990).

87. Ibid.

88. Wald, Matthew L., "Agency Grounds Scores of 737's to Check Wiring," *New York Times* (May 11, 1998).

89. Wald, Matthew L., "Checks of 737's Show More Damaged Wiring," *New York Times* (May 12, 1998).

90. Petroski, op. cit., pp. 95-96.

91. Nather, David, "Safety Issues Plague Aging Airline Fleets," *Dallas Morning News* (July 22, 1990).

92. Marshall, Eliot, "The Salem Case: A Failure of Nuclear Logic," *Science* (April 15, 1983), pp. 280-281.

93. Ibid., p. 281.

94. Pollard, Robert, "Showing their Age," *Nucleus* (Union of Concerned Scientists, Summer 1995), p. 7. Pollard is a senior nuclear safety engineer.

95. Utroska, Daniel, "Holes in the U.S. Nuclear Safety Net," *Bulletin of the Atomic Scientists* (July-August 1987), p. 40.

96. Spinney, op. cit., pp. 46-48.

97. Ibid.

CHAPTER TEN

1. Associated Press, "Pilot's Wrong Keystroke Led to Crash, Airline Says," *New York Times* (August 24, 1996).

2. See, for example, Ornstein, S. M., "Deadly Bloopers," unpublished paper (June 16, 1986), p. 5.

3. Ibid., p. 7.

4. Salpukas, A., "Computer Chaos for Air Travelers," *New York Times* (May 13, 1989).

5. Broad, William J., "New Computer Failure Imperils U.S. Shuttle Astronomy Mission," *New York Times* (December 9, 1990).

6. Clark, D., "Some Scientists Are Angry Over Flaw in Pentium Chip, and Intel's Response," *Wall Street Journal* (November 25, 1994).

7. Communication to the author from an Intel employee in the fall of 1997.

8. Gilpin, K.N., "Stray Rodent Halts NASDAQ Computers," *New York Times* (December 10, 1988).

9. Ibid.

10. Ornstein, op. cit., p. 12.

11. Heath, James R., Keukes, Philip J., Snider, Gregory S. and Williams, R. Stanley, "A Defect-Tolerant Computer Architecture: Opportunities for Nanotechnology," *Science* (June 12, 1998), p. 1716.

12. For further discussion of TMR in this context, see Randell, B., Lee, P. A. and Treleaven, P. C., "Reliability Issues in Computing System Design," Association for Computing Machinery, *Computing Surveys* 10 (no. 2, June 1978), p. 129.

13. Ornstein, op. cit., pp. 9-10.

14. Waldrop, op. cit., March 23, 1984, p. 1280.

15. Ornstein, op. cit., p. 10.

16. Peterson, I., "Phone Glitches and Other Computer Faults," *Science News* (July 6, 1991).

17. Broder, John M. and Zuckerman, Laurence, "Computers Are the Future But Remain Unready for It," *New York Times* (April 7, 1997).

18. Peterson, Ivar, "Year-2000 Chip Danger Looms Large," *Science News* (January 2, 1999), p. 4.

19. Ibid. See also General Accounting Office, Accounting and Information Management Division, *Year 2000 Computing Crisis: An Assessment Guide* (February 1997).

20. Parnas, D.L., "Software Aspects of Strategic Defense Systems," *American Scientist* (September-October 1985), pp. 433 and 436.

21. Ornstein, op. cit., p. 10.

22. Brooks, op. cit., p. 121.

23. Ibid., chapter 11, footnote 5.

24. Ibid, p. 122.

25. Parnas, op. cit., p. 433.

26. Ornstein, op. cit., p. 11.

27. See, for example, "Software for Reliable Networks," *Scientific American* (May 1996).

28. Glanz, James, "Chain of Errors Hauled Probe into Spin," *Science News* (July 24, 1998), p. 449. SOHO is an acronym for Solar and Heliospheric Observatory.

29. Waldrop, M.M., "The Necessity of Knowledge," *Science* (March 23, 1984), p. 1280.

30. In 1927, Werner Heisenberg stated the "uncertainty principle," which holds that it is impossible to specify simultaneously with infinite precision both the linear momentum and the corresponding position of a subatomic particle or photon (a "packet" of energy). The effect of the principle is to change the interpretation of some of the laws of physics from statements of absolute certainty to statements of relative probabilities. That there is some inherent inaccuracy, some essential "unknowability" at the root of the physical universe, has interesting and profound philosophical implications.

31. Peterson, Ivars, "The Soul of a Chess Machine," *Science News* (March 30, 1996), p. 201.

32. Pollack, A., "Pentagon Sought Smart Truck But Found Something Else," *New York Times* (May 30, 1989).

33. Ibid.

34. The best and most comprehensive exposition I have seen of the distinction between what computers do and what the human mind does when it thinks is Theodore Roszak, *The Cult of Information: A Neo-Luddite Treatise on High-Tech, Artificial Intelligence, and the True Art of Thinking* (Berkeley: University of California Press, 1994).

35. In 1996, a computer program "designed to reason" rather than solve a particular problem developed "a major mathematical proof that would have been called creative if a human thought of it". The feat accomplished by the researchers at the Argonne National Laboratories that wrote the program is impressive and interesting, but is very far from proving that the functioning of the human mind can be effectively duplicated by electronic means. See Kolata, Gina, "With Major Math Proof, Brute Computers Show Flash of Reasoning Power," *New York Times* (December 10, 1996).

36. Ornstein, op. cit., p. 38.

37. Parnas, op. cit., p. 438.

38. Cohen, F., "Computer Viruses: Theory and Experiments," *Computers and Security* (February 1987), p. 24.

39. Ibid., p. 29.

40. Markoff, John, "A New Method of Internet Sabotage Is Spreading," *New York Times* (September 19, 1996).

41. Elmer-De-Witt, P. , "Invasion of the Data Snatchers," *Time,* September 26, 1988, p. 67.

42. Marshall, E., "The Scourge of Computer Viruses," *Science,* April 8,1988, p. 133.

43. Elmer-De-Witt, op. cit., pp. 65-66.

44. Dewdney, A.K., "Computer Recreations: In the Game called Core War Hostile Programs Engage in a Battle of Bits," *Scientific American* (May 1984), p. 14.

45. Ibid., p. 18.

46. Cohen, op. cit., p. 31.

47. Markoff, J., "Top Secret, and Vulnerable," *New York Times* (September 25, 1988).

48. Elmer-De-Witt, op. cit., p. 62.

49. The author has not attempted to compile a comprehensive list. However, the instances described in the text are intended to illustrate a variety of different types and sources of attack. In addition to these, there were reported attacks on the National Aeronautics and Space Administration's semipublic "Space Physics Analysis Network" by "hackers" belonging to a group called the Hamburg Chaos Computer Club in West Germany (Highland, H.J., "Case History of A Virus Attack," *Computers & Security,* February 1988,p. 3); on Apple computers at Dallas-based EDS corporation, Boeing, ARCO, NASA, the Internal Revenue Service and the U.S. House of Representatives by a malicious virus called Scores believed to have been written by a disgruntled Apple employee (Elmer-De-Witt, op. cit., pp. 63-64); and on Amiga microcomputers produced by Commodore Corporation by a malicious virus somewhat similar to Scores (Highland, H.J., "Computer Viruses: A Post Mortem," *Computers and Security,* April 1988, p. 120).

50. Higher-security military communications were split off from Arpanet into a second network called Milnet in 1983.

51. Markoff, J., "'Virus' in Military Computers Disrupts Systems Nationwide," *New York Times* (November 4, 1988).

52. Shoch, J.F. and Hupp, J.A., "The 'Worm' Programs—Early Experience with a Distributed Computation," *Communications of the ACM* 25 (no. 3, March 1982), p. 172.

53. Markoff, J., "Invasion of Computer: 'Back Door' Left Ajar," *New York Times* (November 4, 1988).

54. 54 Holden, C., "The Worm's Aftermath," *Science* (November 25, 1988), pp. 1121-1122.

55. Markoff, J., op. cit.

56. Markoff, J., "'Virus' in Military Computers Disrupts System Nationwide," *New York Times* (November 4, 1988).

57. Elmer-De-Witt, op. cit., p. 65. During the mid to late 1990s, hoax virus messages were repeatedly spread widely over the Internet. They warned users that if they opened certain e-mail messages their computers would become instantly infected. One such hoax involved a message headed AOL4FREE.COM, another, Good Times. In order for a virus to infect a computer, the machine must load an executable program. Reading an e-mail message alone does not load a program, although opening an e-mail attachment can.

58. Markoff, John, "Computer System Intruder Plucks Passwords and Avoids Detection," *New York Times* (March 19,1990).

59. Markoff, John, "Data Network Is Found Open to New Threat," *New York Times* (January 23, 1995).
60. Markoff, John, "New Netscape Software Flaw Is Discovered," *New York Times* (May 18, 1996).
61. Lohr, Steve, "A Virus Got You Down? Who You Gonna Call?" *New York Times* (August 12, 1996).
62. Cohen, F., "On the Implications of Computer Viruses and Methods of Defense," *Computers and Security* (April 1988), p. 167.
63. Elmer-De-Witt, op. cit., p. 67.
64. Davis, F.G.F., and Gantenbein, R.E., "Recovering from a Virus Attack," *Journal of Systems and Software* (no. 7, 1987), pp. 255-256.
65. Glath, R.M., "A Practical Guide to Curing a Virus," *Texas Computing* (November 11, 1988), p. 14. See also Davis and Gantenbein, op. cit.; and Gibson, S., "Tech Talk: What Were Simple Viruses May Fast Become a Plague," *Infoworld* (May 2,1988).
66. Cohen, op. cit., February 1987, p. 28.
67. Marshall, E., "The Scourge of Computer Viruses," *Science* (April 8, 1988), p. 134.
68. Cohen, op. cit., February 1987, p. 30.
69. Davis and Gantenbein, op. cit., p. 254.
70. Ibid., p. 256.
71. Cohen, op. cit., April 1988, p. 168.
72. Ibid., p. 168.
73. Gibson, S., "Tech Talk: Effective and Inexpensive Methods for Controlling Software Viruses," *Infoworld* (May 9, 1988), p. 51.
74. "EPA Gets Computer Virus," *News and Observer* (Raleigh, North Carolina) (August 6, 1988).
75. For an interesting nontechnical discussion of how this type of security works, see "Special Report: Computer Security and the Internet," *Scientific American* (October 1998), pp. 95-115.
76. Weiss, P., "Lock-on-a-Chip May Close Hackers Out," *Science News* (November 14, 1998), p. 309.
77. Lewis, Peter H., "Threat to Corporate Computers Is Often the Enemy Within," *New York Times* (March 2, 1998).
78. "Could a Virus Infect Military Computers?" *U.S. News & World Report* (November 14, 1988), p. 13.
79. Markoff, John, "Secure Digital Transactions Just Got a Little Less Secure," *New York Times* (November 11, 1995).
80. Peterson, Ivar, "Power Cracking of Cash Card Codes," *Science News* (June 20, 1998), p. 388; see also Wayner, Peter, "Code Breaker Cracks Smart Cards' Digital Safe," *New York Times* (June 22, 1998).
81. Markoff, John, "Researchers Crack Code in Cell Phones," *New York Times* (April 14, 1998).
82. "'Hackers' Score a New Pentagon Hit," *U.S. News & World Report* (July 29, 1985), p. 7.
83. Ford, Daniel, *The Button: America's Nuclear Warning System—Does It Work?* (New York: Simon and Shuster, 1985), p. 79.
84. Ibid.
85. Easterbrook, G., "Divad," *Atlantic Monthly* (October 1982), p. 37.
86. Ibid., p. 34.
87. Ibid.
88. Dumas, L.J., "The Economics of Warfare," in Barnaby, F., editor, *Future War: Armed Conflict in the Next Decade* (London: Michael Joseph, 1984), pp. 133-134.
89. "Simulated Mobilization for War Produces Major Problems," *New York Times* (December 20, 1980).
90. Ibid.
91. Doherty, R., "Shuttle Security Lapse," *Electronic Engineering Times* (June 6, 1988)
92. Private communication with an informed source at IBM. Fred Cohen has argued that this program was not technically a virus (in Cohen, op. cit., April 1988, p. 170).
93. Wines, M., "Some Point Out Chinks in US Computer Armor," *New York Times* (November 6, 1988), p. 30.
94. Sandza, R., "Spying Through Computers: The Defense Department's Computer System Is Vulnerable," *Newsweek* (June 10, 1985), p. 39.
95. General Accounting Office, *Information Security: Computer Attacks at Department of Defense Pose Increasing Risks,* (Washington, D.C.: May 1996), pp. 2-4.
96. Shenon, Philip, "Defense Dept. Computers Face a Hacker Threat, Report Says," *New York Times* (May 23, 1996).
97. Markoff, J., "'Virus' in Military Computers Disrupts Systems Nationwide," *New York Times* (November 4, 1988).
98. The Brain, Lehigh, Flu-Shot 4, Hebrew University and Fort Worth viruses all destroyed data in the systems they infected. For descriptions of these viruses, see Elmer-De-Witt, op. cit.; and Highland, H.J., "Computer Viruses: A Post Mortem," *Computers and Security* (April 1988).

99.   "'Hackers' Score a New Pentagon Hit," op. cit., p. 7.
100. Elmer-De-Witt, P. , "A Bold Raid on Computer Security," *Time*, (May 2, 1988), p. 58.
101. Wines, op. cit., p. 30.
102. Cohen, op. cit., February 1987, p. 23.

CHAPTER ELEVEN

1.   The likelihood of any event is mathematically measured by its "probability," expressed as a number ranging from zero (the event is impossible) to one (the event is certain). This can be thought of as reflecting the fraction of trials of repeated performance of the chance experience that result in that particular outcome. For example, the probability of randomly drawing a king from an ordinary deck of 52 playing cards is .077. This may be interpreted as meaning that if the random drawing of a card (with replacement after each draw) were repeated, a king would be drawn about 77 times for every 1,000 draws.

2.   Contingency theorists of organizational structure see the desire to reduce or eliminate risk as a driving force in determining the shape and behavior of formal organizations. See Tausky, C., *Work Organizations: Major Theoretical Perspectives* (Itaska, Illinois: Peacock Publishers, 2nd edition, 1980), pp. 61 and 64-82. Experts in child development admonish parents not to keep changing the rules they expect their children to follow because of the anxiety and confusion generated by the unpredictability that constantly changing rules create. See Hetherington, E.M. and Parke, R.D., *Child Psychology: A Contemporary Viewpoint* (New York: McGraw-Hill, Second Edition, 1979), pp. 434-436.

3.   Fromm, E., "Paranoia and Policy," *New York Times* (December 11, 1975).

4.   In 1939, B.F. Skinner analyzed the question of whether or not Shakespeare intentionally used alliteration in his sonnets. He compared the frequency with which certain sounds appear in the English language with the frequency with which they are repeated in a given line of the sonnets. His conclusion was that, although Shakespeare may have used alliteration purposefully, chance was capable of explaining the observed degree of alliteration all by itself. "So far as this aspect of poetry is concerned, Shakespeare might as well have drawn his words out of a hat". See Skinner, B. F., "The Alliteration in Shakespeare's Sonnets: A Study in Literary Behavior," *The Psychological Record* 3 (no. 15), p. 191.

5.   McElheny, V.K., "Improbable Strikes by Lightning Tripped Its System, Con Ed Says," *New York Times* (July 15, 1977). See also Sullivan, W., "Con Ed Delay Cited As A Possible Cause," *New York Times* (July 16, 1977).

6.   "Investigators Agree New York Blackout of 1977 Could Have been Avoided," *Science* (September 15, 1978), pp. 994-996.

7.   McFadden, R.D., "'Disaster Status Given New York and Westchester to Speed Loans; Services Resume After Blackout," *New York Times,* (July 16, 1977).

8.   Tomasson, R.E.,"Meteorite Crashes Into House in Connecticut," *New York Times* (November 10, 1982).

9.   Ibid. See also, Clarke, R.S., Jarosewich, E. and Noonan, A.F., "Preliminary Data on Eight Observed-Fall Chondritic Meteorites," in G.S. Switzer, editor, *Smithsonian: Contributions to the Earth Sciences* (no. 14, 1975), p. 69.

10. *International Herald Tribune* (October 31, 1968).

11. Anderson, D., "Four! A Hole-In-One Record at Open," *New York Times* (June 17, 1989).

12. "Doctors Say Gun Wound to Brain Eliminated A Man's Mental Illness," *New York Times* (February 25, 1988).

13. McFadden, R.D. , "Odds-Defying Jersey Woman Hits Lottery Jackpot 2nd Time," *New York Times* (February 14, 1986).

14. Boffey, P. M., "Fifth US Rocket Failure Is Acknowledged by NASA," *New York Times* (May 10, 1986).

15. As noted in Chapter 4, *Challenger's* next scheduled mission was to be on May 16, when it was due to carry a Centaur rocket in its cargo bay, topped by a satellite powered by plutonium. The Department of Energy (DoE) has been investigating "deep space disposal . . . for concentrated separated fission products from nuclear waste" at least since the late 1970s. Any economically feasible approach to space disposal would have to involve much larger quantities of highly toxic nuclear waste per launch than are likely to have been on the *Challenger's* next payload. How was the waste to be reliably lofted into space? According to DoE, "*NASA's space shuttle* could be used to lift the concentrated waste package" (emphasis added). See Cary, P. , "Challenger Was to Carry Plutonium Next," *Miami Herald* (March

6, 1986); and "Can Nuclear Wastes Be Launched into Space?" a one-page flyer issued by the Office of Nuclear Waste Isolation, Department of Energy (505 King Ave, Columbus, Ohio 43201). For further information, see also "Draft Environmental Impact Statement, Management of Commercially Generated Radioactive Waste," U.S. Department of Energy (Washington: April 1979), vol.1.

16. Manfredi, A.F. and DiMaggio, C., "Effect of U.S. Space Launch Vehicle Problems on the SDI Decision Timeline" (Washington, D.C.: Congressional Research Service, 86-783 SPR, July 17, 1986), p. 1.

17. Boffey, op. cit.

18. Manfredi and Di Maggio, op. cit., p. 1.

19. "Navy Accidents: What Has Gone Wrong," *New York Times* (November 15, 1989), p. 15.

20. Associated Press, "Hospital Fumes That Hurt 6 Are Tied to Nerve Gas," *New York Times* (November 5, 1994).

21. Diaconis, P. and Mosteller, F., "Methods for Studying Coincidences," *Journal of the American Statistical Association* (December 1989), p. 859.

22. Ibid.

23. Ibid.

24. Ibid., p. 858. Diaconis and Mosteller give a formula for calculating the approximate number of people (N) that need to be in the room to have a 50-50 chance that at least two of them will have a birthday within k days of each other:

$$N = 1.2 \ [365/(2k+1)]^{1/2}$$

25. Morgan, K.Z., "Radiation-Induced Health Effects," letter to *Science* (January 28, 1977), p. 346.

26. Perrow, C., *Normal Accidents* (New York: Basic Books, 1984), pp. 137-9. See also National Transportation Safety Board, document AAR-82-3, April 6, 1982, pp. 14-17.

27. Lipkin, R., "Risky Business of Assessing Danger—Part I," *Insight* (May 23, 1988), p. 12.

28. Fischhoff, B., Slovic, P. and Lichtenstein, S., "Fault Trees: Sensitivity of Estimated Failure Probabilities to Problem Representation," *Journal of Experimental Psychology: Human Perception and Performance* 4 (no. 2, 1978), pp. 330-344.

29. Ibid., p. 334.

30. Ibid., p. 335.

31. Ibid., p. 340.

32. Ibid., p. 342.

33. Clarke., L., "The Origins of Nuclear Power: A Case of Institutional Conflict," *Social Problems* 32 (1985), pp. 474-487.

34. Clarke, L., "Politics and Bias in Risk Assessment," *Social Science Journal* 25 (no. 2, 1988), pp. 158-159.

35. Ibid., p. 159. See also Primack, J., "Nuclear Reactor Safety: An Introduction to the Issues," *Bulletin of the Atomic Scientists* (September 1975), p. 16.

36. U.S. Atomic Energy Commission, *Reactor Safety Study: An Assessment of Accident Risks in United States Commercial Nuclear Power Plants* (Washington, D.C.: U.S. Nuclear Regulatory Commission, 1975); Ford, D.F., "A Reporter at Large: The Cult of the Atom—Part II," *New Yorker* (November 1, 1982), p. 58.

37. Clarke, op. cit., 1988, p. 159.

38. Slovic, P. , Fischhoff, B. and Lichtenstein, S., "Facts and Fears: Understanding Perceived Risk," in Schwing, R.C. and Albers, W.A. editors, *Societal Risk Assessment: How Safe Is Enough?* (New York: Plenum Press, 1980), p. 195.

39. Slovic, P. , "Perceptions of Risk," *Science* (April 17, 1987), p. 283.

40. Ibid., p. 281.

41. Ibid.

42. Plous, S., "Biases in the Assimilation of Technological Breakdowns: Do Accidents Make Us Safer?" *Journal of Applied Social Psychology* (1991, 21, 13), p. 1058.

43. Weatherwax, R.K., "Virtues and Limitations of Risk Analysis," *Bulletin of the Atomic Scientists* (September 1975), p. 31.

44. Fischhoff, B., "Cost Benefit Analysis and the Art of Motorcycle Maintenance," *Policy Sciences* (no. 8, 1977), pp. 181-183.

45. National Research Council, National Academy of Sciences, *Long-Term Worldwide Effects of Multiple Nuclear-Weapons Detonations* (Washington, D.C.: National Academy of Sciences, 1975).

46. Slovic, Fischhoff and Lichtenstein, op. cit., 1980, p. 185.

47. Fischhoff, B., Slovic, P. , and Lichtenstein, S., "Knowing with Certainty: The Appropriateness of Extreme Confidence," *Journal of Experimental Psychology: Human Perception and Performance* 3 (no. 4, 1977), p. 555.

48. Ibid., p. 557.
49. Ibid., p. 558.
50. Ibid., p. 554 and 561.
51. Slovic, Fischhoff and Lichtenstein, 1980, op. cit., p. 187.
52. Lichtenstein, S., Fischhoff, B., and Phillips, L.D., "Calibration of Probabilities: the State of the Art to 1980," in Kahneman, D., Slovic, P. and Tversky, A., editors, *Judgement Under Uncertainty: Heuristics and Biases* (Cambridge: Cambridge University Press, 1982), pp. 315-318.
53. Hynes, M. and Vanmarcke, E., "Reliability of Embankment Performance Predictions," *Proceedings of the ASCE Engineering Mechanics Division Specialty Conference* (Waterloo, Ontario: University of Waterloo Press, 1976), pp. 31-33.
54. Slovic, Fischhoff and Lichtenstein, op. cit., 1980, p. 187.
55. Ibid., p. 185.
56. Lichtenstein, Fischhoff and Phillips, op. cit., p. 333. See also Plous, S., "Thinking the Unthinkable: The Effects of Anchoring on Likelihood Estimates of Nuclear War," *Journal of Applied Social Psychology* (1989, 19, 1), pp. 67-91.
57. Ibid.
58. Slovic, Fischhoff and Lichtenstein, op. cit., 1980, p. 188.
59. Bazelon, D.L., "Risk and Responsibility," *Science* (July 20, 1979), p. 279.

CHAPTER TWELVE

1. Canberra Commission, *Report on the Elimination of Nuclear Weapons,* p. 2. At this writing, the full report is only generally available on the World Wide Web at the address: http//www.dfat.gov.au/dfat/cc/cchome.html.
2. Ibid., pp. 2 and 1.
3. Diamond, J., "Air Force General Calls for End to Atomic Arms," *Boston Globe* (July 16, 1994), as cited in Forrow, Lachlan and Sidel, Victor W., "Medicine and Nuclear War," *Journal of the American Medical Association* (August 5, 1998), p. 459.
4. ECAAR Newsletter, volume 9, nos.1 and 2 (May-June 1996), p. 4.
5. Butler, Lee, "A Voice of Reason," *Bulletin of the Atomic Scientists* (May-June 1998), pp. 59 and 61.
6. Epstein, William, *The Last Chance: Nuclear Proliferation and Arms Control* (New York: The Free Press, 1976), appendix 4, p. 319.
7. Weiner, Tim, "Iraq Spies on UN So It Can Predict Arms Inspections," *New York Times* (November 25, 1997).
8. Office of Technology Assessment, Congress of the United States, *Environmental Monitoring for Nuclear Safeguards* (Washington, D.C.: U.S. Government Printing Office, September 1995), pp. 1 and 5-6.
9. Skindrud, E., "Bomb Testers Beware: Trace Gases Linger," *Science News* (August 10, 1996).
10. Annan, Kofi, "The Unpaid Bill That's Crippling the UN," *New York Times* (March 9, 1998).
11. Because biological and chemical weapons of mass destruction are much simpler, cheaper and easier to create, it is harder to achieve this degree of assurance with them than it is with nuclear weapons. But at least in the case of prohibiting the buildup of national arsenals, there is no reason to believe it cannot be done. Terrorist manufacture of one or a few of these weapons, on the other hand, is much harder to prevent.
12. One of the more interesting and potentially effective proposals for verifying disarmament is "inspection by the people," developed by Seymour Melman and his colleagues more than 40 years ago. Based on the principle that "a common feature of any organized production effort to evade a disarmament inspection system . . . is the participation of a large number of people," inspection by the people would give substantial rewards to anyone with knowledge of illicit activities that revealed them to international authorities, along with safe passage out of the country and protection against future retaliation. See Seymour Melman, editor, *Inspection for Disarmament* (New York: Columbia University Press, 1958), p. 38.
13. A number of other processes can also be used to destroy chemical weapons. In a study for the U.S. Army, the National Research Council (of the National Academy of Sciences and National Academy of Engineering) compared three other nonincineration options with chemical neutralization: chemical oxidation using electricity and silver compounds; exposure to high-temperature hydrogen and steam;

and a process involving a high-temperature molten metal bath. National Research Council, *Review and Evaluation of Alternative Chemical Disposal Technologies* (Washington, D.C.: National Academy Press, 1996).

14. The Johnston Atoll Chemical Agent Disposal System (JACADS), as the incinerator is known, released VX or GB nerve gas to the environment once in December 1990, once in March 1994, twice in March 1995 and once on the first day of April 1995. Information on the last two of these nerve gas releases is available in Program Manager of Chemical Demilitarization, U.S. Department of Defense, *JACADS 1995 Annual Report of RCRA [Resource Conservation and Recovery Act] Noncompliances* (February 25, 1996). The earlier releases were mentioned in Environmental Protection Agency (Region 9), *Report of JACADS Operational Problems* (August 1994). Information on the explosions and other operational problems at JACADS can be found in Mitre Corporation, *Summary Evaluation of the JACADS Operational Verification Testing* (May 1993). Documents concerning the problems at JACADS have been closely followed by the Chemical Weapons Working Group of the Kentucky Environmental Foundation (P. O. Box 467, Berea, Kentucky 40403).

15. Brooke, James, "Incineration of Poison Gas Loses Support," *New York Times* (February 7, 1997).

16. Brooke, James, "So Far So Good as Chemical Weapons Are Burned in Utah, Officials Say," *New York Times* (April 13, 1998).

17. National Research Council, *Review and Evaluation of Alternative Chemical Disposal Technologies* (Washington, D.C.: National Academy Press, 1996).

18. Chepesiuk, Ron, "A Sea of Trouble," *Bulletin of the Atomic Scientists* (September-October 1997).

19. Weiner, Tim, "Soviet Defector Warns of Biological Weapons," *New York Times* (February 25, 1998).

20. Alibek, Ken, "Russia's Deadly Expertise," *New York Times* (March 27, 1998).

21. U.S. General Accounting Office, *Department of Energy: Problems and Progress in Managing Plutonium* (GAO/RCED-98-68), p. 4.

22. As of 1996, access to these bunkers was obstructed by very large concrete blocks placed in front of each bunker's steel entry door. A huge crane powerful enough to move these concrete blocks aside was stored in a guarded area nearby. Conversation with Professor William J. Weida of Colorado College (formerly of the U.S. Air Force Academy), author of an extensive study of the nuclear weapons production complex (February 17, 1996).

23. Wald, Matthew L., "Energy Dept. Announces Dual Plan for Disposal of Plutonium," *New York Times* (January 15, 1997).

24. Weapons-grade plutonium is normally about 93 percent Pu-239, 6 percent Pu-240 and 0.5 percent Pu-241, while typical reactor-grade plutonium is only about 64 percent Pu-239, with 24 percent Pu-240 and 11.5 percent Pu-241. Although neither Pu-240 or Pu-241 is nearly as suitable for weapons purposes as Pu-239, is it a mistake to believe that reactor-grade plutonium could not be used to make a powerful nuclear weapon. The critical mass for reactor-grade plutonium is 6.6 kilograms, only 40 percent more than for weapons-grade. See Chow, Brian G. and Solomon, Kenneth A., *Limiting the Spread of Weapon-Usable Fissile Materials* (Santa Monica, California: National Defense Research Institute, RAND Corporation, 1993), pp. 62-63.

25. Ibid.

26. William J. Weida, "MOX Use and Subsidies" (January 28, 1997, unpublished), p. 4. Weida's calculations are based on Office of Fissile Materials Disposition, Department of Energy, *Technical Summary Report for Surplus Weapons-Usable Plutonium Disposition* (Washington, D.C.: DOE/MD-003, 1996) and Revision 1 of that document (issued October 10, 1996). As of early 1997, the Department of Energy was still not being clear about how many reactors would be required to burn all the plutonium coming out of dismantled weapons. It will take at least one, and maybe a few.

An earlier and apparently less comprehensive estimate by the National Defense Research Institute (RAND) put the cost of the MOX option at between $7,600 and $18,000 per kilogram of weapons-grade plutonium used. For 50 metric tons, the cost would thus be $380 to $900 million. This study found underground disposal to be a safer and cheaper alternative. (Chow and Solomon, op. cit., p. 66.)

27. Toevs, James and Beard, Carl A., "Gallium in Weapons-Grade Plutonium and MOX Fuel Fabrication," Los Alamos National Laboratories document LA-UR-96-4674 (1996).

28. Wald, Matthew L., "Factory Is Set to Process Dangerous Nuclear Waste," *New York Times* (March 13, 1996).

29. For an interesting summary debate between advocates of the two-track approach and advocates of vitrification only, see Moore, Mike, "Plutonium: the Disposal Decision," Holdren, John P. , "Work with Russia" and Lyman, Edwin S. and Leventhal, Paul, "Bury the Stuff," *Bulletin of the Atomic Scientists* (March-April 1997).

30. Hall, Brian, "Overkill Is Not Dead," *New York Times Magazine* (March 15, 1998).

31. Myers, Steven Lee, "Pentagon Ready to Shrink Arsenal of Nuclear Bombs," *New York Times* (November 23, 1998).

32. "First Nuclear Triggers in 40 Years to Be Made," *Palm Beach Post* (April 5, 1998); see also Carlson, Scott and Mello, Greg, "DOE Poised to Throw Money into Pits at LANL," *The Nuclear Examiner* (November 1997), p. 2.

33. Wald, Matthew L., "U.S. to Put a Civilian Reactor to Military Use," *New York Times* (August 11, 1997).

34. Ibid., p. 44.

35. Ibid.

36. Ibid., pp. 42 and 44.

37. For a more complete discussion, see, for example, Fischer, Dietrich, *Preventing War in the Nuclear Age* (Totowa, New Jersey: Rowman and Allanheld, 1984), pp. 47-62.

38. Political scientist Gene Sharp has argued persuasively for a completely nonmilitary form he calls "civilian-based defense." See, for example, Sharp, Gene, *Civilian-Based Defense: A Post-Military Weapons System* (Princeton University Press, 1990). Dietrich Fischer has also made major contributions to the debate over nonoffensive defense. For an especially clear exposition, see Fischer, Dietrich, *Preventing War in the Nuclear Age* (Rowman and Allenheld, 1984).

39. Hufbauer, Gary C., Schott, Jeffrey J. and Elliott, Kimberly, A., *Economic Sanctions Reconsidered: History and Current Policy* (Washington, D.C.: Institute for International Economics, 1990).

40. Ibid., pp. 94-105.

41. For a more detailed analysis of this proposal and the reasoning behind it, see Dumas, L.J., "A Proposal for a New United Nations Council on Economic Sanctions," in Cortright, David and Lopez, George, editors, *Economic Sanctions: Panacea or Peacebuilding in a Post–Cold War World* (Boulder, Colorado: Westview Press, 1995), pp. 187-200.

42. Elsewhere, I have analyzed some of the principles and institutions that could make international economic relations more successful as a means to this end. See Dumas, Lloyd J., "Economics and Alternative Security: Toward a Peacekeeping International Economy," in Weston, Burns, editor, *Alternative Security: Living Without Nuclear Deterrence* (Boulder, Colorado: Westview Press, 1990), pp. 137-175.

43. Boulding, Kenneth, *Stable Peace* (Austin: University of Texas Press, 1978).

44. For a more comprehensive analysis, see Dumas, op.cit., 1990.

45. Sivard, Ruth L., *World Military and Social Expenditures, 1993* (Washington, D.C.: World Priorities, 1993), pp. 20-21. For total wars and war-related deaths from 1900 through 1995, see Sivard, Ruth L., *World Military and Social Expenditures, 1996* (Washington, D.C.: World Priorities, 1996), pp. 17-19.

46. Though these limits of liability have been increased over the years, they remain a small fraction of estimated damages. For example, in 1982, the Sandia National Laboratories prepared a study for the Nuclear Regulatory Commission (NRC), called *Calculation of Reactor Accident Consequences (CRAC2) for U.S. Nuclear Power Plants (Health Effects and Costs)*. The NRC transmitted the CRAC2 study results to Congress later that year. These included estimates of potential damages of $135 billion for a major accident at the Salem Nuclear Plant, Unit 1 reactor (New Jersey), $158 billion for Diablo Canyon, Unit 2 reactor (California), $186 billion for San Onofre, Unit 2 reactor (California) and $314 billion for Indian Point, Unit 3 reactor (New York). Six years later, the 1988 update of Price-Anderson raised the liability limit to $7 billion, only 2 to 5 percent of these damage estimates.

47. Solar power can use the heat of the sun directly, generate electricity with sunlight via solar cells, or even be used to liberate hydrogen that can be cleanly burned from water; geothermal energy make use of the earth's internal heat; biomass conversion turns plant or animal waste into clean burning methane gas; turbines that generate electricity can be driven by waterfalls (hydropower), the force of the incoming and outgoing tides (tidal power), or the wind (windpower); and energy can be extracted from the sea by making use of ocean thermal gradients, the differences in temperature between different layers of seawater.

48. Dumas, L.J., *The Conservation Response: Strategies for the Design and Operation of Energy-Using Systems* (Lexington, Massachusetts: D.C. Heath and Company, 1976).

49. It is theoretically possible to transmute plutonium, uranium and other dangerous radioactive substances into more benign elements by bombarding them with subatomic particles in particle accelerators. The technology to do this in a practical and economically feasible way with large quantities

of radioactive material is nowhere near at hand. While it would be useful to pursue research in this area, whether or not this can be done must be regarded as a speculative matter for the foreseeable future.

50. Some of the more dangerous radionuclides, like strontium-90 and cesium-137 decay to very low levels within a few centuries. Others, like plutonium-239, americium-241 and neptunium-237 take many thousands to millions of years to become essentially harmless. For an interesting, brief discussion of this issue, see Whipple, Chris G., "Living with High-Level Radioactive Waste," *Scientific American* (June 1996), p. 78.

51. Brooke, James, "Underground Haven, Or a Nuclear Hazard?" *New York Times* (February 6, 1997).

52. Wald, Matthew L., "Admitting Error at a Weapons Plant," *New York Times* (March 23, 1998); Zorpette, Glenn, "Hanford's Nuclear Wasteland," *Scientific American* (May 1996), pp. 88-97.

53. U.S. General Accounting Office, *Nuclear Waste: Management and Technical Problems Continue to Delay Characterizing Hanford's Tank Waste* (Washington, D.C.: January 1996), pp. 1-2, 4 and 16.

54. Ibid., p. 8.

55. Hoffman, David, "Rotting Nuclear Subs Pose Threat in Russia: Moscow Lacks Funds for Disposal," *Washington Post* (November 16, 1998).

56. Warf, James C. and Plotkin, Sheldon C., "Disposal of High Level Nuclear Waste," Global Security Study no. 23 (Santa Barbara, California: Nuclear Age Peace Foundation, September 1996), p. 6.

57. Hollister, Charles D. and Nadis, Steven, "Burial of Radioactive Waste Under the Seabed," *Scientific American* (January 1998), pp. 61-62.

58. Ibid., pp. 62-65.

59. Whipple, Chris G., "Can Nuclear Waste Be Stored Safely at Yucca Mountain?" *Scientific American* (June 1996), pp. 72-76.

60. U.S. General Accounting Office, *Nuclear Waste: Impediments to Completing the Yucca Mountain Repository Project* (Washington, D.C.: January 1997), p. 4.

61. Wald, Matthew L., "Doubt Cast on Prime Site as Nuclear Waste Dump," *New York Times* (June 20, 1997).

62. Wernicke, Brian, et al., "Anomalous Strain Accumulation in the Yucca Mountain Area, Nevada"; and Kerr, Richard A., "A Hint of Unrest at Yucca Mountain," *Science* (March 27, 1998), pp. 2096-2100 and pp. 2040-2041, respectively.

63. Kerr, Richard A., "Yucca Mountain Panel Says DoE lacks Data," *Science* (February 26, 1999).

64. James Brooke, op. cit.

65. For a particularly interesting and erudite analysis of the problems with the WIPP site, see Citizens for Alternatives to Radioactive Dumping, "Greetings from WIPP, the Hot Spot: Everything You Always Wanted to Know About WIPP" (1997: CARD, 144 Harvard SE, Albuquerque, NM 87106).

66. Associated Press, "Nation's First Nuclear Waste Depository Opens" (March 26, 1999).

67. Lawler, Andrew, "Researchers Vie for Role in Nuclear-Waste Cleanup," *Science* (March 21, 1997).

68. The F/A-18 is estimated to cost $73 million per plane. See Eisman, Dale, "Pentagon Commits $2 billion to Super Hornet Program, Declares 'Wing-Drop' Problem Resolved," *The Virginian-Pilot* (Norfolk, Virginia) (April 6, 1998).

69. One interesting alternative is bioremediation, though much careful study is necessary before living organisms can be seriously considered as agents for degrading nuclear waste. See Travis, John, "Meet the Superbug: Radiation-Resistant Bacteria May Clean Up the Nation's Worst Waste Sites," *Science News* (December 12, 1998).

70. Environmental Protection Agency, "National Priorities List: Final and Proposed [General Superfund and Federal Facilities] Sites" (December 1996), p. 22. This unpublished list was sent to the author by the EPA Region 6 Office (Dallas, Texas) in March 1997.

    According to EPA Headquarters in Washington, an "operable unit" is a component of the cleanup process, such as groundwater treatment or land cleanup. There have been an average of 1.8 operable units per site. (Correspondence and telephone conversations between the author and Program Analyst Ed Ziomkoski at the Program Analysis and Research Management Center of U.S. EPA, March 27 and April 4, 1997).

71. Correspondence and telephone conversations between the author and Program Analyst Ed Ziomkoski at the Program Analysis and Research Management Center of U.S. EPA; March 27 and April 4, 1997.

72. Federal Facilities Profile, OSWER Directive 9200.3-14-1C, Exhibit D.1, pp. D-2 and D-3, sent to the author by Reneé P. Wynn, Federal Facilities Restoration and Reuse Office, U.S. Environmental Protection Agency, Washington, D.C. (April 11, 1997).

73. Guerrero, Peter F., "Superfund: Times to Assess and Clean Up Hazardous Waste Sites Exceed Program Goals," Testimony before the Subcommittee on National Economic Growth, Natural Resources and

Regulatory Affairs of the Committee on Government Reform and Oversight, U.S. House of Representatives (Washington, D.C.: February 13, 1997), p. 1. See also U.S. General Accounting Office, *Superfund: Times to Complete the Assessment and Cleanup of Hazardous Waste Sites* (Washington, D.C.: March 1997).

74. U.S. General Accounting Office, *Superfund: EPA Could Further Ensure the Safe Operation of On-Site Incinerators* (Washington, D.C.: March 1997).

INDEX